Student's Solution

for use with

Basic College Mathematics

Second Edition

Ignacio Bello
Hillsborough Community College

McGraw Hill **Higher Education**

Boston   Burr Ridge, IL   Dubuque, IA   Madison, WI   New York   San Francisco   St. Louis
Bangkok   Bogotá   Caracas   Kuala Lumpur   Lisbon   London   Madrid   Mexico City
Milan   Montreal   New Delhi   Santiago   Seoul   Singapore   Sydney   Taipei   Toronto

The McGraw·Hill Companies

Student's Solutions Manual for use with
BASIC COLLEGE MATHEMATICS, SECOND EDITION
IGNACIO BELLO

Published by McGraw-Hill Higher Education, an imprint of The McGraw-Hill Companies, Inc., 1221 Avenue of the Americas, New York, NY 10020. Copyright © 2006 by The McGraw-Hill Companies, Inc. All rights reserved.

1 2 3 4 5 6 7 8 9 0 QPD/QPD 0 9 8 7 6 5

ISBN 0-07-294556-7

www.mhhe.com

Student's Solutions Manual

Bello, *Basic College Mathematics*, 2e

Table of Contents

Preface

Dear Student,

This Student's Solution Manual for use with Bello, *Basic College Mathematics, 2nd Edition*, provides complete, worked-out solutions to the following:

- Odd-numbered items in each section's set of exercises.
- All problems paired with examples.
- Odd- and even-numbered items in the chapter Review Exercises.
- Odd- and even-numbered items in the Cumulative Review exercises.

How to use this manual:

If you want to get good grades on your tests, then this manual can help you. <u>Treat every homework assignment as a practice test</u>. Resist the urge to look up all the solutions just to get it over with, because you're not allowed to do that on a test.

So, <u>first try every exercise on your own</u>, then check your work against the solution. If you're stumped, then do as much of the exercise as you can. Next, <u>look up the rest of the steps one at a time</u>, and try to figure out each step before you look at it. (It's handy to keep a piece of paper covering the solution steps so you don't see them all at once.) This is a study habit that will help you improve your test scores and build your confidence in math.

Seeing some solutions will give you that "aha!" sense that you knew how to do it all along and just needed a refresher. If you see other solutions that still appear too complex to follow, then it will help you to <u>review the textbook's explanations and examples</u> to see how similar problems were discussed and solved.

You can also go to the <u>text's MathZone website</u> to view lecture demonstrations, practice problems, and videos by the authors explaining how to work through sample exercises. With all these resources, you'll find that this is a course that you can do well in.

Chapter 1

Whole Numbers

Section 1.1 – Standard Numerals

Problems

1. $9000 + 200 + 40 + 1$

2. $100 + 90 + 7$

3. $90,000 + 8000 + 700 + 3$

4. $80,000 + 9000 + 700 + 80 + 9$

5. 9225

6. 11,111

7. 50,206

8. **a.** Ninety-three
   b. Two hundred nine thousand, three hundred seventy-six
   c. Seventy-five billion, one hundred forty-two million, six hundred forty-two thousand, eight hundred ninety-three

9. 310,692,712

10. 1,125,000,000

11. 1784

Exercises 1.1

1. $30 + 4$

3. $100 + 8$

5. $2000 + 500$

7. $7000 + 40$

9. $20,000 + 3000 + 10 + 8$

11. $600,000 + 4000$

13. $90,000 + 1000 + 300 + 80 + 7$

15. $60,000 + 8000 + 20$

17. $80,000 + 80 + 2$

19. $70,000 + 100 + 90 + 8$

21. 78

23. 308

25. 822

27. 701

29. 3473

31. 5250

33. 2030

35. 8090

37. 7001

39. 6600

41. Fifty-seven

43. Three thousand, four hundred eight

45. One hundred eighty-one thousand, three hundred sixty-two

47. Forty-one million, three hundred thousand

49. One billion, two hundred thirty-one million, three hundred forty-one thousand

1

51. 809

53. 4897

55. 2003

57. 2,023,045

59. 345,033,894

61. One hundred seventy-three thousand, eight hundred eighty

63. Thirteen million, five hundred thirty-seven thousand

65. 14,979,000

67. Annual Cost Public: Fourteen thousand, eight hundred seventy-two dollars

    Four Year Public: Sixty-three thousand, six hundred twenty-seven dollars

    Annual Cost Private: Thirty-two thousand, forty-four dollars

    Four Year Private: One hundred thirty-seven thousand, ninety-one dollars

69. Annual Cost Public: Sixteen thousand, two hundred forty-one dollars

    Four Year Public: Sixty-nine thousand, four hundred eighty-two dollars

    Annual Cost Private: Thirty-four thousand, nine hundred ninety-three dollars

    Four Year Private: One hundred forty-nine thousand, seven hundred seven dollars

71. Annual Cost Public: Seventeen thousand, seven hundred thirty-six dollars

    Four Year Public: Seventy-five thousand, eight hundred seventy-six dollars

    Annual Cost Private: Thirty-eight thousand, two hundred thirteen dollars

    Four Year Private: One hundred sixty-three thousand, four hundred eighty-four dollars

73. Eight hundred twenty-three thousand, five hundred

75. Nine hundred fifty-four thousand

77. Forty-six thousand, three hundred dollars

79. Thirty-three thousand, six hundred dollars

81. 5182

83. 7001

85. 9799.00

87. Twenty-eight thousand, nine hundred

89. Ninety thousand, eight hundred

91. Answers may vary.

93. The value of the 8 is 8.

95. The second digit from the right (the number 7) tells the number of tens.

97. Let $10 = x$. Then
$4 \times 10^2 + 8 \times 10 + 6$ becomes
$4x^2 + 8x + 6$.

99. Answer may vary.

Section 1.2 – Ordering and Rounding Whole Numbers

Problems

1. **a.** $23 < 25$
   b. $31 > 27$

2. **a.** $1081 > 1048 > 980$
   b. Phoenix

3. 300

4. 7000

5. 6000

6. **a.** 1,013,090
   b. 1,013,100
   c. 1,013,000

7. **a.** The GU package estimate is $1000.
   b. The SR package estimate is $700.
   c. The XV package estimate is $100.

Exercises 1.2

1. $8 < 10$

3. $8 > 0$

5. $102 < 120$

7. $999 > 990$

8. $777 > 770$

9. $1001 < 1010$

11. The number to right of the underlined number is 3 (less than 5), so do not change the underlined number. Change the number to the right of the underlined number to zero. Thus, $\underline{7}3 \rightarrow 70$.

13. The number to right of the underlined number is 6 (more than 5), so add one to the underlined number (obtaining 9). Change the number to the right of the underlined number to zero. Thus, $\underline{8}6 \rightarrow 90$.

15. The number to right of the underlined number is 8 (more than 5), so add one to the underlined number (obtaining 10). Change the number to the right of the underlined number to zero. Thus, $\underline{8}8 \rightarrow 100$.

17. The number to right of the underlined number is 0 (less than 5), so do not change the underlined number. Change all the numbers to the right of the underlined number to zeros. Thus, $\underline{1}03 \rightarrow 100$.

19. The number to right of the underlined number is 8 (more than 5), so add one to the underlined number (obtaining 4). Change all the numbers to the right of the underlined number to zeros. Thus, $\underline{3}86 \rightarrow 400$.

21. The number to right of the underlined number is 5, so add one to the underlined number (obtaining 10). Change all the numbers to the right of the underlined number to zeros. Thus, $\underline{9}50 \rightarrow 1000$.

23. The number to right of the underlined number is 3 (less than 5), so do not change the underlined number. Change all the numbers to the right of the underlined number to zeros. Thus, $\underline{2}308 \rightarrow 2000$.

25. The number to right of the underlined number is 9 (more than 5), so add one to the underlined number (obtaining 7). Change all the numbers to the right of the underlined number to zeros. Thus, $\underline{6}999 \rightarrow 7000$.

27. The number to right of the underlined number is 9 (more than 5), so add one to the underlined number (obtaining 10). Change all the numbers to the right of the underlined number to zeros. Thus, $9999 \to 10,000$.

29. The number to right of the underlined number is 0 (less than 5), so do not change the underlined number. Change all the numbers to the right of the underlined number to zeros. Thus, $9099 \to 9000$.

		TEN	HUNDRED	THOUSAND
31.	586	590	600	1000
33.	29,450	29,450	29,500	29,000
35.	49,992	49,990	50,000	50,000
37.	259,906	259,910	259,900	260,000
39.	289,000	289,000	289,000	289,000

41. Underline the place to which we are rounding: $1\underline{4}7$. The first number to the right of the underlined number is 7 (more than 5), so we add one to the underlined number, obtaining 5. Change all the numbers to the right of the underlined number to zeros. Thus, $1\underline{4}7 \to 150$.

43. Underline the place to which we are rounding: $1\underline{0}69$. The first number to the right of the underlined number is 6 (more than 5), so we add one to the underlined number, obtaining 1. Change all the numbers to the right of the underlined number to zeros. Thus, $1\underline{0}69 \to 1100$.

45. Underline the place to which we are rounding: $1\underline{0},950$. The first number to the right of the underlined number is 9 (more than 5), so we add one to the underlined number, obtaining 1. Change all the numbers to the right of the underlined number to zeros. Thus, $1\underline{0},950 \to 11,000$.

47. Underline the place to which we are rounding: $7,\underline{8}95,563$. The first number to the right of the underlined number is 9 (more than 5), so we add one to the underlined number, obtaining 9. Change all the numbers to the right of the underlined number to zeros. Thus, $7,\underline{8}95,563 \to 7,900,000$.

49. Underline the place to which we are rounding: $\$8\underline{5},747,184$. The first number to the right of the underlined number is 7 (more than 5), so we add one to the underlined number, obtaining 6. Change all the numbers to the right of the underlined number to zeros. Thus, we have $\$8\underline{5},747,184 \to \$86,000,000$.

51.
a. $\$689 < \$968 < \$1019$
b. $\$1019 > \$968 > \$689$
c. The Dimension 8250 is the most expensive model.
d. The most economical model is the Dimension 4550.

53.
a. $<$
b. $<$
c. $>$

55. Nine thousand, four hundred ninety-four dollars

57. Nine thousand, six hundred ninety-four dollars

59. Eleven thousand, thirty-five dollars

61. $23,899.\underline{5}6 \rightarrow \$23,900$

63. $349.\underline{4}8 \rightarrow \$349$

65. $25,675.\underline{6}3 \rightarrow \$25,676$

67. Answers will vary.

Mastery Test 1.2

69. a. $7\underline{6}5 \rightarrow 770$

   b. $3\underline{6}4 \rightarrow 360$

   c. $\underline{8}62 \rightarrow 900$

71. $4\underline{9},773 \rightarrow \$50,000$

Section 1.3 – Addition

Problems

1.
$$\begin{array}{r} 45 \\ +32 \\ \hline 77 \end{array}$$

2.
$$\begin{array}{r} 236 \\ +741 \\ \hline 977 \end{array}$$

3.
$$\begin{array}{r} 1 \\ 243 \\ +\ 29 \\ \hline 272 \end{array}$$

4.
$$\begin{array}{r} 1 \\ 263 \\ +475 \\ \hline 738 \end{array}$$

5.
$$\begin{array}{r} 632 \\ +754 \\ \hline 1386 \end{array}$$

6.
$$\begin{array}{r} 1 \\ 813 \\ +1702 \\ \hline 2515 \end{array}$$

7.
$$\begin{array}{r} 1\ 1 \\ 3943 \\ +4672 \\ \hline 8615 \end{array}$$

8.
$$\begin{array}{r} 2\ 2\ 2\ 1 \\ 2451 \\ 4741 \\ 7879 \\ +6563 \\ \hline 21,634 \end{array}$$

9.
$$\begin{array}{r} 2 \\ 13 \\ 6 \\ 4 \\ 7 \\ +8 \\ \hline 38 \end{array}$$

10.
$$\begin{array}{r} 1\underline{2},240 \rightarrow 12,000 \\ \underline{6}450 \rightarrow \ 6000 \\ 800 \rightarrow \ 1000 \\ \hline \$19,000 \end{array}$$

11. $(8+6)+10 = 14+10 = 24$ miles

Exercises 1.3

1.
$$\begin{array}{r} 8 \\ +3 \\ \hline 11 \end{array}$$

3.
$$\begin{array}{r} 4 \\ +7 \\ \hline 11 \end{array}$$

5.
$$\begin{array}{r} 6 \\ +6 \\ \hline 16 \end{array}$$

7.
$$\begin{array}{r} 10 \\ +20 \\ \hline 30 \end{array}$$

9.
$$\begin{array}{r} 82 \\ +83 \\ \hline 165 \end{array}$$

11.
$$\begin{array}{r} 4 \\ 53 \\ +72 \\ \hline 129 \end{array}$$

13.
$$\begin{array}{r} 1 \\ 9 \\ +51 \\ \hline 60 \end{array}$$

15.
$$\begin{array}{r} 0 \\ +11 \\ \hline 11 \end{array}$$

17.
$$\begin{array}{r} 1 \\ 28 \\ +\ 6 \\ \hline 34 \end{array}$$

19.
$$\begin{array}{r} 1 \\ 26 \\ +\ 9 \\ \hline 35 \end{array}$$

21.
$$\begin{array}{r} 1 \\ 4 \\ 13 \\ +\ 6 \\ \hline 23 \end{array}$$

23.
$$\begin{array}{r} 1 \\ 3 \\ 31 \\ +47 \\ \hline 81 \end{array}$$

25.
$$\begin{array}{r} 1 \\ 21 \\ 19 \\ +87 \\ \hline 127 \end{array}$$

27.
$$\begin{array}{r} 1 \\ 67 \\ +58 \\ \hline 125 \end{array}$$

29.
$$\begin{array}{r} 1 \\ 97 \\ +35 \\ \hline 132 \end{array}$$

31.
$$\begin{array}{r} 1 \\ 85 \\ +67 \\ \hline 152 \end{array}$$

33.
$$\begin{array}{r} 1\ 1 \\ 386 \\ +\ 14 \\ \hline 400 \end{array}$$

35.
$$\begin{array}{r} 1 \\ 6347 \\ +\ 426 \\ \hline 6773 \end{array}$$

37.
$$\begin{array}{r} 1 \\ 432 \\ +1381 \\ \hline 1813 \end{array}$$

39.
$$\begin{array}{r} 1\ 1 \\ 136 \\ +3587 \\ \hline 3723 \end{array}$$

41.
$$\begin{array}{r} 1 \\ 4605 \\ +\ \ 39 \\ \hline 4644 \end{array}$$

43.
$$\begin{array}{r} 1\ 2 \\ 108 \\ 2134 \\ +\ \ 98 \\ \hline 2340 \end{array}$$

45.
$$\begin{array}{r} \overset{1\ \ 1}{} \\ \overset{305}{} \\ 6312 \\ +8573 \\ \hline 15,190 \end{array}$$

47.
$$\begin{array}{r} \overset{11\ \ 1}{} \\ 82,583 \\ +\ 8,692 \\ \hline 91,275 \end{array}$$

49.
$$\begin{array}{r} \overset{1\ \ 11}{} \\ 63,126 \\ 77,684 \\ +80,000 \\ \hline 220,810 \end{array}$$

51. $11+65+201+305 = 582$ descendants

53. $37+5+5+17 = 64$ million tons of material

55. $118+60+17 = 195$ gallons

57. $1250+222 = 1472$ feet

59. $2248+32 = 2280$ ounces

61. $90+90+90+90 = 360$ feet

63. $160+360+160+360 = 1040$ feet

65. **a.** $\underline{1}040 \rightarrow 1000$ **b.** $\underline{2}25 \rightarrow 200$
   c. $\underline{7}2 \rightarrow 70$

67. $98+83+93+209 = \$483$

69. Family A:
$107+98+75+150+90+100 = \$620$
Family B:
$258+175+130+190+105+150 = \$1008$

71. **a.** $2+0+0+2+4+2 = 10$ (average)
   b. $4+2+4+2+4+2 = 18$ (high)
   c. $0+2+2+0+0+2 = 6$ (below average)

73. **a.**

Basic Cow	$\$499.\underline{9}5 \rightarrow \$500$
Shipping and handling	$3\underline{5}.75 \rightarrow 36$
Extra Stomach	$7\underline{9}.25 \rightarrow 79$
Two tone exterior	$142.10 \rightarrow 142$
Produce storage compartment	$12\underline{6}.50 \rightarrow 127$
Heavy duty straw chopper	$18\underline{9}.60 \rightarrow 190$
4 spigot/high output drain system	$14\underline{9}.20 \rightarrow 149$
Automatic fly swatter	$8\underline{8}.50 \rightarrow 89$
Genuine cowhide upholstery	$17\underline{9}.90 \rightarrow 180$
Deluxe dual horns	$5\underline{9}.25 \rightarrow 59$
Automatic fertilizer attachment	$33\underline{9}.40 \rightarrow 339$
4 x 4 traction drive assembly	$88\underline{4}.16 \rightarrow 884$
Pre-delivery wash and comb (Farmer Prep)	$6\underline{9}.80 \rightarrow 70$
Additional Farmer Markup and hay fees	$30\underline{0}.00 \rightarrow 300$

b. FARMERS SUGGESTED LIST PRICE $\$2843.36 \rightarrow \$2844.00$

c. TOTAL LIST PRICE (including options) $\$3143.36$

d. 4 x 4 traction drive assembly ($884.16)

e. Shipping and handling ($35.75)

75. a. $11 + 63 = 74$ miles
   b. $140 + 101 + 96 = 337$ miles
   c. via Salinas, Paso Robles, and
     Bakersfield

77. $73 + 96 + 95 + 147 + 71 = 482$
   The perimeter is 482 miles.

79. Answers will vary.

81. Answers will vary.

83. a. $(101 + 96) + 62 = 197 + 62 = 259$ mi.
   b. $101 + (96 + 62) = 101 + 158 = 259$ mi.
   c. The Associative law of Addition

85.
$$\begin{array}{r} 3x + 9 \\ 8x + 7 \\ \hline 11x + 16 \end{array}$$

87.
$$\begin{array}{r} 3x^3 + 8x^2 + 7x + 9 \\ 9x^3 + 8x^2 + 4x + 5 \\ \hline 12x^3 + 16x^2 + 11x + 14 \end{array}$$

Mastery Test 1.3

89. 25,549

91 1347

93.
$$\begin{array}{r} 2538 \\ 3120 \\ +3330 \\ \hline 8988 \end{array}$$
It will cost $9000. The exact answer is $8988.

Section 1.4 – Subtraction

Problems

1.
$$\begin{array}{r} 15 \\ -\ 9 \\ \hline 6 \end{array}$$

2.
$$\begin{array}{r} 95 \\ -62 \\ \hline 33 \end{array}$$

3.
$$\begin{array}{r} {}^{4\ 14}6\cancel{5}4 \\ -239 \\ \hline 415 \end{array}$$

4.
$$\begin{array}{r} {}^{7\ 14}\cancel{8}\cancel{4}6 \\ -\ 72 \\ \hline 774 \end{array}$$

5.
$$\begin{array}{r} {}^{7\ 10}6\cancel{8}0 \\ -295 \\ \hline 5 \end{array} \rightarrow \begin{array}{r} {}^{5\ 17}\cancel{6}\cancel{8}0 \\ -295 \\ \hline \boxed{385} \end{array}$$

6.
$$\begin{array}{r} {}^{4\ 10}52\cancel{5}0 \\ -1478 \\ \hline 2 \end{array} \rightarrow \begin{array}{r} {}^{1\ 14}5\cancel{2}\cancel{5}0 \\ -1478 \\ \hline 72 \end{array} \rightarrow \begin{array}{r} {}^{4\ 11}5\cancel{2}\cancel{5}0 \\ -1478 \\ \hline \boxed{3772} \end{array}$$

7.
$$\begin{array}{r} {}^{5\ 9\ 15}\cancel{6}\cancel{0}5 \\ -247 \\ \hline 358 \end{array}$$

8.
$$\begin{array}{r} {}^{5\ 9\ 9\ 12}\cancel{6}\cancel{0}\cancel{0}2 \\ -5843 \\ \hline 159 \end{array}$$

9.
$$\begin{array}{r} {}^{2\ 10\ 17}\cancel{3}\cancel{1}7 \\ -\ 59 \\ \hline 258 \end{array}$$
The balance carried forward is $258.

10. $14,880 = $10,999 + $3882
$14,880 = $14,881 is a false statement.
The price should be $10,998.

Exercises 1.4

1.
$$\begin{array}{r} 14 \\ -\ 6 \\ \hline 8 \end{array}$$

3.
$$\begin{array}{r} 13 \\ -\ 6 \\ \hline 7 \end{array}$$

5.
$$\begin{array}{r} 17 \\ -\ 0 \\ \hline 17 \end{array}$$

7.
$$\begin{array}{r} 17 \\ -\ 9 \\ \hline 8 \end{array}$$

9.
$$\begin{array}{r} 66 \\ -32 \\ \hline 34 \end{array}$$

11.
$$\begin{array}{r} 83 \\ -20 \\ \hline 63 \end{array}$$

13.
$$\begin{array}{r} {}^{8\ 10} \\ \cancel{9}0 \\ -56 \\ \hline 34 \end{array}$$

15.
$$\begin{array}{r} {}^{6\ 13} \\ \cancel{7}3 \\ -58 \\ \hline 15 \end{array}$$

17.
$$\begin{array}{r} {}^{4\ 10} \\ \cancel{5}03 \\ -291 \\ \hline 212 \end{array}$$

19.
$$\begin{array}{r} {}^{6\ 18} \\ 5\cancel{7}8 \\ -499 \\ \hline 9 \end{array} \rightarrow \begin{array}{r} {}^{4\ 16} \\ \cancel{5}\cancel{7}8 \\ -499 \\ \hline \boxed{79} \end{array}$$

21.
$$\begin{array}{r} {}^{71\ 15} \\ \cancel{7}25 \\ -318 \\ \hline 407 \end{array}$$

23.
$$\begin{array}{r} {}^{86\ 12} \\ \cancel{8}\cancel{7}2 \\ -657 \\ \hline 215 \end{array}$$

25.
$$\begin{array}{r} {}^{5\ 10} \\ 5\cancel{6}0 \\ -278 \\ \hline 2 \end{array} \rightarrow \begin{array}{r} {}^{4\ 15} \\ \cancel{5}\cancel{6}0 \\ -278 \\ \hline \boxed{282} \end{array}$$

27.
$$\begin{array}{r} {}^{89\ 15} \\ 9\cancel{0}5 \\ -726 \\ \hline 179 \end{array}$$

29.
$$\begin{array}{r} {}^{59\ 17} \\ 6\cancel{0}7 \\ -398 \\ \hline 209 \end{array}$$

31.
$$\begin{array}{r} 5837 \\ -3216 \\ \hline 2621 \end{array}$$

33.
$$\begin{array}{r} {}^{477\ 13} \\ 4\cancel{7}83 \\ -1278 \\ \hline 3505 \end{array}$$

35.
$$\begin{array}{r} {}^{684\ 13} \\ 68\cancel{5}3 \\ -2765 \\ \hline 8 \end{array} \rightarrow \begin{array}{r} {}^{67\ 14} \\ 6\cancel{8}\cancel{5}3 \\ -2765 \\ \hline \boxed{4088} \end{array}$$

37.
$$\begin{array}{r} {}^{2\ 12} \\ 5\cancel{3}25 \\ -2432 \\ \hline 93 \end{array} \rightarrow \begin{array}{r} {}^{4\ 12} \\ 5\cancel{3}25 \\ -2432 \\ \hline \boxed{2893} \end{array}$$

39.
$$
\begin{array}{c}
\overset{5\ 10}{38\cancel{6}0} \\
-2971 \\
\hline
9
\end{array}
\rightarrow
\begin{array}{c}
\overset{7\ 15}{3\cancel{8}\cancel{6}0} \\
-2971 \\
\hline
89
\end{array}
\rightarrow
\begin{array}{c}
\overset{2\ 17}{\cancel{3}\cancel{8}\cancel{6}0} \\
-2971 \\
\hline
\boxed{889}
\end{array}
$$

41.
$$
\begin{array}{c}
\overset{2\ 14}{76\cancel{3}4} \\
-\ 388 \\
\hline
6
\end{array}
\rightarrow
\begin{array}{c}
\overset{5\ 12}{7\cancel{6}\cancel{3}4} \\
-\ \ 388 \\
\hline
\boxed{7246}
\end{array}
$$

43.
$$
\begin{array}{c}
\overset{49\ 17}{\cancel{5}\cancel{0}73} \\
-\ 782 \\
\hline
4291
\end{array}
$$

45.
$$
\begin{array}{c}
\overset{599\ 13}{\cancel{6}\cancel{0}\cancel{0}3} \\
-\ 289 \\
\hline
5714
\end{array}
$$

47.
$$
\begin{array}{c}
\overset{3\ 15}{13,\cancel{4}56} \\
-\ 7,576 \\
\hline
80
\end{array}
\rightarrow
\begin{array}{c}
\overset{12\ 13}{\cancel{1}\cancel{3},\cancel{4}56} \\
-\ 7,576 \\
\hline
\boxed{5,880}
\end{array}
$$

49.
$$
\begin{array}{c}
\overset{4999\ 10}{5\cancel{0},\cancel{0}\cancel{0}0} \\
-23,569 \\
\hline
26,431
\end{array}
$$

51.
$$
\begin{array}{c}
\overset{20\ 10}{\cancel{2}\cancel{1}17} \\
-1250 \\
\hline
857
\end{array}
$$
The difference is 867 feet.

53.
$$
\begin{array}{c}
\overset{195\ 12}{\cancel{1}\cancel{9}\cancel{6}2} \\
-1958 \\
\hline
4
\end{array}
$$
She was 4 years old.

55.
$$
\begin{array}{c}
\overset{0\ 12}{22\cancel{1}2} \\
-\ 193 \\
\hline
9
\end{array}
\rightarrow
\begin{array}{c}
\overset{1\ 10}{2\cancel{2}\cancel{1}2} \\
-\ \ 193 \\
\hline
2019
\end{array}
$$
The difference in heights is 2019 feet.

57.
$$
\begin{array}{c}
\overset{19\ 10}{3\cancel{2}00} \\
-2999 \\
\hline
01
\end{array}
\rightarrow
\begin{array}{c}
\overset{2\ 11}{\cancel{3}\cancel{2}00} \\
-2999 \\
\hline
201
\end{array}
$$
You can save \$201.

59. $6300 - 4500 - 787 = 1013$
She made \$1013 profit on the car.

61.
$$
\begin{array}{r}
20,000 \\
-\ 4,672 \\
\hline
15,328 \\
-\ 1,500 \\
\hline
13,828 \\
-\ 1,383 \\
\hline
12,445
\end{array}
$$
The real buying power is \$12,445.

63. a.
$$
\begin{array}{r}
19,922 \\
-\ \ \ 500 \\
\hline
19,422
\end{array}
\rightarrow
\begin{array}{r}
19,422 \\
-\ \ \ 447 \\
\hline
18,975
\end{array}
$$
The dealer's cost is \$18,975.

b.
$$
\begin{array}{r}
22,000 \\
-18,975 \\
\hline
3,025
\end{array}
$$
The dealer is making \$3025.

65.
$$
\begin{array}{c}
\overset{9\ 11}{\cancel{1}\cancel{0}19} \\
-\ 968 \\
\hline
51
\end{array}
$$
The difference in price is \$51.

67.
$$
\begin{array}{c}
\overset{5\ 18}{9\cancel{6}8} \\
-689 \\
\hline
9
\end{array}
\rightarrow
\begin{array}{c}
\overset{8\ 15}{\cancel{9}\cancel{6}8} \\
-689 \\
\hline
279
\end{array}
$$
The difference in price is \$279.

69.
$$
\begin{array}{c}
\overset{9\ 11}{\cancel{1}\cancel{0}19} \\
-\ 689 \\
\hline
330
\end{array}
$$
You need \$330 more.

71.
$$\begin{array}{r} \overset{2\ 13}{\cancel{3}3,600} \\ -29,500 \\ \hline 4,100 \end{array}$$
The difference is \$4100.

73.
$$\begin{array}{r} \overset{5\ 13}{4\cancel{6},300} \\ -2\ 9,500 \\ \hline 800 \end{array} \rightarrow \begin{array}{r} \overset{3\ 15}{\cancel{4}\cancel{6},300} \\ -2\ 9,500 \\ \hline 16,800 \end{array}$$
The difference is \$16,800.

75. Non-Hispanic White and White are closest in their household income.

77. $345 + 5400 = 5745$

79. $2348 = 2000 + 300 + 40 + 8$

81.

	Dollars	Cents
BAL BROTFORD	260	
AMT DEPOSITED		
——	——	
TOTAL		
AMT THIS CHECK	50	
BAL CARDFORD	210	

83.

	Dollars	Cents
BAL BROTFORD	150	
AMT DEPOSITED		
——	——	
TOTAL		
AMT THIS CHECK	28	
BAL CARDFORD	122	

85.
$$\begin{array}{r} 442 \\ -363 \\ \hline 79°F \end{array}$$

87.
$$\begin{array}{r} 800 \\ -468 \\ \hline 332°F \end{array}$$

89.
$$\begin{array}{r} 800 \\ -442 \\ \hline 358°F \end{array}$$

91. Answers may vary.

93. Answers may vary.

95.
$$\begin{array}{r} 5x^2 + 8x + 7 \\ -(3x^2 + 6x + 2) \\ \hline \end{array} \rightarrow \begin{array}{r} 5x^2 + 8x + 7 \\ -3x^2 - 6x - 2 \\ \hline \boxed{2x^2 + 2x + 5} \end{array}$$

97.
$$\begin{array}{r} 3x^3 + 5x^2 + 7x + 9 \\ - \quad (2x^2 + 5x + 3) \\ \hline \end{array} \rightarrow$$
$$\begin{array}{r} 3x^3 + 5x^2 + 7x + 9 \\ - \quad (2x^2 + 5x + 3) \\ \hline \boxed{3x^3 + 3x^2 + 2x + 6} \end{array}$$

Mastery Test 1.4

99. $857 - 62 = 795$

101. $703 - 257 = 446$

103. $\$347 - \$59 = \$288$

Section 1.5 – Multiplication

Problems

1.
$$\begin{array}{r} 23 \\ \times\ 6 \\ \hline 18 \\ 120 \\ \hline 138 \end{array}$$

2.
$$\begin{array}{r} 53 \\ \times\ 6 \\ \hline 18 \\ 300 \\ \hline 318 \end{array}$$

$$\begin{array}{r} \overset{1}{}52 \\ \times\ 38 \\ \hline 416 \\ 156 \\ \hline 1976 \end{array}$$

3.

$$\begin{array}{r} \overset{1}{\overset{\diagup}{213}} \\ \times\ \ 514 \\ \hline 852 \\ 213 \\ 1065 \\ \hline 109,482 \end{array}$$

4.

$$\begin{array}{r} \overset{2}{304} \\ \times\ \ 512 \\ \hline 608 \\ 304 \\ 1520 \\ \hline 155,648 \end{array}$$

5.

$$\begin{array}{r} \overset{2}{\overset{\diagup}{290}} \\ \times\ \ 134 \\ \hline 1160 \\ 870 \\ 290 \\ \hline 38,860 \end{array}$$

6.

$$\begin{array}{r} \overset{1}{620} \\ \times\ \ 318 \\ \hline 4960 \\ 620 \\ 1860 \\ \hline 197,160 \end{array}$$

7.

8. a. $1000 \times 5 = 5000$
   b. $40 \times 90 = 3600$
   c. $700 \times 80 = 56,000$

$$\begin{array}{r} \overset{1}{120} \\ \times\ \ \ 4 \\ \hline 480 \\ \times\ \ 12 \\ \hline 960 \\ 480 \\ \hline 5760 \end{array}$$

9.

10. a. 1 quart = 2 pints
        = 2(15 berries) = 30 berries
    b. 16 quarts = 2(8 quarts)
           = 2(1 flat)
           = 2(12 pounds) = 24 pounds
    c. 8000 acres × 2000 flats × \$10
        = \$160,000,000 or \$160 million

11. $A = 40 \text{ ft} \times 38 \text{ ft}$
      $= 8 \times 40 + 30 \times 40$
      $= 320 + 1200$
      $= 1520 \text{ ft}^2$

Exercises 1.5

1. a. $\begin{array}{r} 3 \\ \times 4 \\ \hline 12 \end{array}$ **b.** $\begin{array}{r} 3 \\ \times 7 \\ \hline 21 \end{array}$

3. a. $\begin{array}{r} 9 \\ \times 1 \\ \hline 9 \end{array}$ **b.** $\begin{array}{r} 8 \\ \times 1 \\ \hline 8 \end{array}$

5. a. $\begin{array}{r} 5 \\ \times 8 \\ \hline 40 \end{array}$ **b.** $\begin{array}{r} 9 \\ \times 8 \\ \hline 72 \end{array}$

7. a. $\begin{array}{r} 0 \\ \times 0 \\ \hline 0 \end{array}$ **b.** $\begin{array}{r} 0 \\ \times 3 \\ \hline 0 \end{array}$

9. a. $\begin{array}{r} 1 \\ \times 8 \\ \hline 8 \end{array}$ **b.** $\begin{array}{r} 1 \\ \times 4 \\ \hline 4 \end{array}$

11. $6 \cdot 8 = 48$

13. $1 \cdot 5 = 5$

15. $9 \cdot 9 = 81$

17. $0 \cdot 1 = 0$

19. $4 \cdot 4 = 16$

21.
$$\begin{array}{r} 10 \\ \times\ 9 \\ \hline 90 \end{array}$$

23.
$$\begin{array}{r} 20 \\ \times\ 8 \\ \hline 160 \end{array}$$

25.
$$\begin{array}{r} {}^{1} \\ 53 \\ \times\ 6 \\ \hline 318 \end{array}$$

27.
$$\begin{array}{r} {}^{5} \\ 48 \\ \times\ 17 \\ \hline 336 \\ 480 \\ \hline 816 \end{array}$$

29.
$$\begin{array}{r} 608 \\ \times\ 32 \\ \hline 1216 \\ 1824 \\ \hline 19,456 \end{array}$$

31.
$$\begin{array}{r} 1234 \\ \times\ 3 \\ \hline 3702 \end{array}$$

33.
$$\begin{array}{r} {}^{3\,1} \\ 35,209 \\ \times\ 16 \\ \hline 211254 \\ 35209 \\ \hline 563,344 \end{array}$$

35.
$$\begin{array}{r} {}^{1} \\ 63 \\ \times\ 40 \\ \hline 2520 \end{array}$$

37.
$$\begin{array}{r} {}^{2\,4} \\ 249 \\ \times\ 50 \\ \hline 12,450 \end{array}$$

39.
$$\begin{array}{r} {}^{1\,2} \\ 346 \\ \times\ 420 \\ \hline 6920 \\ 1384 \\ \hline 145,320 \end{array}$$

41.
$$\begin{array}{r} {}^{1} \\ 2260 \\ \times\ 200 \\ \hline 452,000 \end{array}$$

43.
$$\begin{array}{r} 3020 \\ \times\ 405 \\ \hline 15100 \\ 12080 \\ \hline 1,223,100 \end{array}$$

45. $20 \cdot 5 = 2 \cdot 5 \times 10 = 10 \times 10 = 100$

47. $700 \times 80 = 7 \cdot 8 \times 100 = 56 \times 100 = 56,000$

49. $300 \cdot 200 = 3 \cdot 2 \times 10,000$
$= 6 \times 10,000$
$= 60,000$

51. $25 \times 8 = 200$ feet

53. a. 20 seconds $= 20 \times 1$ second
$= 20 \times 300,000$ km
$= 2 \cdot 3 \times 1,000,000$ km
$= 6 \times 1,000,000$ km
$= 6,000,000$ km

b. 25 sec $= 25 \times 1$ second
$= 25 \times 300,000$ km
$= 25 \cdot 3 \times 100,000$ km
$= 75 \times 100,000$ km
$= 7,500,000$ km

c. 30 seconds $= 30 \times 1$ second
$$= 30 \times 300{,}000 \text{ km}$$
$$= 3 \cdot 3 \times 1{,}000{,}000 \text{ km}$$
$$= 9 \times 1{,}000{,}000 \text{ km}$$
$$= 9{,}000{,}000 \text{ km}$$

55. $206 \times 4 \times 12 \rightarrow$

$$
\begin{array}{r}
206 \\
\times\ \ 4 \\
\hline
824 \\
\times\ 12 \\
\hline
1648 \\
824 \\
\hline
9888
\end{array}
$$

You end up paying \$9888 for the \$8000 you borrowed.

57. a. 5 yrs $= 5 \times 52$ weeks $= 260$ weeks

    b. 10 yrs $= 10 \times 52$ weeks $= 520$ weeks

59. 8 yrs $= 8 \times 8$ billion barrels
$$= 64 \text{ billion barrels}$$

61. $A = 30 \text{ ft} \times 10 \text{ ft} = 300 \text{ ft}^2$
The area is 300 ft^2.

63. $A = 90 \text{ ft} \times 90 \text{ ft} = 8100 \text{ ft}^2$
The area is 8100 ft^2.

65. $A = 360 \text{ ft} \times 160 \text{ ft} =$

$$
\begin{array}{r}
360 \\
\times\ 160 \\
\hline
21600 \\
360 \\
\hline
57{,}600 \text{ ft}^2
\end{array}
$$

67. $234 = 200 + 30 + 4$

69.
$$
\begin{array}{r}
^{1\ 1} \\
349 \\
+\ 786 \\
\hline
1135
\end{array}
$$

71.
$$
\begin{array}{r}
^{49\ 10}\ \ ^{2\ 14} \\
3{,}5\!\!\!/0\!\!\!/0 \\
-\ 728 \\
\hline
72
\end{array}
\rightarrow
\begin{array}{r}
\!\!\!/3\ 5\!\!\!/0\!\!\!/0 \\
-\ 7\ 2\ 8 \\
\hline
\boxed{2\ 7\ 7\ 2}
\end{array}
$$

73. Ideal weight $= 72 \times 4 - 130$
$$= 288 - 130$$
$$= 158 \text{ lb}$$

75. a. Calories needed $= 150 \times 15$
$$= 15 \cdot 15 \times 10$$
$$= 225 \times 10$$
$$= 2250$$

    b. Calories needed $= 150 \times 17$
$$= 15 \cdot 17 \times 10$$
$$= 255 \times 10$$
$$= 2550$$

77. Answers may vary.

79. Answers may vary.

81.
$$
\begin{array}{r}
2x + 7 \\
\times\ \ \ \ \ 5 \\
\hline
10x + 35
\end{array}
$$

83.
$$
\begin{array}{r}
3x + 1 \\
\times\ 2x + 2 \\
\hline
6x + 2 \\
6x^2 + 2x \\
\hline
6x^2 + 8x + 2
\end{array}
$$

Mastery Test 1.5

85. $210 \times 5 \times 12 =$

$$
\begin{array}{r}
210 \\
\times\ \ \ 5 \\
\hline
1050 \\
\times\ \ 12 \\
\hline
2100 \\
1050 \\
\hline
12{,}600
\end{array}
$$

87.
$$
\begin{array}{r}
^{3} \\
3\!\!\!/19 \\
\times\ \ 450 \\
\hline
15950 \\
1276 \\
\hline
143{,}550
\end{array}
$$

89.
$$\begin{array}{r} \overset{2}{} \\ \overset{2}{\cancel{7}} \\ 129 \\ \times\ 318 \\ \hline 1032 \\ 129 \\ 387 \\ \hline 41,022 \end{array}$$

91.
$$\begin{array}{r} \overset{4}{56} \\ \times\ 7 \\ \hline 392 \end{array}$$

93. 1 quart = 2 pints
$$= 2(15 \text{ berries}) = 30 \text{ berries}$$

Section 1.6 – Division

Problems

1. a. $63 \div 9 = \square \Rightarrow 63 = 9 \times \square$

Since $9 \times 7 = 63$, $63 \div 9 = \boxed{7}$

b. $56 \div 8 = \square \Rightarrow 56 = 8 \times \square$

Since $8 \times 7 = 56$, $56 \div 8 = \boxed{7}$

2. a. $9 \div 9 = \square \Rightarrow 9 = 9 \times \square$

Since $9 \times 1 = 9$, $9 \div 9 = \boxed{1}$

b. $9 \div 1 = \square \Rightarrow 9 = 1 \times \square$

Since $1 \times 9 = 9$, $9 \div 1 = \boxed{9}$

c. $0 \div 9 = \square \Rightarrow 0 = 9 \times \square$

Since $9 \times 0 = 0$, $0 \div 9 = \boxed{0}$

d. $9 \div 0 = \square \Rightarrow 9 = 0 \times \square$

Since there is *no* number such that $9 = 0 \times \square$, $9 \div 0$ is not defined.

3.
$$\begin{array}{r} 131 \\ 7\overline{)917} \\ -7 \\ \hline 21 \\ -21 \\ \hline 07 \\ -7 \\ \hline 0 \end{array}$$
so $917 \div 7 = 131$

4.
$$\begin{array}{r} 341 \\ 4\overline{)1367} \\ -12 \\ \hline 16 \\ -16 \\ \hline 7 \\ -4 \\ \hline 3 \end{array}$$
so $1367 \div 4 = 341 \text{ r } 3$

5.
$$\begin{array}{r} 101 \\ 7\overline{)709} \\ -7 \\ \hline 009 \\ -7 \\ \hline 2 \end{array}$$
so $709 \div 7 = 101 \text{ r } 2$

6.
$$\begin{array}{r} 22 \\ 45\overline{)1029} \\ -90 \\ \hline 129 \\ -90 \\ \hline 39 \end{array}$$
so $1029 \div 45 = 22 \text{ r } 39$

7.
$$\begin{array}{r} 750 \\ 26\overline{)19,500} \\ -182 \\ \hline 130 \\ -130 \\ \hline 00 \end{array}$$

The biweekly salary will be $750.

Exercises 1.6

1. $30 \div 5 = 6$

3. $28 \div 4 = 7$

5. $21 \div 7 = 3$

7. $0 \div 2 = 0$

9. $7 \div 7 = 1$

11. $\dfrac{36}{9} = 4$

13. $\dfrac{3}{0}$ is not defined

15. $\dfrac{32}{8} = 4$

17. $\dfrac{24}{1} = 24$

19. $\dfrac{62}{6} = 10 \text{ r } 2$

21.
$$
\begin{array}{r}
61 \\
6{\overline{)366}} \\
-36 \\
\hline
06 \\
-\;06 \\
\hline
0
\end{array}
$$
answer: 61

23.
$$
\begin{array}{r}
631 \\
8{\overline{)5048}} \\
-48 \\
\hline
24 \\
-24 \\
\hline
08 \\
-\;8 \\
\hline
0
\end{array}
$$
answer: 631

25.
$$
\begin{array}{r}
513 \\
4{\overline{)2055}} \\
-20 \\
\hline
05 \\
-\;4 \\
\hline
15 \\
-12 \\
\hline
3
\end{array}
$$
answer: 513 r 3

27.
$$
\begin{array}{r}
24 \\
14{\overline{)336}} \\
-28 \\
\hline
56 \\
-56 \\
\hline
0
\end{array}
$$
answer: 24

29.
$$
\begin{array}{r}
21 \\
19{\overline{)399}} \\
-38 \\
\hline
19 \\
-19 \\
\hline
0
\end{array}
$$
answer: 21

31.
$$
\begin{array}{r}
60 \\
10{\overline{)605}} \\
-60 \\
\hline
05 \\
-0 \\
\hline
5
\end{array}
$$
answer: 60 r 5

33.
$$
\begin{array}{r}
44 \\
16{\overline{)704}} \\
-64 \\
\hline
64 \\
-64 \\
\hline
0
\end{array}
$$
answer: 44

35.
$$
\begin{array}{r}
9 \\
81{\overline{)805}} \\
-729 \\
\hline
76
\end{array}
$$
answer: 9 r 76

37.
$$
\begin{array}{r}
42 \\
12{\overline{)505}} \\
-48 \\
\hline
25 \\
-24 \\
\hline
1
\end{array}
$$
answer: 42 r 1

39.
$$
\begin{array}{r}
59 \\
22{\overline{)1305}} \\
-110 \\
\hline
205 \\
-198 \\
\hline
7
\end{array}
$$
answer: 59 r 7

41.
$$
\begin{array}{r}
214 \\
42{\overline{)9013}} \\
-84 \\
\hline
61 \\
-42 \\
\hline
193 \\
-168 \\
\hline
25
\end{array}
$$
answer: 214 r 25

43.
$$\begin{array}{r} 45 \\ 123\overline{)5583} \\ \underline{-492} \\ 663 \\ \underline{-615} \\ 48 \end{array}$$
answer: 45 r 48

45.
$$\begin{array}{r} 87 \\ 417\overline{)36,279} \\ \underline{-3336} \\ 2919 \\ \underline{-2919} \\ 0 \end{array}$$
answer: 87

47.
$$\begin{array}{r} 630 \\ 50\overline{)31,500} \\ \underline{-300} \\ 150 \\ \underline{-150} \\ 00 \\ \underline{-\ 0} \\ 0 \end{array}$$
answer: 630

49.
$$\begin{array}{r} 934 \\ 654\overline{)611,302} \\ \underline{-5886} \\ 2270 \\ \underline{-1962} \\ 3082 \\ \underline{-2616} \\ 466 \end{array}$$
answer: 934 r 466

51.
$$\begin{array}{r} 504 \\ 25\overline{)12,600} \\ \underline{-125} \\ 10 \\ \underline{-\ 0} \\ 100 \\ \underline{-100} \\ 0 \end{array}$$
There were 504 shares sold.

53. $\dfrac{42}{14} = 3$; Three loads were washed.

55.
$$\begin{array}{r} 600 \\ 52\overline{)31,200} \\ \underline{-312} \\ 000 \end{array}$$
The weekly salary is $600.

57. $\dfrac{11,600}{10} = 1160$ Btu per hour

59. $\dfrac{1500}{5} = 300$ words per minute

61. $345 < 354$

63.
$$\begin{array}{r} 1003 \\ \times\quad 305 \\ \hline 5015 \\ 3009\quad \\ \hline 305,915 \end{array}$$

65. $\dfrac{2,100,000}{40} = 52,500$

The average income is $52,500 per year.

67. $\dfrac{1,500,000}{40} = 37,500$

The average income is $37,500 per year.

69. $\dfrac{1,000,000}{40} = 25,000$

The average income is $25,000 per year.

71. $\dfrac{250 + 210 + 200 + 240}{4} = \dfrac{900}{4} = 225$

The average salary was $225 per week.

73. $\dfrac{24 + 25 + 24 + 26 + 26}{5} = \dfrac{125}{5} = 25$

The average price was $25 per share.

75. Answers may vary.

77. Answers may vary.

79.
$$\begin{array}{r} x+3 \\ x+1\overline{)x^2 + 4x + 5} \\ \underline{-x^2 - 3x} \\ x+5 \\ \underline{-x-3} \\ 2 \end{array}$$
answer: $(x+3)$ r 2

81.

$$x+3\overline{)x^2+4x+6}\quad\text{answer: }(x+1)\text{ r }3$$

$$\begin{array}{r}x+1\\[-2pt]\underline{-x^2-3x}\\x+6\\\underline{-x-3}\\3\end{array}$$

Mastery Test 1.6

83. a. $48\div6=\square$ becomes $48=6\times\square$; $\square=8$

    b. $37\div1=\square$ becomes $37=1\times\square$; $\square=37$

85. a. $9\div9=\square$ becomes $9=9\times\square$; $\square=1$

    b. $7\div1=\square$ becomes $7=1\times\square$; $\square=7$

87. a.

$$\begin{array}{r}132\\[-2pt]6\overline{)792}\\\underline{-6}\\19\\\underline{-18}\\12\\\underline{-12}\\0\end{array}$$

    b.

$$\begin{array}{r}192\\[-2pt]9\overline{)1728}\\\underline{-9}\\82\\\underline{-81}\\18\\\underline{-18}\\0\end{array}$$

    answer: 132　　　　answer: 192

Section 1.7 – Primes, Factors, and Exponents

Problems

1. a. $19=1\times19$ has exactly two factors so 19 is prime.

    b. $15=1\times15=3\times5$ has more than two factors so 15 is composite.

2. a. $13=1\times13$; 13 is the prime factor

    b. $12=1\times12=2\times6=3\times4$ so 1, 2, 3, 4, 6, and 12 are the factors of 12. Of these, 2 and 3 are prime factors.

3. a. $5\underline{|35}$ so $35=5\times7$
         7

    b. $2\underline{|16}$ so $16=2\times2\times2\times2$
       $2\underline{|8}$
       $2\underline{|4}$
         2

    c. 97 is not divisible by 2, 3, or 5.

    $\dfrac{97}{7}=13\text{ r }6$ and $\dfrac{97}{11}=8\text{ r }9$ so 97 is not divisible by 7 or 11. Thus, 97 is prime.

4. a. $3\underline{|27}$ so $27=3\times3\times3=3^3$
       $3\underline{|9}$
         3

    b. $2\underline{|98}$ so $98=2\times7\times7=2\times7^2$
       $7\underline{|49}$
         7

5. a. $2^2\times3^3=2\times2\times3\times3\times3=4\times27=108$

    b. $2^3\times5^2=2\times2\times2\times5\times5=8\times25=200$

6. a. $3^1\times4^2=3\times16=48$

    b. $2^2\times3^0\times5^3=4\times1\times125=500$

Exercises 1.7

1. $7=1\times7$; prime

3. $6=1\times6$ or 2×3. 6 is composite with factors 1, 2, 3, and 6.

5. $24=1\times24$ or 2×12 or 3×8 or 4×6. 24 is composite with factors 1, 2, 3, 4, 6, 8, 12 and 24.

7. $25=1\times25$ or 5×5. 25 is composite with factors 1, 5, and 25.

9. $23=1\times23$; prime

11. $14=1\times14$ or 2×7. The prime factors are 2 and 7.

13. $18=1\times18$ or 2×9 or 3×6. The prime factors are 2 and 3.

15. $29=1\times29$. The prime factor is 29.

17. $22 = 1 \times 22$ or 2×11. The prime factors are 2 and 11.

19. $21 = 1 \times 21$ or 3×7. The prime factors are 3 and 7.

21. $34 = 2 \times 17$

23. 41 is prime.

25.

$$\begin{array}{r} 2\overline{)64} \\ 2\overline{)32} \\ 2\overline{)16} \\ 2\overline{)8} \\ 2\overline{)4} \\ 2 \end{array}$$

so $64 = 2^6$

27. $\dfrac{91}{7} = 13$ so $91 = 7 \times 13$

29.

$$\begin{array}{r} 2\overline{)190} \\ 5\overline{)95} \\ 19 \end{array}$$

so $190 = 2 \times 5 \times 19$

31. $3^0 \times 2^2 = 1 \times 4 = 4$

33. $2^0 \times 10^0 = 1 \times 1 = 1$

35. $5^2 \times 2^2 = 25 \times 4 = 100$

37. $4^3 \times 2^1 \times 4^0 = 64 \times 2 \times 1 = 128$

39. $5^2 \times 2^3 \times 11^0 = 25 \times 8 \times 1 = 200$

41. The prime numbers less than 50 are 2, 3, 5, 7, 11, 13, 17, 19, 23, 29, 31, 37, 41, 43, and 47. Thus, there are 15 prime numbers less than 50.

43. $100,000,000 = 10^8$

45.

Number of folds	Number of Pieces Thick
4	$2^{\boxed{4}} = \boxed{16}$
5	$2^{\boxed{5}} = \boxed{32}$

47. $1,000,000,000 = 10^9$

49. $3 \times 10^4 = 3 \times 10,000 = 30,000$

51. $138 = 100 + 30 + 8$
$\quad = 1 \times 10^2 + 3 \times 10 + 8$

53. $1208 = 1000 + 200 + 8$
$\quad = 1 \times 10^3 + 2 \times 10^2 + 8$

55. a. 12; yes   **b.** 13; no   **c.** 20; yes

57. a. 125; yes   **b.** 301; no   **c.** 240; yes

59. a. 420: yes, since 20 is divisible by 4
b. 308: yes, since 08 is divisible by 4
c. 234: no, since 34 is not divisible by 4
d. 1236: yes, since 36 is divisible by 4

61. a. $\dfrac{424}{8} = 53$; yes, 1424 is divisible by 8

b. $\dfrac{630}{8} = 78 \text{ r } 6$; no, 1630 is not divisible by 8

c. $\dfrac{360}{8} = 45$; yes, 2360 is divisible by 8

d. $\dfrac{148}{8} = 18 \text{ r } 4$; no, 2148 is not divisible by 8

63. a. 450: yes   **b.** 432: no
c. 567: no   **d.** 980: yes

65. a. Odd. Reasons may vary. Sample answer: Every even number greater than two divisible by 2 and is therefore composite.
b. Answers may vary.
c. Answers may vary.

67. Answers may vary. Sample answer: The "number of factors" refer to all factors whereas the "number of prime" factors refer to only those factors that are prime. For example, 12 has 6 factors (1, 2, 3, 4, 6, and 12) but only 2 prime factors (2 and 3).

Mastery Test 1.7

69. a. $41 = 1 \times 41$; prime
   b. 39 is divisible by 3; composite

71. a. $3600 = 36 \times 100$
$= 4 \times 9 \times 4 \times 25$
$= 16 \times 9 \times 25 = 2^4 \times 3^2 \times 5^2$

b. $360 = 36 \times 10$
$= 4 \times 9 \times 2 \times 5$
$= 8 \times 9 \times 5 = 2^3 \times 3^2 \times 5$

73. $5^2 \times 2^2 \times 7^0 = 25 \times 4 \times 1 = 100$

Section 1.8 – Order of Operations and Grouping Symbols

Problems

1. a. $7 \cdot 2^3 - 7 = 7 \cdot 8 - 7$
$= 56 - 7$
$= 49$

b. $23 + 2^2 \cdot 5 = 23 + 4 \cdot 5$
$= 23 + 20$
$= 43$

2. a. $48 \div 6 - (3+1) = 48 \div 6 - 2$
$= 6 - 2$
$= 4$

b. $10 \div 2 \cdot 2 \cdot 2 + 2 - 1 = 5 \cdot 2 \cdot 2 + 2 - 1$
$= 20 + 2 - 1$
$= 22 - 1$
$= 21$

3. $6 \div 3 \cdot 2 + 2(5-3) - 2^2 \cdot 1$
$= 6 \div 3 \cdot 2 + 2(2) - 2^2 \cdot 1$
$= 6 \div 3 \cdot 2 + 2(2) - 4 \cdot 1$
$= 2 \cdot 2 + 2(2) - 4 \cdot 1$
$= 4 + 2(2) - 4 \cdot 1$
$= 4 + 4 - 4 \cdot 1$
$= 4 + 4 - 4$
$= 8 - 4$
$= 4$

4. $25 \div 5 + \{3 \cdot 2^2 - [5 + (4-1)]\}$
$= 25 \div 5 + \{3 \cdot 2^2 - [5 + 3]\}$
$= 25 \div 5 + \{3 \cdot 2^2 - 8\}$
$= 25 \div 5 + \{3 \cdot 4 - 8\}$
$= 25 \div 5 + \{12 - 8\}$
$= 25 \div 5 + 4$
$= 5 + 4$
$= 9$

5. Ideal rate $= [(205 - 35) \cdot 7] \div 10$
$= [170 \cdot 7] \div 10$
$= 1190 \div 10$
$= 119$

6. a. rental fee + cost for 40 people – coupon
150 + 40(25) – 100
b. $150 + 40(25) - 100 = 150 + 1000 - 100$
$= 1150 - 100$
$= 1050$
The total cost of the party is $1050.

Exercises 1.8

1. $4 \cdot 5 + 6 = 20 + 6$
$= 26$

3. $7 + 3 \cdot 2 = 7 + 6$
$= 13$

5. $7 \cdot 8 - 3 = 56 - 3$
$= 53$

7. $20 - 3 \cdot 5 = 20 - 15$
$= 5$

9. $48 \div 6 - (3+2) = 48 \div 6 - 5$
$= 8 - 5$
$= 3$

11. $3 \cdot 4 \div 2 + (6-2) = 3 \cdot 4 \div 2 + 4$
$= 12 \div 2 + 4$
$= 6 + 4$
$= 10$

13. $6 \div 3 \cdot 3 \cdot 3 + 4 - 1 = 2 \cdot 3 \cdot 3 + 4 - 1$
$$= 6 \cdot 3 + 4 - 1$$
$$= 18 + 4 - 1$$
$$= 22 - 1$$
$$= 21$$

15. $8 \div 2 \cdot 2 \cdot 2 - 3 + 5 = 4 \cdot 2 \cdot 2 - 3 + 5$
$$= 8 \cdot 2 - 3 + 5$$
$$= 16 - 3 + 5$$
$$= 13 + 5$$
$$= 18$$

17. $10 \div 5 \cdot 2 + 8 \cdot (6 - 4) - 3 \cdot 4$
$$= 10 \div 5 \cdot 2 + 8 \cdot 2 - 3 \cdot 4$$
$$= 2 \cdot 2 + 8 \cdot 2 - 3 \cdot 4$$
$$= 4 + 8 \cdot 2 - 3 \cdot 4$$
$$= 4 + 16 - 3 \cdot 4$$
$$= 4 + 16 - 12$$
$$= 20 - 12$$
$$= 8$$

19. $4 \cdot 8 \div 2 - 3(4 - 1) + 9 \div 3$
$$= 4 \cdot 8 \div 2 - 3(3) + 9 \div 3$$
$$= 32 \div 2 - 3(3) + 9 \div 3$$
$$= 16 - 3(3) + 9 \div 3$$
$$= 16 - 9 + 9 \div 3$$
$$= 16 - 9 + 3$$
$$= 7 + 3$$
$$= 10$$

21. $20 \div 5 + \{3 \cdot 4 - [4 + (5 - 3)]\}$
$$= 20 \div 5 + \{3 \cdot 4 - [4 + 2]\}$$
$$= 20 \div 5 + \{3 \cdot 4 - 6\}$$
$$= 20 \div 5 + \{12 - 6\}$$
$$= 20 \div 5 + 6$$
$$= 4 + 6$$
$$= 10$$

23. $(20 - 15) \cdot [20 \div 2 - (2 \cdot 2 + 2)]$
$$= 5 \cdot [20 \div 2 - (4 + 2)]$$
$$= 5 \cdot [20 \div 2 - 6]$$
$$= 5 \cdot [10 - 6]$$
$$= 5 \cdot 4$$
$$= 20$$

25. $\{4 \div 2 \cdot 6 - (3 + 2 \cdot 3) + [5(3 + 2) - 1]\}$
$$= \{4 \div 2 \cdot 6 - (3 + 2 \cdot 3) + [5(5) - 1]\}$$
$$= \{4 \div 2 \cdot 6 - (3 + 2 \cdot 3) + [25 - 1]\}$$
$$= \{4 \div 2 \cdot 6 - (3 + 2 \cdot 3) + 24\}$$
$$= \{4 \div 2 \cdot 6 - (3 + 6) + 24\}$$
$$= \{4 \div 2 \cdot 6 - 9 + 24\}$$
$$= \{2 \cdot 6 - 9 + 24\}$$
$$= \{12 - 9 + 24\}$$
$$= 3 + 24$$
$$= 27$$

27. **a.** pay for 3 hours + cost for gas and oil
$$3(10) \quad + \quad 2$$
b. $3(10) + 2 = 30 + 2$
$$= 32$$
The total earnings is $32.

29. $9 \cdot 15 = 135$
$135 \div 510 \cdot 100 = 26.4706$
About 26% of the calories in this meal come from fat.

31. $9 \cdot 19 = 171$
$171 \div 380 \cdot 100 = 45$
About 45% of the calories in this meal come from fat.

33. $4 \cdot 8 = 32$
$32 \div 36 \cdot 100 = 88.889$
About 89% of the calories in this dish come from carbohydrates.

35. $4 \cdot 150 = 600$
$600 \div 1380 \cdot 100 = 43.478$
About 43% of the calories in this meal come from carbohydrates.

37. $2^4 \cdot 3^2 = 16 \cdot 9 = 144$

39. $2^3 \cdot 3^2 \cdot 10^2 = 8 \cdot 9 \cdot 100 = 7200$;
Seven thousand, two hundred

41. child's dose $= (10 \cdot 75) \div 150$
$$= 750 \div 150$$
$$= 5 \text{ mg}$$

43. child's dose $= (6 \cdot 4) \div (6 + 2)$
$$= 24 \div 8$$
$$= 3 \text{ tablets every 12 hours}$$

45. Here is one: $(8 \div 2)(3 + 3)$

47. Here is one: $(8 \div 2)(9 - 3)$

49. Multiplying first gives
$$4/2 \cdot 2 + 3 = 4/4 + 3$$
$$= 1 + 3$$
$$= 4.$$
Performing the operation from left to right gives
$$4/2 \cdot 2 + 3 = 2 \cdot 2 + 3$$
$$= 4 + 3$$
$$= 7.$$

51. $(x + y)(x + z) = (3 + 2)(3 + 4)$
$$= (5)(7)$$
$$= 35$$

53. $x \cdot z \div y - z = 3 \cdot 4 \div 2 - 4$
$$= 12 \div 2 - 4$$
$$= 6 - 4$$
$$= 2$$

55. $8 \cdot (z \div y + x - z) = 8 \cdot (4 \div 2 + 3 - 4)$
$$= 8 \cdot (2 + 3 - 4)$$
$$= 8 \cdot (5 - 4)$$
$$= 8 \cdot 1$$
$$= 8$$

Mastery Test 1.8

57. $2^3 \div 8 \cdot 2 + 4(6 - 1) - 2 \cdot 3$
$$= 2^3 \div 8 \cdot 2 + 4(5) - 2 \cdot 3$$
$$= 8 \div 8 \cdot 2 + 4(5) - 2 \cdot 3$$
$$= 1 \cdot 2 + 4(5) - 2 \cdot 3$$
$$= 2 + 20 - 6$$
$$= 16$$

59. $27 \div 3 \cdot 3 \cdot 3 + 4 - 1 = 9 \cdot 3 \cdot 3 + 4 - 1$
$$= 27 \cdot 3 + 4 - 1$$
$$= 81 + 4 - 1$$
$$= 85 - 1$$
$$= 84$$

61. $2^3 + 4 \cdot 5 = 8 + 4 \cdot 5$
$$= 8 + 20$$
$$= 28$$

63. a. set up fee + cost for 10 people – discount
$$50 + 10(500) - 100$$
b. $50 + 10(500) - 100 = 50 + 5000 - 100$
$$= 5050 - 100$$
$$= 4950$$
The total cost of the cruise is $4950.

Section 1.9 – Equations and Problem Solving

Problems

1. a. $x + 6 = 15$
If we replace x with 9, we get $9 + 6 = 15$, which is true. Thus, the solution is $x = 9$.

b. $13 - x = 4$
If we replace x with 9, we get $13 - 9 = 4$, which is true. Thus, the solution is $x = 9$.

c. $24 = 8x$. We need a number that multiplied by 8 would give 24. The number is 3. Thus, the solution is $x = 3$.

d. $36 \div x = 9$. We need a number x so when 36 is divided by this number it gives 9. The number is 4. Thus, the solution is $x = 4$.

2. a.
$$n - 13 = 17$$
$$n - 13 + 13 = 17 + 13$$
$$n = 30$$
The solution is $n = 30$.

b.
$$20 = m - 3$$
$$20 + 3 = m - 3 + 3$$
$$23 = m$$
The solution is $m = 23$.

3. a.
$$n + 10 = 13$$
$$n + 10 - 10 = 13 - 10$$
$$n = 3$$
The solution is $n = 3$.

b.　　$39 = 18 + m$
　　　　$39 - 18 = 18 + m - 18$
　　　　　　$21 = m$
　　The solution is $m = 21$.

4. a.　　$4x = 36$
　　　　$4x \div 4 = 36 \div 4$
　　　　　　$x = 9$
　　The solution is $x = 9$.
b.　　　$42 = 7x$
　　　　$42 \div 7 = 7x \div 7$
　　　　　　$6 = x$
　　The solution is $x = 6$.

5. Let h = height of the antenna.
Tower ht. + antenna ht. reached 1710
　　　$1559 + h = 1710$
　$1559 + h - 1559 = 1710 - 1559$
　　　　　　$h = 151$
The height of the antenna is 151 feet.
Check: Tower ht. + antenna ht. = 1710
　　　　$1559 + 151 = 1710$ ✓

Exercises 1.9

1. $m + 9 = 17$. What number added to 9 gives 17? The number is 8; $m = 8$.

3. $13 - x = 9$. What number subtracted from 13 gives 9? The number is 4; $x = 4$.

5. $20 = 4x$. What number multiplied by 4 gives 20? The number is 5; $x = 5$.

7. $9x = 54$. What number multiplied by 9 gives 54? The number is 6; $x = 6$.

9. $30 \div y = 6$. What number divided into 30 gives 6? The number is 5; $x = 5$.

11. $9 = 63 \div t$. What number divided into 63 gives 9? The number is 7; $x = 7$.

13.　　$z - 18 = 30$
　　$z - 18 + 18 = 30 + 18$
　　　　$z = 48$

15.　　$40 = p - 12$
　　$40 + 12 = p - 12 + 12$
　　　　$52 = p$

17.　　$x + 17 = 31$
　　$x + 17 - 17 = 31 - 17$
　　　　$x = 14$

19.　　$30 = 17 + m$
　　$30 - 17 = 17 + m - 17$
　　　　$13 = m$

21.　　$4x = 8$
　　$4x \div 4 = 8 \div 4$
　　　　$x = 2$

23.　　$9x = 36$
　　$9x \div 9 = 36 \div 9$
　　　　$x = 4$

25. s = speed of Thrust 2.
"Th" speed is 130 more than "Th 2" speed
　　　$763 = s + 130$
　$763 - 130 = s + 130 - 130$
　　　　$633 = s$
The speed of the Thrust 2 is 633 mph.

27. d = distance flown by Glenn
Bryan's dist. is 15 miles less than Glenn's
　　　$22 = d - 15$
　$22 + 15 = d - 15 + 15$
　　　$37 = d$
Glenn flew 37 miles.

29. w = weight of an African elephant
brontos wt. is 4 times African elephant wt.
　　　$60,000 = 4w$
　$60,000 \div 4 = 4w \div 4$
　　　$15,000 = w$
The weight of an African elephant is 15,000 lb.

31. Let c = number of calories in Coke
Q.P. calories + coke calories reaches 570
　　　$420 + c = 570$
　$420 + c - 420 = 570 - 420$
　　　　$c = 150$
There are 150 calories in the Coke.

33. Let c = number of calories in Coke
Whopper cal. + coke cal. reaches 920
$$640 + c = 920$$
$$640 + c - 640 = 920 - 640$$
$$c = 280$$
There are 280 calories in the Coke.

35. Let f = number of calories in the fries;
then $f + 120$ = no. of calories in the CB
fries cal. + cheeseburger cal. is 540
$$f + (f + 120) = 540$$
$$f + f + 120 = 540$$
$$2f + 120 = 540$$
$$2f + 120 - 120 = 540 - 120$$
$$2f = 420$$
$$2f \div 2 = 420 \div 2$$
$$f = 210$$
and so $f + 120 = 210 + 120 = 330$
There are 210 calories in the fries and 330
calories in the cheeseburger.

37. Let t = cost of tuition
books & supplies + tuition totals 2272
$$700 + t = 2272$$
$$700 + t - 700 = 2272 - 700$$
$$t = 1572$$
The cost of tuition is $1572.

39. Let g = grant money awarded; then
$g - 400$ = scholarships awarded
scholarships + grants = 3600
$$(g - 400) + g = 3600$$
$$2g - 400 = 3600$$
$$2g = 4000$$
$$g = 2000$$
and so $g - 400 = 2000 - 400 = 1600$
The average awards for scholarships is
$1600 and for grants is $2000.

41. Balance $= 1000 + 5p$.
$$2115 = 1000 + 5p$$
$$1115 = 5p$$
$$223 = p$$
The direct deposit checks were $223 each.

43. Let x = old balance.
old bal. – checks written = new bal.
$$x - (50 + 120 + 70 + 65) = 907$$
$$x - 305 = 907$$
$$x = 1212$$
His old balance is $1212.

45. no. of installments × amount = 1350
$$3t = 1350$$
$$t = \frac{1350}{3} = 450$$
Each installment was $450.

47. Let m = miles from Yeehaw to Miami
dist. from T to Y + dist. from Y to M = 264
$$106 + x = 264$$
$$x = 158$$
It is 158 miles from Yeehaw Junction to
Miami.

49. Let c = number of calories
$$c = 3500 \times 15 = 52{,}500 \text{ calories}$$

Let e = number of excess calories
$$e = 1800 - 1300 = 500 \text{ calories}$$

Let d = number of days
$$500d = 52{,}500$$
$$d = 105 \text{ days}$$

51. a. $1\underline{5}7 \to 160$
   b. $\underline{1}57 \to 200$

53. $d = 50t$
$$300 = 50t$$
$$6 = t$$
It would take 6 hours to travel 300 miles.

55. Answers may vary.

57. $x + 1 = 2$ Check: $x + 1 = 2$
$$x + 1 - 1 = 2 - 1 \qquad\qquad (1) + 1 = 2$$
$$x = 1 \qquad\qquad\qquad 2 = 2$$
The solution is 1. Yes, the answer does
satisfy the definition of a solution.

Mastery Test 1.9

59. $x + 8 = 17$
$x + 8 - 8 = 17 - 8$
$x = 9$

61. $20 = 4x$
$20 \div 4 = 4x \div 4$
$5 = x$

63. $40 = 38 + m$
$40 - 38 = 38 + m - 38$
$2 = m$

65. $x - 5 = 10$
$x - 5 + 5 = 10 + 5$
$x = 15$

67. $20 = x + 5$
$20 - 5 = x + 5 - 5$
$15 = x$

Review Exercises – Chapter 1

1. a. $127 = 100 + 20 + 7$
  b. $189 = 100 + 80 + 9$
  c. $380 = 300 + 80$
  d. $1490 = 1000 + 400 + 90$
  e. $2559 = 2000 + 500 + 50 + 9$

2. a. $40 + 9 = 49$
  b. $500 + 80 + 6 = 586$
  c. $500 + 3 = 503$
  d. $800 + 10 = 810$
  e. $1000 + 4 = 1004$

3. a. $79 \rightarrow$ Seventy-nine
  b. $143 \rightarrow$ One hundred forty-three
  c. $1249 \rightarrow$ One thousand, two hundred forty-nine
  d. $5659 \rightarrow$ Five thousand, six hundred fifty-nine
  e. $12,347 \rightarrow$ Twelve thousand, three hundred forty-seven

4. a. 26   **b.** 192
  c. 468   **d.** 1644
  e. 42,801

5. a. $27 < 29$ **b.** $30 > 28$
  c. $23 < 25$ **d.** $19 < 39$
  e. $39 > 19$

6. a. $2\underline{8}48 \rightarrow 2800$ **b.** $9\underline{7}46 \rightarrow 9700$
  c. $3\underline{5}50 \rightarrow 3600$ **d.** $4\underline{4}44 \rightarrow 4400$
  e. $5\underline{5}55 \rightarrow 5600$

7. a. $\$21,\underline{0}90 \rightarrow \$21,100$
  b. $\$27,\underline{2}70 \rightarrow \$27,300$
  c. $\$35,\underline{5}40 \rightarrow \$35,500$
  d. $\$26,\underline{4}60 \rightarrow \$26,500$
  e. $\$22,\underline{9}90 \rightarrow \$23,000$

8. a.
$$\begin{array}{r} 3402 \\ + 8576 \\ \hline 11,978 \end{array}$$

  b.
$$\begin{array}{r} {}^{1\,1} \\ 2098 \\ + 2383 \\ \hline 4481 \end{array}$$

  c.
$$\begin{array}{r} {}^{1\,1} \\ 3099 \\ + 6547 \\ \hline 9646 \end{array}$$

  d.
$$\begin{array}{r} {}^{1} \\ 4563 \\ + 8603 \\ \hline 13,166 \end{array}$$

  e.
$$\begin{array}{r} {}^{1\,1} \\ 3480 \\ + 9769 \\ \hline 13,249 \end{array}$$

9. a. Perimeter $= 3 + 3 + 1 = 7$ ft
  b. Perimeter $= 4 + 4 + 3 = 11$ ft
  c. Perimeter $= 5 + 5 + 4 = 14$ ft
  d. Perimeter $= 4 + 5 + 3 = 12$ ft
  e. Perimeter $= 8 + 10 + 6 = 24$ ft

10. a.
$$\begin{array}{r} {}^{3\,\,17} \\ \not{4}7 \\ - 18 \\ \hline 29 \end{array}$$

  b.
$$\begin{array}{r} {}^{2\,\,16} \\ \not{3}6 \\ - 19 \\ \hline 17 \end{array}$$

  c.
$$\begin{array}{r} {}^{4\,\,15} \\ \not{5}5 \\ - 26 \\ \hline 29 \end{array}$$

  d.
$$\begin{array}{r} {}^{3\,\,16} \\ \not{4}6 \\ - 37 \\ \hline 9 \end{array}$$

  e.
$$\begin{array}{r} {}^{8\,\,13} \\ \not{9}3 \\ - 44 \\ \hline 49 \end{array}$$

11. a.

$$
\begin{array}{r}
{}^{4\;14}\\
6\,\cancel{5}\,4\\
-4\,6\,7\\
\hline
7
\end{array}
\rightarrow
\begin{array}{r}
{}^{5\;14}\\
\cancel{6}\,\cancel{5}\,4\\
-4\,6\,7\\
\hline
\boxed{1\,8\,7}
\end{array}
$$

b.

$$
\begin{array}{r}
{}^{3\;17}\\
5\,\cancel{4}\,7\\
-4\,5\,8\\
\hline
9
\end{array}
\rightarrow
\begin{array}{r}
{}^{4\;13}\\
\cancel{5}\,\cancel{4}\,7\\
-4\,5\,8\\
\hline
\boxed{8\,9}
\end{array}
$$

c.

$$
\begin{array}{r}
{}^{4\;12}\\
9\,\cancel{5}\,2\\
-8\,6\,3\\
\hline
9
\end{array}
\rightarrow
\begin{array}{r}
{}^{8\;14}\\
\cancel{9}\,\cancel{5}\,2\\
-8\,6\,3\\
\hline
\boxed{8\,9}
\end{array}
$$

d.

$$
\begin{array}{r}
{}^{4\;11}\\
8\,\cancel{5}\,1\\
-6\,7\,3\\
\hline
8
\end{array}
\rightarrow
\begin{array}{r}
{}^{7\;14}\\
\cancel{8}\,\cancel{5}\,1\\
-6\,7\,3\\
\hline
\boxed{1\,7\,8}
\end{array}
$$

e.

$$
\begin{array}{r}
{}^{2\;12}\\
4\,\cancel{3}\,2\\
-2\,4\,6\\
\hline
6
\end{array}
\rightarrow
\begin{array}{r}
{}^{3\;12}\\
\cancel{4}\,\cancel{3}\,2\\
-2\,4\,6\\
\hline
\boxed{1\,8\,6}
\end{array}
$$

12. a.

$$
\begin{array}{r}
{}^{3\,9\;13}\\
5\,\cancel{4}\cancel{0}3\\
-\;8\,6\,9\\
\hline
3\,4
\end{array}
\rightarrow
\begin{array}{r}
{}^{4\;13}\\
\cancel{5}\,\cancel{4}\cancel{0}3\\
-\;8\,6\,9\\
\hline
\boxed{\$4\,5\,3\,4}
\end{array}
$$

b.

$$
\begin{array}{r}
{}^{3\,9\;13}\\
5\,\cancel{4}\cancel{0}3\\
-\;7\,7\,8\\
\hline
2\,5
\end{array}
\rightarrow
\begin{array}{r}
{}^{4\;13}\\
\cancel{5}\,\cancel{4}\cancel{0}3\\
-\;7\,7\,8\\
\hline
\boxed{\$4\,6\,2\,5}
\end{array}
$$

c.

$$
\begin{array}{r}
{}^{3\,9\;13}\\
5\,\cancel{4}\cancel{0}3\\
-\;9\,8\,9\\
\hline
1\,4
\end{array}
\rightarrow
\begin{array}{r}
{}^{4\;13}\\
\cancel{5}\,\cancel{4}\cancel{0}3\\
-\;9\,8\,9\\
\hline
\boxed{\$4\,4\,1\,4}
\end{array}
$$

d.

$$
\begin{array}{r}
{}^{3\,9\;13}\\
5\,\cancel{4}\cancel{0}3\\
-\;6\,7\,6\\
\hline
2\,7
\end{array}
\rightarrow
\begin{array}{r}
{}^{4\;13}\\
\cancel{5}\,\cancel{4}\cancel{0}3\\
-\;6\,7\,6\\
\hline
\boxed{\$4\,7\,2\,7}
\end{array}
$$

e.

$$
\begin{array}{r}
{}^{3\,9\;13}\\
5\,\cancel{4}\cancel{0}3\\
-\;7\,6\,5\\
\hline
3\,8
\end{array}
\rightarrow
\begin{array}{r}
{}^{4\;13}\\
\cancel{5}\,\cancel{4}\cancel{0}3\\
-\;7\,6\,5\\
\hline
\boxed{\$4\,6\,3\,8}
\end{array}
$$

13. a.

$$
\begin{array}{r}
{}^{2}\\
{}^{3}\\
36\\
\times\,45\\
\hline
180\\
144\\
\hline
1620
\end{array}
$$

b.

$$
\begin{array}{r}
{}^{3}\\
{}^{7}\\
28\\
\times\,49\\
\hline
252\\
112\\
\hline
1372
\end{array}
$$

c.

$$
\begin{array}{r}
{}^{2}\\
{}^{6}\\
47\\
\times\,39\\
\hline
423\\
141\\
\hline
1833
\end{array}
$$

d.

$$
\begin{array}{r}
{}^{1}\\
{}^{2}\\
56\\
\times\,24\\
\hline
224\\
112\\
\hline
1344
\end{array}
$$

e.

$$
\begin{array}{r}
{}^{7}\\
{}^{1}\\
48\\
\times\,92\\
\hline
96\\
432\\
\hline
4416
\end{array}
$$

14. a.

$$
\begin{array}{r}
{}^{1\,1}\\
123\\
\times\,216\\
\hline
738\\
123\\
246\\
\hline
26{,}568
\end{array}
$$

b.

$$
\begin{array}{r}
{}^{1}\\
231\\
\times\,413\\
\hline
691\\
231\\
924\\
\hline
95{,}403
\end{array}
$$

c.

$$
\begin{array}{r}
{}^{2\,3}\\
{}^{2\,2}\\
{}^{1\,2}\\
345\\
\times\,654\\
\hline
1380\\
1725\\
2070\\
\hline
225{,}630
\end{array}
$$

d.

$$
\begin{array}{r}
{}^{2}\\
{}^{1}\\
231\\
\times\,843\\
\hline
691\\
924\\
1848\\
\hline
194{,}733
\end{array}
$$

e.

$$
\begin{array}{r}
{}^{3\;4}\\
{}^{4\;5}\\
{}^{7\;8}\\
879\\
\times\,569\\
\hline
7911\\
5274\\
4395\\
\hline
500{,}151
\end{array}
$$

15. a.
$$\begin{array}{r} \overset{1\,1}{} \\ \overset{1\,1}{} \\ 234 \\ \times\, 330 \\ \hline 7020 \\ 702 \\ \hline 77{,}220 \end{array}$$

b.
$$\begin{array}{r} \overset{1}{} \\ \overset{1}{} \\ 546 \\ \times\, 220 \\ \hline 10920 \\ 1092 \\ \hline 120{,}120 \end{array}$$

c.
$$\begin{array}{r} \overset{1\,2}{} \\ \overset{1\,2}{} \\ 324 \\ \times\, 550 \\ \hline 16200 \\ 1620 \\ \hline 178{,}200 \end{array}$$

d.
$$\begin{array}{r} \overset{1\,2}{} \\ \overset{1\,2}{} \\ 124 \\ \times\, 450 \\ \hline 6200 \\ 620 \\ \hline 55{,}800 \end{array}$$

e.
$$\begin{array}{r} \overset{3}{} \\ \overset{8\,1}{} \\ 892 \\ \times\, 490 \\ \hline 80280 \\ 3568 \\ \hline 437{,}080 \end{array}$$

16. a. $36 \times \$220 = \$7920$
b. $24 \times \$220 = \$5280$
c. $48 \times \$220 = \$10{,}560$
d. $30 \times \$220 = \$6600$
e. $60 \times \$220 = \$13{,}200$

17. a. $36 \text{ in.} \times 10 \text{ in.} = 360 \text{ in.}^2$
b. $24 \text{ in.} \times 10 \text{ in.} = 240 \text{ in.}^2$
c. $30 \text{ in.} \times 12 \text{ in.} = 360 \text{ in.}^2$
d. $18 \text{ in.} \times 10 \text{ in.} = 180 \text{ in.}^2$
e. $24 \text{ in.} \times 12 \text{ in.} = 288 \text{ in.}^2$

18. a. $\dfrac{0}{2} = 0$
b. $\dfrac{0}{5} = 0$
c. $\dfrac{0}{12} = 0$

19. a. $\dfrac{2}{0}$ is not defined
b. $\dfrac{5}{0}$ is not defined
c. $\dfrac{12}{0}$ is not defined

20. a.
$$\begin{array}{r} 15 \\ 5\,\overline{)75} \\ \underline{-5} \\ 25 \\ \underline{-25} \\ 0 \end{array}$$
answer: 15

b.
$$\begin{array}{r} 12 \\ 7\,\overline{)84} \\ \underline{-7} \\ 14 \\ \underline{-14} \\ 0 \end{array}$$
answer: 12

c.
$$\begin{array}{r} 15 \\ 6\,\overline{)90} \\ \underline{-6} \\ 30 \\ \underline{-30} \\ 0 \end{array}$$
answer: 15

d.
$$\begin{array}{r} 11 \\ 8\,\overline{)88} \\ \underline{-8} \\ 8 \\ -8 \\ \hline 0 \end{array}$$
answer: 11

e.
$$\begin{array}{r} 17 \\ 4\,\overline{)68} \\ \underline{-4} \\ 28 \\ \underline{-28} \\ 0 \end{array}$$
answer: 17

21. a.
$$\begin{array}{r} 31 \\ 9\,\overline{)279} \\ \underline{-27} \\ 9 \\ \underline{-9} \\ 0 \end{array}$$
answer: 31

b.
$$\begin{array}{r} 42 \\ 9\,\overline{)378} \\ \underline{-36} \\ 18 \\ \underline{-18} \\ 0 \end{array}$$
answer: 42

c.
$$\begin{array}{r} 103 \\ 8\,\overline{)824} \\ \underline{-8} \\ 2 \\ -0 \\ \hline 24 \\ \underline{-24} \\ 0 \end{array}$$
answer: 103

d. $\quad \begin{array}{r} 21 \\ 6\overline{)126} \\ \underline{-12} \\ 6 \\ \underline{-6} \\ 0 \end{array}$ answer: 21

e. $\quad \begin{array}{r} 65 \\ 7\overline{)455} \\ \underline{-42} \\ 35 \\ \underline{-35} \\ 0 \end{array}$ answer: 65

22. a. $\begin{array}{r} 46 \\ 21\overline{)967} \\ \underline{-84} \\ 127 \\ \underline{-126} \\ 1 \end{array}$ answer: 46 r 1

b. $\begin{array}{r} 42 \\ 24\overline{)1009} \\ \underline{-96} \\ 49 \\ \underline{-48} \\ 1 \end{array}$ answer: 42 r 1

c. $\begin{array}{r} 25 \\ 35\overline{)876} \\ \underline{-70} \\ 176 \\ \underline{-175} \\ 1 \end{array}$ answer: 25 r 1

d. $\begin{array}{r} 37 \\ 29\overline{)1074} \\ \underline{-87} \\ 204 \\ \underline{-203} \\ 1 \end{array}$ answer: 37 r 1

e. $\begin{array}{r} 48 \\ 51\overline{)2450} \\ \underline{-204} \\ 410 \\ \underline{-408} \\ 2 \end{array}$ answer: 48 r 2

23. a. $\$11{,}232 \div 9 = \$1248$ per period
b. $\$11{,}232 \div 12 = \$936$ per period
c. $\$11{,}232 \div 24 = \$468$ per period
d. $\$11{,}232 \div 26 = \$432$ per period
e. $\$11{,}232 \div 52 = \$216$ per period

24. a. $41 = 1 \times 41$; prime
b. $26 = 1 \times 26$ or 2×13; composite
c. $37 = 1 \times 37$; prime
d. $81 = 1 \times 81$ or 9×9; composite
e. $2 = 1 \times 2$; prime

25. a. $40 = 1 \times 40$ or 2×20 or 4×10 or 5×8; the prime factors are 2 and 5.
b. $25 = 1 \times 25$ or 5×5; the prime factor is 5.
c. $75 = 1 \times 75$ or 3×25 or 5×15; the prime factors are 3 and 5.
d. $128 = 1 \times 128$ or 2×64 or 4×32 or 8×16; the prime factor is 2.
e. $68 = 1 \times 68$ or 2×34 or 4×17; the prime factors are 2 and 17.

26. a. $\begin{array}{l} 2\overline{)50} \\ 5\overline{)25} \\ \quad 5 \end{array}$ Thus, $50 = 2 \times 5 \times 5$

b. $\begin{array}{l} 2\overline{)34} \\ \quad 17 \end{array}$ Thus, $34 = 2 \times 17$

c. $\begin{array}{l} 2\overline{)76} \\ 2\overline{)38} \\ \quad 19 \end{array}$ Thus, $76 = 2 \times 2 \times 19$

d. $\begin{array}{l} 3\overline{)39} \\ \quad 13 \end{array}$ Thus, $39 = 3 \times 13$

e. $\begin{array}{l} 3\overline{)81} \\ 3\overline{)27} \\ 3\overline{)9} \\ \quad 3 \end{array}$ Thus, $81 = 3 \times 3 \times 3 \times 3$

27. a. $2^2 = 2 \times 2 = 4$
b. $3^2 = 3 \times 3 = 9$
c. $5^3 = 5 \times 5 \times 5 = 125$
d. $2^7 = 2 \times 2 \times 2 \times 2 \times 2 \times 2 \times 2 = 128$
e. $3^5 = 3 \times 3 \times 3 \times 3 \times 3 = 243$

28. a. $3^2 \times 5^3 = 9 \times 125 = 1125$
b. $3^3 \times 8^0 = 27 \times 1 = 27$
c. $3^2 \times 5^2 \times 2^0 = 9 \times 25 \times 1 = 225$
d. $5^0 \times 2^3 \times 5^2 = 1 \times 8 \times 25 = 200$
e. $4^2 \times 9^0 \times 5^1 = 16 \times 1 \times 5 = 80$

29. a. $2^2 \times 3 \times 8^0 = 4 \times 3 \times 1 = 12$

b. $5^2 \times 7^0 \times 2^1 = 25 \times 1 \times 2 = 50$

c. $3^3 \times 5^2 \times 6^0 = 27 \times 25 \times 1 = 675$

d. $5^2 \times 3^0 \times 2^0 = 25 \times 1 \times 1 = 25$

e. $5^0 \times 3^0 \times 2^0 = 1 \times 1 \times 1 = 1$

30. a. $7 \cdot 8 - 2 = 56 - 2 = 54$

b. $6 \cdot 8 - 3 = 48 - 3 = 45$

c. $5 \cdot 8 - 4 = 40 - 4 = 36$

d. $4 \cdot 8 - 5 = 32 - 5 = 27$

e. $3 \cdot 8 - 5 = 24 - 5 = 19$

31. a. $30 + 4 \cdot 5 = 30 + 20 = 50$

b. $31 + 5 \cdot 5 = 31 + 25 = 56$

c. $32 + 6 \cdot 5 = 32 + 30 = 62$

d. $33 + 7 \cdot 5 = 33 + 35 = 68$

e. $34 + 8 \cdot 5 = 34 + 40 = 74$

32. a. $48 \div 6 - (1 + 2) = 48 \div 6 - (3)$
$$= 8 - 3$$
$$= 5$$

b. $48 \div 8 - (2 + 2) = 48 \div 8 - (4)$
$$= 6 - 4$$
$$= 2$$

c. $48 \div 4 - (2 + 3) = 48 \div 4 - (5)$
$$= 12 - 5$$
$$= 7$$

d. $48 \div 3 - (2 + 4) = 48 \div 3 - (6)$
$$= 16 - 6$$
$$= 10$$

e. $48 \div 2 - (2 + 5) = 48 \div 2 - (7)$
$$= 24 - 7$$
$$= 17$$

33. a. $9 \div 3 \cdot 3 \cdot 3 + 3 - 1 = 3 \cdot 3 \cdot 3 + 3 - 1$
$$= 9 \cdot 3 + 3 - 1$$
$$= 27 + 3 - 1$$
$$= 30 - 1$$
$$= 29$$

b. $9 \div 3 \cdot 3 + 3 - 1 = 3 \cdot 3 + 3 - 1$
$$= 9 + 3 - 1$$
$$= 12 - 1$$
$$= 11$$

c. $8 \div 2 \cdot 2 \cdot 2 + 2 - 1 = 4 \cdot 2 \cdot 2 + 2 - 1$
$$= 8 \cdot 2 + 2 - 1$$
$$= 16 + 2 - 1$$
$$= 18 - 1$$
$$= 17$$

d. $8 \div 2 \cdot 2 + 2 - 1 = 4 \cdot 2 + 2 - 1$
$$= 8 + 2 - 1$$
$$= 10 - 1$$
$$= 9$$

e. $8 \div 4 \cdot 4 + 4 - 1 = 2 \cdot 4 + 4 - 1$
$$= 8 + 4 - 1$$
$$= 12 - 1$$
$$= 11$$

34. a. $20 \div 5 + \{3 \cdot 9 - [3 + (5 - 2)]\}$
$$= 20 \div 5 + \{3 \cdot 9 - [3 + (3)]\}$$
$$= 20 \div 5 + \{3 \cdot 9 - [6]\}$$
$$= 20 \div 5 + \{27 - [6]\}$$
$$= 20 \div 5 + \{21\}$$
$$= 4 + 21$$
$$= 25$$

b. $20 \div 5 + \{4 \cdot 9 - [3 + (5 - 3)]\}$
$$= 20 \div 5 + \{4 \cdot 9 - [3 + (2)]\}$$
$$= 20 \div 5 + \{4 \cdot 9 - [5]\}$$
$$= 20 \div 5 + \{36 - [5]\}$$
$$= 20 \div 5 + \{31\}$$
$$= 4 + 31$$
$$= 35$$

c. $24 \div 6 + \{5 \cdot 9 - [3 + (5 - 4)]\}$
$$= 24 \div 6 + \{5 \cdot 9 - [3 + (1)]\}$$
$$= 24 \div 6 + \{5 \cdot 9 - [4]\}$$
$$= 24 \div 6 + \{45 - [4]\}$$
$$= 24 \div 6 + \{41\}$$
$$= 4 + 41$$
$$= 45$$

d. $24 \div 4 + \{6 \cdot 9 - [3 + (5 - 5)]\}$
$$= 24 \div 4 + \{6 \cdot 9 - [3 + (0)]\}$$
$$= 24 \div 4 + \{6 \cdot 9 - [3]\}$$
$$= 24 \div 4 + \{54 - [3]\}$$
$$= 24 \div 4 + \{51\}$$
$$= 6 + 51$$
$$= 57$$

e. $24 \div 3 + \{7 \cdot 9 - [3 + (5 - 1)]\}$
$$= 24 \div 3 + \{7 \cdot 9 - [3 + (4)]\}$$
$$= 24 \div 3 + \{7 \cdot 9 - [7]\}$$
$$= 24 \div 3 + \{63 - [7]\}$$
$$= 24 \div 3 + \{56\}$$
$$= 8 + 56$$
$$= 64$$

35. amount for work + parts = total cost
  a. $30(3) + 80 = 90 + 80 = 170$
    The total cost is \$170.
  b. $30(5) + 80 = 150 + 80 = 230$
    The total cost is \$230.
  c. $30(2) + 80 = 60 + 80 = 140$
    The total cost is \$140.
  d. $30(4) + 80 = 120 + 80 = 200$
    The total cost is \$200.
  e. $30(6) + 80 = 180 + 80 = 260$
    The total cost is \$260.

36. **a.** $x + 6 = 18$
    What number added to 6 gives 18?
    The solution is 12; $x = 12$.
  b. $x + 7 = 18$
    What number added to 7 gives 18?
    The solution is 11; $x = 11$.
  c. $x + 8 = 18$
    What number added to 8 gives 18?
    The solution is 10; $x = 10$.
  d. $x + 9 = 18$
    What number added to 9 gives 18?
    The solution is 9; $x = 9$.
  e. $x + 10 = 18$
    What number added to 10 gives 18?
    The solution is 8; $x = 8$.

37. **a.** $10 - x = 3$
    What number subtracted from 10 gives 3? The solution is 7; $x = 7$.
  b. $10 - x = 4$
    What number subtracted from 10 gives 4? The solution is 6; $x = 6$.
  c. $10 - x = 5$
    What number subtracted from 10 gives 5? The solution is 5; $x = 5$.
  d. $10 - x = 6$
    What number subtracted from 10 gives 6? The solution is 4; $x = 4$.
  e. $10 - x = 7$
    What number subtracted from 10 gives 7? The solution is 3; $x = 3$.

38. **a.** $20 = 4x$
    What number multiplied by 4 gives 20?
    The solution is 5; $x = 5$.
  b. $20 = 5x$
    What number multiplied by 5 gives 20?
    The solution is 4; $x = 4$.
  c. $20 = 10x$
    What number multiplied by 10 gives 20? The solution is 2; $x = 2$.
  d. $20 = 20x$
    What number multiplied by 20 gives 20? The solution is 1; $x = 1$.
  e. $20 = 2x$
    What number multiplied by 2 gives 20? The solution is 10; $x = 10$.

39. **a.** $28 \div x = 4$
    28 divided by what number gives 4? The solution is 7; $x = 7$.
  b. $24 \div x = 4$
    24 divided by what number gives 4? The solution is 6; $x = 6$.
  c. $20 \div x = 4$
    20 divided by what number gives 4? The solution is 5; $x = 5$.
  d. $16 \div x = 4$
    28 divided by what number gives 4? The solution is 4; $x = 4$.
  e. $12 \div x = 4$
    12 divided by what number gives 4? The solution is 3; $x = 3$.

40. **a.** $n - 10 = 11$
    $n - 10 + 10 = 11 + 10$
    $n = 21$
  b. $n - 4 = 12$
    $n - 4 + 4 = 12 + 4$
    $n = 16$
  c. $n - 27 = 13$
    $n - 27 + 27 = 13 + 27$
    $n = 40$
  d. $n - 48 = 14$
    $n - 48 + 48 = 14 + 48$
    $n = 62$
  e. $n - 18 = 15$
    $n - 18 + 18 = 15 + 18$
    $n = 33$

41. a. $20 = m - 12$
$20 + 12 = m - 12 + 12$
$32 = m$

b. $20 = m - 38$
$20 + 38 = m - 38 + 38$
$58 = m$

c. $11 = m - 14$
$11 + 14 = m - 14 + 14$
$25 = m$

d. $42 = m - 15$
$42 + 15 = m - 15 + 15$
$57 = m$

e. $49 = m - 16$
$49 + 16 = m - 16 + 16$
$65 = m$

42. a. $33 = 18 + m$
$33 - 18 = 18 + m - 18$
$15 = m$

b. $32 = 19 + m$
$32 - 18 = 19 + m - 19$
$13 = m$

c. $37 = 19 + m$
$37 - 19 = 19 + m - 19$
$18 = m$

d. $39 = 17 + m$
$39 - 17 = 17 + m - 17$
$22 = m$

e. $46 = 17 + m$
$46 - 17 = 17 + m - 17$
$29 = m$

43. a. $3x = 36$
$3x \div 3 = 36 \div 3$
$x = 12$

b. $4x = 52$
$4x \div 4 = 52 \div 4$
$x = 13$

c. $6x = 72$
$6x \div 6 = 72 \div 6$
$x = 12$

d. $7x = 63$
$7x \div 7 = 63 \div 7$
$x = 9$

e. $9x = 108$
$9x \div 9 = 108 \div 9$
$x = 12$

44. a. $10 = 2x$
$10 \div 2 = 2x \div 2$
$5 = x$

b. $16 = 4x$
$16 \div 4 = 4x \div 4$
$4 = x$

c. $20 = 5x$
$20 \div 5 = 5x \div 5$
$4 = x$

d. $36 = 6x$
$36 \div 6 = 6x \div 6$
$6 = x$

e. $48 = 8x$
$48 \div 8 = 8x \div 8$
$6 = x$

45. Let h = height of the antenna

a. $1430 + h = 1520$
$1430 + h - 1430 = 1520 - 1430$
$h = 90$
The height of the antenna is 90 feet.

b. $1430 + h = 1530$
$1430 + h - 1430 = 1530 - 1430$
$h = 100$
The height of the antenna is 100 feet.

c. $1430 + h = 1540$
$1430 + h - 1430 = 1540 - 1430$
$h = 110$
The height of the antenna is 110 feet.

d. $1430 + h = 1515$
$1430 + h - 1430 = 1515 - 1430$
$h = 85$
The height of the antenna is 85 feet.

e. $1430 + h = 1505$
$1430 + h - 1430 = 1505 - 1430$
$h = 75$
The height of the antenna is 75 feet.

Chapter 2

Fractions and Mixed Numbers

Section 2.1 – Fractions and Mixed Numbers

Problems

1. $\dfrac{2}{5} = \dfrac{2 \text{ shaded parts}}{5 \text{ total parts}}$

$\dfrac{1}{7} = \dfrac{1 \text{ shaded part}}{7 \text{ total parts}}$

2. **a.** $\dfrac{6}{6} = 1$; thus $\dfrac{6}{6}$ is an improper fraction.

b. $\dfrac{3}{19}$ is less than 1; thus $\dfrac{3}{19}$ is a proper fraction.

c. $\dfrac{19}{3}$ is greater than 1; thus $\dfrac{19}{3}$ is an improper fraction.

d. $\dfrac{0}{3} = 0$, which is less than 1; thus $\dfrac{0}{19}$ is a proper fraction.

c. $\dfrac{7}{1}$ is greater than 1; thus $\dfrac{7}{1}$ is an improper fraction.

3. **a.** $\dfrac{26}{5} = 5$ with a remainder of 1. Thus $\dfrac{26}{5} = 5\dfrac{1}{5}$.

b. $\dfrac{47}{6} = 7$ with a remainder of 5. Thus $\dfrac{47}{6} = 7\dfrac{5}{6}$.

4. **a.** $5\dfrac{3}{4} = \dfrac{4 \times 5 + 3}{4} = \dfrac{23}{4}$

b. $8\dfrac{2}{7} = \dfrac{7 \times 8 + 2}{7} = \dfrac{58}{7}$

5. **a.** 1 week $= \dfrac{7}{30}$ month

b. 30 days $= \dfrac{30}{30} = 1$ month

c. 60 days $= \dfrac{60}{30} = 2$ months

6. **a.** Safflower oil

b. $\dfrac{8}{10}$

7. $\dfrac{28}{4} = 7$

Exercises 2.1

1. $\dfrac{1 \text{ part shaded}}{2 \text{ equal parts}} = \dfrac{1}{2}$

3. $\dfrac{1 \text{ part shaded}}{3 \text{ equal parts}} = \dfrac{1}{3}$

5. $\dfrac{5 \text{ parts shaded}}{12 \text{ equal parts}} = \dfrac{5}{12}$

7. $\dfrac{1 \text{ part shaded}}{4 \text{ equal parts}} = \dfrac{1}{4}$

9. $\dfrac{3 \text{ parts shaded}}{4 \text{ equal parts}} = \dfrac{3}{4}$

11. $\dfrac{2}{4}$

13. $\dfrac{3}{4}$

15. $\frac{4}{4}$ or 1

17. $\frac{2}{3}$

19. 1

21. $\frac{9}{61}$ is less than 1 and is therefore a proper fraction.

23. $\frac{4}{17}$ is less than 1 and is therefore a proper fraction.

25. $\frac{8}{41}$ is less than 1 and is therefore a proper fraction.

27. $\frac{8}{16}$ is less than 1 and is therefore a proper fraction.

29. $\frac{3}{100}$ is less than 1 and is therefore a proper fraction.

31. $\frac{31}{10} = 3$ with remainder 1; thus $\frac{31}{10} = 3\frac{1}{10}$.

33. $\frac{8}{7} = 1$ with remainder 1; thus $\frac{8}{7} = 1\frac{1}{7}$.

35. $\frac{29}{8} = 3$ with remainder 5; thus $\frac{29}{8} = 3\frac{5}{8}$.

37. $\frac{69}{9} = 7$ with remainder 6; thus

$\frac{69}{9} = 7\frac{6}{9} = 7\frac{2}{3}$.

39. $\frac{101}{10} = 10$ with remainder 1; thus

$\frac{101}{10} = 10\frac{1}{10}$.

41. $5\frac{1}{7} = \frac{7 \times 5 + 1}{7} = \frac{35 + 1}{7} = \frac{36}{7}$

43. $4\frac{1}{10} = \frac{10 \times 4 + 1}{10} = \frac{40 + 1}{10} = \frac{41}{10}$

45. $1\frac{2}{11} = \frac{11 \times 1 + 2}{11} = \frac{11 + 2}{11} = \frac{13}{11}$

47. $8\frac{3}{10} = \frac{10 \times 8 + 3}{10} = \frac{80 + 3}{10} = \frac{83}{10}$

49. $2\frac{1}{6} = \frac{6 \times 2 + 1}{6} = \frac{12 + 1}{6} = \frac{13}{6}$

51. 7 hours $= \frac{7}{24}$ day

53. 7 ounces $= \frac{7}{16}$ pound

55. $\frac{5}{8}$

57. $\frac{51}{100}$

59. a. Dog food: $\frac{60}{60} = 1$ minute

    b. Toothpaste: $\frac{90}{60} = 1\frac{1}{2}$ minutes

    c. Soap: $\frac{45}{60} = \frac{3}{4}$ minute

    d. Cereal: $\frac{15}{60} = \frac{1}{4}$ minute

61. Single person: $\frac{25}{98}$

63. Five persons: $\frac{6}{98} = \frac{3}{49}$

65. Six persons or more: $\frac{2+1}{98} = \frac{3}{98}$

67. Mexico: $\frac{37}{99}$

69. Spain: $\frac{2}{99}$

71. 23 $\underline{<}$ 27

73. 2|28
2|14
7   Thus, $28 = 2^2 \cdot 7$

75. 2|180
2|90
3|45
3|15
5   Thus, $200 = 2^2 \cdot 3^2 \cdot 5$

77. $\frac{0}{4} = 0$ was used; $\frac{4}{4} = 1$ remains

79. $\frac{2}{4} = \frac{1}{2}$ was used; $\frac{2}{4} = \frac{1}{2}$ remains

81. $\frac{180}{10} = 18$ miles per gallon

83. $\frac{210}{10} = 21$ miles per gallon

85. $\frac{25 \text{ miles}}{1 \text{ gallon}} \cdot 14 \text{ gallons} = 350$ miles

87. $340 \text{ miles} \cdot \frac{1 \text{ gallon}}{20 \text{ miles}} = \frac{340}{20} = 17$
gallons needed. Thus, no you can't make it since you need 17 gallons and you only have 14 gallons.

89. $240 \text{ miles} \cdot \frac{1 \text{ gallon}}{20 \text{ miles}} = \frac{240}{20} = 12$ gal.

91. Answers may vary.

93. Answers may vary.

Mastery Test 2.1

95. 5 months = $\frac{5}{12}$ year

97. $\frac{25}{3} = 8$ with remainder 1; thus $\frac{25}{3} = 8\frac{1}{3}$.

99. $\frac{17}{17} = 1$ and is therefore an improper fraction.

Section 2.2 – Equivalent Fractions

Problems

1. a. $\frac{2}{7} = \frac{?}{28}$. The denominator 7 has to be multiplied by 4 to get the denominator 28 so the numerator 2 has to be multiplied by 4. Thus $\frac{2}{7} = \frac{2 \times 4}{7 \times 4} = \frac{8}{28}$.

b. $\frac{5}{6} = \frac{20}{?}$. The numerator 5 has to be multiplied by 4 to get the numerator 20 so the denominator 6 has to be multiplied by 4. Thus $\frac{5}{6} = \frac{5 \times 4}{6 \times 4} = \frac{20}{24}$.

2. a. $\frac{42}{54} = \frac{?}{18}$. The denominator 54 has to be divided by 3 to get the denominator 18 so the numerator 42 has to be divided by 3. Thus $\frac{42}{54} = \frac{42 \div 3}{54 \div 3} = \frac{14}{18}$.

b. $\frac{6}{20} = \frac{3}{?}$. The numerator 6 has to be divided by 2 to get the numerator 3 so the denominator 20 has to be divided by 2. Thus $\frac{6}{20} = \frac{6 \div 2}{20 \div 2} = \frac{3}{10}$.

3. a. $\frac{16}{80} = \frac{\overset{1}{\cancel{2}} \times \overset{1}{\cancel{2}} \times \overset{1}{\cancel{2}} \times \overset{1}{\cancel{2}}}{\underset{1}{\cancel{2}} \times \underset{1}{\cancel{2}} \times \underset{1}{\cancel{2}} \times \underset{1}{\cancel{2}} \times 5} = \frac{1}{5}$

b. $\frac{70}{155} = \frac{2 \times \overset{1}{\cancel{5}} \times 7}{\underset{1}{\cancel{5}} \times 31} = \frac{14}{31}$

4. a. $\frac{3}{17} > \frac{2}{17}$ since they have the same

denominator and $3 > 2$.

b. $\frac{1}{5} = \frac{1 \times 9}{5 \times 9} = \frac{9}{45}$ and $\frac{2}{9} = \frac{2 \times 5}{9 \times 5} = \frac{10}{45}$

Since $\frac{9}{45} < \frac{10}{45}$, we have $\frac{1}{5} < \frac{2}{9}$.

5. a. $\frac{4}{7}$ and $\frac{5}{8}$; $4 \times 8 = 32 < 7 \times 5 = 35$ so

$\frac{4}{7} < \frac{5}{8}$.

b. $\frac{2}{7}$ and $\frac{3}{11}$; $2 \times 11 = 22 > 7 \times 3 = 21$ so

$\frac{2}{7} > \frac{3}{11}$.

6. $\frac{28\cancel{0}}{12\cancel{0}} = \frac{28}{12} = \frac{\overset{1}{\cancel{4}} \times 7}{\underset{1}{\cancel{4}} \times 3} = \frac{7}{3}$

Exercises 2.2

1. $\frac{3}{5} = \frac{?}{50}$; $\frac{3}{5} = \frac{3 \times 10}{5 \times 10} = \frac{30}{50}$. The missing

number is 30.

3. $\frac{1}{6} = \frac{5}{?}$; $\frac{1}{6} = \frac{1 \times 5}{6 \times 5} = \frac{5}{30}$. The missing

number is 30.

5. $\frac{3}{5} = \frac{27}{?}$; $\frac{3}{5} = \frac{3 \times 9}{5 \times 9} = \frac{27}{45}$. The missing

number is 45.

7. $1\frac{2}{3} = \frac{?}{9}$; $\frac{5}{3} = \frac{5 \times 3}{3 \times 3} = \frac{15}{9}$. The missing

number is 15.

9. $4\frac{1}{2} = \frac{?}{16}$; $\frac{9}{2} = \frac{9 \times 8}{2 \times 8} = \frac{72}{16}$. The missing

number is 72.

11. $\frac{12}{15} = \frac{?}{5}$; $\frac{12}{15} = \frac{12 \div 3}{15 \div 3} = \frac{4}{5}$. The missing

number is 4.

13. $\frac{8}{24} = \frac{4}{?}$; $\frac{8}{24} = \frac{8 \div 2}{24 \div 2} = \frac{4}{12}$. The missing

number is 12.

15. $\frac{21}{56} = \frac{?}{8}$; $\frac{21}{56} = \frac{21 \div 7}{56 \div 7} = \frac{3}{8}$. The missing

number is 3.

17. $\frac{28}{30} = \frac{\overset{1}{\cancel{2}} \times 14}{\underset{1}{\cancel{2}} \times 15} = \frac{14}{15}$

19. $\frac{13}{52} = \frac{13}{2 \times 26} = \frac{\overset{1}{\cancel{13}}}{2 \times 2 \times \underset{1}{\cancel{13}}} = \frac{1}{4}$

21. $\frac{56}{24} = \frac{2 \times 28}{2 \times 12}$

$= \frac{2 \times 2 \times 14}{2 \times 2 \times 6} = \frac{\overset{1}{\cancel{2}} \times \overset{1}{\cancel{2}} \times \overset{1}{\cancel{2}} \times 7}{\underset{1}{\cancel{2}} \times \underset{1}{\cancel{2}} \times \underset{1}{\cancel{2}} \times 3} = \frac{7}{3}$

23. $\frac{21}{28} = \frac{3 \times \overset{1}{\cancel{7}}}{4 \times \underset{1}{\cancel{7}}} = \frac{3}{4}$

25. $\frac{22}{33} = \frac{2 \times \overset{1}{\cancel{11}}}{3 \times \underset{1}{\cancel{11}}} = \frac{2}{3}$

27. $\frac{45}{210} = \frac{5 \times 9}{5 \times 42} = \frac{\overset{1}{\cancel{5}} \times \overset{1}{\cancel{3}} \times 3}{\underset{1}{\cancel{5}} \times \underset{1}{\cancel{3}} \times 14} = \frac{3}{14}$

29. $\frac{231}{1001} = \frac{3 \times 77}{7 \times 143} = \frac{3 \times \overset{1}{\cancel{7}} \times \overset{1}{\cancel{11}}}{\underset{1}{\cancel{7}} \times \underset{1}{\cancel{11}} \times 13} = \frac{3}{13}$

31. $\frac{7}{8}$ is the greater of the two numbers, since

the denominators are the same and $7 > 5$.

33. $\frac{5}{11}$ is the greater of the two numbers,

since the denominators are the same and $5 > 4$.

35. $\frac{2}{3}$ and $\frac{4}{5}$; $2 \times 5 = 10 < 3 \times 4 = 12$ so $\frac{2}{3} < \frac{4}{5}$.

37. $1\frac{4}{7} < 1\frac{5}{7}$ since the whole number parts and the denominators are the same and $4 < 5$.

39. $11\frac{2}{7} = 11\frac{2 \times 8}{7 \times 8} = 11\frac{16}{56}$ and $11\frac{3}{8} = 11\frac{3 \times 7}{8 \times 7} = 11\frac{21}{56}$. Thus, $11\frac{2}{7} < 11\frac{3}{8}$ since the whole number parts and denominators are the same and $16 < 21$.

41. $\frac{46}{100} = \frac{2 \times 23}{2 \times 50} = \frac{23}{50}$. This is $\frac{23}{50}$ of the personal income tax revenues.

43. $\frac{46\cancel{0}}{276\cancel{0}} = \frac{2 \times 23}{2 \times 6 \times 23} = \frac{1}{6}$. The fraction of the budget to be spent on defense is $\frac{1}{6}$.

45. $\frac{100}{365} = \frac{5 \times 20}{5 \times 73} = \frac{20}{73}$. This is $\frac{20}{73}$ of the days.

47. a. $\frac{26}{52} = \frac{1 \times 26}{2 \times 26} = \frac{1}{2}$ of the deck is red.

   b. $\frac{13}{52} = \frac{1 \times 13}{2 \times 2 \times 13} = \frac{1}{4}$ of the deck is hearts.

   c. $\frac{4}{52} = \frac{1 \times 4}{2 \times 2 \times 13} = \frac{1}{13}$ of the deck is Kings.

49. $\frac{3}{5}$ and $\frac{1}{2}$; $3 \times 2 = 6 > 5 \times 1 = 5$ so $\frac{3}{5} > \frac{1}{2}$. The recipe calling for $\frac{3}{5}$ cup takes more sugar.

51. $\frac{24}{96} = \frac{\cancel{2} \times \cancel{2} \times \cancel{2} \times \cancel{3}}{\cancel{2} \times \cancel{2} \times \cancel{2} \times \cancel{3} \times 4} = \frac{1}{4}$. He uses $\frac{1}{4}$ of his time in shopping and paperwork.

53. $\frac{12}{96} = \frac{\cancel{2} \times \cancel{2} \times \cancel{3}}{\cancel{2} \times \cancel{2} \times 2 \times \cancel{3} \times 4} = \frac{1}{8}$. He uses $\frac{1}{8}$ of his time doing kitchen work.

55. $\frac{25}{200} = \frac{\cancel{5} \times \cancel{5}}{\cancel{5} \times \cancel{5} \times 8} = \frac{1}{8}$. The fraction of a Big Mac that is protein is $\frac{1}{8}$.

57. $\frac{35}{200} = \frac{\cancel{5} \times 7}{\cancel{5} \times 40} = \frac{7}{40}$. The fraction of a Big Mac that is fat is $\frac{7}{40}$.

59. $\frac{52}{140} = \frac{\cancel{2} \times \cancel{2} \times 13}{\cancel{2} \times \cancel{2} \times 35} = \frac{13}{35}$. The fraction of the 2 slices that is carbohydrates is $\frac{13}{35}$.

61. $3\frac{3}{8} = \frac{8 \times 3 + 3}{8} = \frac{24 + 3}{8} = \frac{27}{8}$

63. $7\frac{9}{10} = \frac{10 \times 7 + 9}{10} = \frac{70 + 9}{10} = \frac{79}{10}$

65. $10\frac{2}{13} = \frac{10 \times 13 + 2}{13} = \frac{130 + 2}{13} = \frac{132}{13}$

67. $\frac{25}{30} = \frac{5}{6}$ or "5 to 6"

69. a. $\frac{2\cancel{00}}{56\cancel{00}} = \frac{\cancel{2}}{\cancel{2} \times 2 \times 14} = \frac{1}{28}$ or "1 to 28"

   b. $\frac{1}{20} = \frac{?}{8000}$. The denominator 20 has to be multiplied by 400 to get the denominator 8000 so the numerator 1 has to be multiplied by 400. Thus $\frac{1}{20} = \frac{1 \times 400}{20 \times 400} = \frac{400}{8000}$ and so 400 teachers are needed.

71. Answers may vary. Sample answer: $\frac{1}{2}$ and $\frac{4}{8}$.

73. Answers may vary.

75. $\frac{6x^3}{8x^2} = \frac{\cancel{2} \cdot 3 \cdot \cancel{x} \cdot \cancel{x} \cdot x}{\cancel{2} \cdot 2 \cdot 2 \cdot \cancel{x} \cdot \cancel{x}} = \frac{3x}{4}$

77. $\dfrac{12x^4}{18x^3} = \dfrac{\cancel{2}\cdot 2\cdot \cancel{3}\cdot \cancel{x}\cdot \cancel{x}\cdot \cancel{x}\cdot x}{\cancel{2}\cdot \cancel{3}\cdot 3\cdot \cancel{x}\cdot \cancel{x}\cdot \cancel{x}} = \dfrac{2x}{3}$

Mastery Test 2.2

79. $\dfrac{3}{5} = \dfrac{?}{25}$. The denominator 5 has to be multiplied by 5 to get the denominator 25 so the numerator 3 has to be multiplied by 5. Thus $\dfrac{3}{5} = \dfrac{3\times 5}{5\times 5} = \dfrac{15}{25}$.

81. $\dfrac{9}{75} = \dfrac{?}{25}$. The denominator 754 has to be divided by 3 to get the denominator 25 so

the numerator 9 has to be divided by 3.
Thus $\dfrac{9}{75} = \dfrac{9\div 3}{75\div 3} = \dfrac{3}{25}$.

83. $\dfrac{3}{8}$ and $\dfrac{2}{5}$; $3\times 5 = 15 < 8\times 2 = 16$ so $\dfrac{3}{8} < \dfrac{2}{5}$.

85. $\dfrac{3}{5}$ and $\dfrac{2}{5}$; the denominators are the same and $3 > 2$ so $\dfrac{3}{5} > \dfrac{2}{5}$.

87. $\dfrac{20}{115} = \dfrac{\cancel{5}\times 4}{\cancel{5}\times 23} = \dfrac{4}{23}$

Section 2.3 – Multiplication and Division of Fractions and Mixed Numbers

Problems

1. a. $\dfrac{2}{5}\cdot \dfrac{4}{7} = \dfrac{2\cdot 4}{5\cdot 7} = \dfrac{8}{35}$

   b. $\dfrac{3}{4}\cdot \dfrac{2}{3} = \dfrac{3\cdot 2}{4\cdot 3} = \dfrac{6}{12} = \dfrac{1}{2}$.

2. a. Ground beef:
   $\dfrac{1}{3}\cdot 1\dfrac{1}{2} = \dfrac{1}{3}\cdot \dfrac{3}{2} = \dfrac{1\cdot 3}{3\cdot 2} = \dfrac{3}{6} = \dfrac{1}{2}$ lb

   b. Veal: $\dfrac{1}{3}\cdot 1 = \dfrac{1}{3}$ lb

   c. Onion: $\dfrac{1}{3}\cdot \dfrac{1}{4} = \dfrac{1\cdot 1}{3\cdot 4} = \dfrac{1}{12}$ cup

   d. Salt: $\dfrac{1}{3}\cdot 2 = \dfrac{1}{3}\cdot \dfrac{2}{1} = \dfrac{2}{3}$ tsp

   e. Garlic salt: $\dfrac{1}{3}\cdot \dfrac{1}{2} = \dfrac{1}{6}$ tsp

   f. Pepper: $\dfrac{1}{3}\cdot \dfrac{1}{3} = \dfrac{1}{9}$ tsp

3. $3\dfrac{1}{4}\cdot \dfrac{4}{3}\cdot \dfrac{3}{13} = \dfrac{\cancel{13}}{\cancel{4}}\cdot \dfrac{\cancel{4}}{\cancel{3}}\cdot \dfrac{\cancel{3}}{\cancel{13}} = 1$

4. a. $\left(\dfrac{2}{5}\right)^3 = \dfrac{2}{5}\cdot \dfrac{2}{5}\cdot \dfrac{2}{5} = \dfrac{8}{125}$ lb

b. $\left(1\dfrac{1}{4}\right)^2 = \left(\dfrac{5}{4}\right)^2 = \dfrac{5}{4}\cdot \dfrac{5}{4} = \dfrac{25}{16}$

c. $\left(\dfrac{2}{3}\right)^2 \cdot \left(2\dfrac{1}{4}\right) = \dfrac{\cancel{2}}{\cancel{3}}\cdot \dfrac{\cancel{2}}{\cancel{3}}\cdot \dfrac{\cancel{9}}{\cancel{4}} = 1$

5. a. $\dfrac{5}{7}\div \dfrac{3}{8} = \dfrac{5}{7}\cdot \dfrac{8}{3} = \dfrac{40}{21}$

   b. $\dfrac{3}{4}\div 7 = \dfrac{3}{4}\cdot \dfrac{1}{7} = \dfrac{3}{28}$

6. a. $5\dfrac{1}{2}\div \dfrac{3}{4} = \dfrac{11}{2}\div \dfrac{3}{4} = \dfrac{11}{\cancel{2}}\cdot \dfrac{\cancel{4}}{3} = \dfrac{22}{3}$

   b. $\dfrac{6}{7}\div 3\dfrac{1}{7} = \dfrac{6}{7}\div \dfrac{22}{7} = \dfrac{\cancel{6}}{\cancel{7}}\cdot \dfrac{\cancel{7}}{\cancel{22}} = \dfrac{3}{11}$

7. $2\dfrac{1}{2}\cdot 41,000 = \dfrac{5}{\cancel{2}}\cdot \dfrac{\cancel{41,000}}{1} = 102,500$

They can afford $102,500 home.

8. $130 \div 1\frac{2}{5} = 130 \div \frac{7}{5} = 130 \cdot \frac{5}{7} = \frac{650}{7} = 92\frac{6}{7}$

About 93 brushings are possible.

9. Area $= 4\frac{1}{3} \cdot 5\frac{2}{3} = \frac{13}{3} \cdot \frac{17}{3} = \frac{221}{9} = 24\frac{5}{9}$ yd^2

Exercises 2.3

1. $\dfrac{3}{4} \cdot \dfrac{7}{8} = \dfrac{3 \cdot 7}{4 \cdot 8} = \dfrac{21}{32}$

3. $\dfrac{1}{\not{6}} \cdot \dfrac{\not{6}}{7} = \dfrac{1}{7}$

5. $\dfrac{2}{\not{6}} \cdot \dfrac{\not{6}}{3} = \dfrac{2}{3}$

7. $3 \cdot \dfrac{2}{5} = \dfrac{3}{1} \cdot \dfrac{2}{5} = \dfrac{6}{5}$ or $1\frac{1}{5}$

9. $\dfrac{\overset{1}{\not{6}}}{\underset{2}{\not{6}}} \cdot \dfrac{\overset{1}{\not{6}}}{\underset{1}{\not{6}}} = \dfrac{1}{2}$

11. $\dfrac{\overset{1}{\not{7}}}{\underset{1}{\not{8}}} \cdot \dfrac{\overset{3}{\not{15}}}{\underset{2}{\not{14}}} = \dfrac{3}{2}$ or $1\frac{1}{2}$

13. $\dfrac{\overset{2}{\not{6}}}{\underset{1}{\not{7}}} \cdot \dfrac{\overset{2}{\not{14}}}{\underset{1}{\not{8}}} = \dfrac{4}{1} = 4$

15. $1\frac{2}{3} \cdot \dfrac{6}{5} = \dfrac{\not{5}}{\not{8}} \cdot \dfrac{\overset{2}{\not{6}}}{\underset{1}{\not{8}}} = \dfrac{2}{1} = 2$

17. $\dfrac{9}{4} \cdot 3\frac{1}{9} = \dfrac{\not{9}}{\underset{1}{\not{4}}} \cdot \dfrac{\overset{7}{\not{28}}}{\not{9}} = \dfrac{7}{1} = 7$

19. $2\frac{1}{3} \cdot 4\frac{1}{2} = \dfrac{7}{\underset{1}{\not{3}}} \cdot \dfrac{\overset{3}{\not{9}}}{2} = \dfrac{21}{2}$ or $10\frac{1}{2}$

21. $3 \cdot 4\frac{1}{3} = \dfrac{\not{3}}{1} \cdot \dfrac{13}{\not{3}} = \dfrac{13}{1} = 13$

23. $5\frac{1}{6} \cdot 12 = \dfrac{31}{\underset{1}{\not{6}}} \cdot \dfrac{\overset{2}{\not{12}}}{1} = \dfrac{62}{1} = 62$

25. $\left(\dfrac{1}{3}\right)^2 = \dfrac{1}{3} \cdot \dfrac{1}{3} = \dfrac{1}{9}$

27. $\left(2\frac{1}{2}\right)^2 = \left(\dfrac{5}{2}\right)^2 = \dfrac{5}{2} \cdot \dfrac{5}{2} = \dfrac{25}{4}$ or $6\frac{1}{4}$

29. a. $\dfrac{\overset{1}{\not{3}}}{\underset{1}{\not{4}}} \times \dfrac{\overset{2}{\not{8}}}{\underset{3}{\not{9}}} \times \dfrac{1}{5} = \dfrac{2}{15}$

b. $\dfrac{\not{3}}{\underset{2}{\not{12}}} \times \dfrac{\overset{1}{\not{6}}}{\not{7}} \times \dfrac{\not{7}}{\not{8}} = \dfrac{1}{2}$

31. $\left(\dfrac{2}{3}\right)^2 \cdot \dfrac{3}{4} = \dfrac{2}{3} \cdot \dfrac{\not{2}}{\not{3}} \cdot \dfrac{\not{3}}{\not{4}} = \dfrac{1}{3}$

33. $\dfrac{14}{27} \cdot \left(\dfrac{3}{7}\right)^2 = \dfrac{\overset{2}{\not{14}}}{\underset{3}{\not{27}}} \cdot \dfrac{\not{3}}{\not{7}} \cdot \dfrac{\not{3}}{7} = \dfrac{2}{21}$

35. $\left(\dfrac{2}{3}\right)^3 = \dfrac{2}{3} \cdot \dfrac{2}{3} \cdot \dfrac{2}{3} = \dfrac{8}{27}$

37. $5 \div \dfrac{2}{3} = \dfrac{5}{1} \cdot \dfrac{3}{2} = \dfrac{15}{2}$ or $7\frac{1}{2}$

39. $\dfrac{4}{5} \div 6 = \dfrac{\overset{2}{\not{4}}}{5} \cdot \dfrac{1}{\underset{3}{\not{6}}} = \dfrac{2}{15}$

41. $\dfrac{2}{3} \div \dfrac{6}{7} = \dfrac{\overset{1}{\not{2}}}{3} \cdot \dfrac{7}{\underset{3}{\not{6}}} = \dfrac{7}{9}$

43. $\dfrac{4}{5} \div \dfrac{8}{15} = \dfrac{\not{4}}{\not{5}} \cdot \dfrac{\overset{3}{\not{15}}}{\underset{2}{\not{8}}} = \dfrac{3}{2}$ or $1\frac{1}{2}$

45. $\dfrac{2}{3} \div \dfrac{5}{12} = \dfrac{2}{\cancel{3}} \cdot \dfrac{\cancel{12}^{4}}{5} = \dfrac{8}{5}$ or $1\dfrac{3}{5}$

47. $\dfrac{3}{4} \div \dfrac{3}{4} = \dfrac{\cancel{3}}{\cancel{4}} \cdot \dfrac{\cancel{4}}{\cancel{3}} = 1$

49. $\dfrac{3}{5} \div 1\dfrac{1}{2} = \dfrac{3}{5} \div \dfrac{3}{2} = \dfrac{\cancel{3}}{5} \cdot \dfrac{2}{\cancel{3}} = \dfrac{2}{5}$

51. $3\dfrac{3}{4} \div \dfrac{3}{8} = \dfrac{15}{4} \div \dfrac{3}{8} = \dfrac{\cancel{15}^{5}}{\cancel{4}_{1}} \cdot \dfrac{\cancel{8}^{2}}{\cancel{3}_{1}} = 10$

53. $6\dfrac{1}{2} \div 2\dfrac{1}{2} = \dfrac{13}{2} \div \dfrac{5}{2} = \dfrac{13}{\cancel{2}} \cdot \dfrac{\cancel{2}}{5} = \dfrac{13}{5}$ or $2\dfrac{3}{5}$

55. $3\dfrac{1}{8} \div 1\dfrac{1}{3} = \dfrac{25}{8} \div \dfrac{4}{3} = \dfrac{25}{8} \cdot \dfrac{3}{4} = \dfrac{75}{32}$ or $2\dfrac{11}{32}$

57. $3\dfrac{1}{8} \div 3\dfrac{1}{8} = \dfrac{25}{8} \div \dfrac{25}{8} = \dfrac{\cancel{25}}{\cancel{8}} \cdot \dfrac{\cancel{8}}{\cancel{25}} = 1$

59. $1\dfrac{2}{3} \div 13\dfrac{3}{4} = \dfrac{5}{3} \div \dfrac{55}{4} = \dfrac{\cancel{5}}{3} \cdot \dfrac{4}{\cancel{55}_{11}} = \dfrac{4}{33}$

61. $\dfrac{\cancel{6}}{7} \cdot \dfrac{2}{\cancel{6}} = \dfrac{2}{7}$ square miles

63. $\dfrac{4}{5} \cdot 90 = \dfrac{4}{\cancel{5}_{1}} \cdot \dfrac{\cancel{90}^{18}}{1} = \dfrac{72}{1} = 72$ people

65. $\dfrac{8}{15} \cdot 30 = \dfrac{8}{\cancel{15}_{1}} \cdot \dfrac{\cancel{30}^{2}}{1} = \dfrac{16}{1} = 16$ days

67. $1\dfrac{1}{2} \div \dfrac{3}{16} = \dfrac{3}{2} \div \dfrac{3}{16} = \dfrac{\cancel{3}}{\cancel{2}_{1}} \cdot \dfrac{\cancel{16}^{8}}{\cancel{3}} = 8$ turns

69. $10\dfrac{1}{2} \div \dfrac{5}{8} = \dfrac{21}{2} \div \dfrac{5}{8} = \dfrac{21}{\cancel{2}_{1}} \cdot \dfrac{\cancel{8}^{4}}{5} = \dfrac{84}{5}$ or $16\dfrac{4}{5}$

vests

71. $40 \cdot 16\dfrac{1}{2} = \dfrac{\cancel{40}^{20}}{1} \cdot \dfrac{33}{\cancel{2}_{1}} = 660$ feet

73. $5 \cdot 2\dfrac{1}{5} = \dfrac{\cancel{5}}{1} \cdot \dfrac{11}{\cancel{5}} = 11$ liters

75. $80\dfrac{3}{5} \div 6\dfrac{1}{5} = \dfrac{403}{5} \div \dfrac{31}{5} = \dfrac{\cancel{403}^{13}}{\cancel{5}} \cdot \dfrac{\cancel{5}}{\cancel{31}_{1}} = 13$ gal

77. $\dfrac{18\cancel{0}}{16\cancel{0}} = \dfrac{9}{8} = 1\dfrac{1}{8}$ grams

79. $\dfrac{13\cancel{0}}{12\cancel{0}} = \dfrac{13}{12} = 1\dfrac{1}{12}$ grams

81. $4\dfrac{3}{8} \cdot 5\dfrac{3}{4} = \dfrac{35}{8} \cdot \dfrac{23}{4} = \dfrac{805}{32} = 25\dfrac{5}{32}$ sq. in.

83. $4\dfrac{1}{8} \cdot 9\dfrac{1}{2} = \dfrac{33}{8} \cdot \dfrac{19}{2} = \dfrac{627}{16} = 39\dfrac{3}{16}$ sq. in.

85. $9\dfrac{1}{4} \cdot 4\dfrac{2}{3} = \dfrac{37}{4} \cdot \dfrac{14}{3} = \dfrac{518}{12} = 43\dfrac{1}{6}$ sq. in.

87. $10\dfrac{1}{2} \cdot 12\dfrac{4}{5} = \dfrac{21}{2} \cdot \dfrac{64}{5} = \dfrac{1344}{10} = 134\dfrac{2}{5}$ sq. in.

89. $15\dfrac{3}{4} \div 3\dfrac{1}{2} = \dfrac{63}{4} \div \dfrac{7}{2} = \dfrac{\cancel{63}^{9}}{\cancel{4}_{2}} \cdot \dfrac{\cancel{2}}{\cancel{7}} = \dfrac{9}{2} = 4\dfrac{1}{2}$ yd

91. $84 = 2 \times 42$
$= 2 \times 2 \times 21$
$= 2 \times 2 \times 3 \times 7$
$= 2^2 \times 3 \times 7$

93. $72 = 9 \times 8$
$= (3 \times 3) \times (2 \times 2 \times 2)$
$= 2^3 \times 3^2$

95. $105 = 5 \times 21 = 5 \times 3 \times 7 = 3 \times 5 \times 7$

97. $36 \times 3\frac{1}{2} = \frac{\overset{18}{\cancel{36}}}{1} \cdot \frac{7}{\cancel{2}} = 126$ miles

99. $36 \times 2\frac{1}{2} = \frac{\overset{18}{\cancel{36}}}{1} \cdot \frac{5}{\cancel{2}} = 90$ miles

101. $2\frac{1}{2} \times 1\frac{1}{4} = \frac{5}{2} \cdot \frac{5}{4} = \frac{25}{8} = 3\frac{1}{8}$ inches

103. Answers may vary.

105. Answers may vary.

Mastery Test 2.3

107. $2 \cdot \frac{1}{4} = \frac{2}{1} \cdot \frac{1}{4} = \frac{1}{2}$ cup

109. $5 \cdot \frac{3}{4} = \frac{5}{1} \cdot \frac{3}{4} = \frac{15}{4}$ or $3\frac{3}{4}$

111. $\left(\frac{3}{4}\right)^2 \cdot \left(1\frac{1}{2}\right) = \frac{3}{4} \cdot \frac{3}{4} \cdot \frac{3}{2} = \frac{27}{32}$

113. $1\frac{3}{5} \div \frac{2}{5} = \frac{8}{5} \div \frac{2}{5} = \frac{8}{\cancel{5}} \cdot \frac{\cancel{5}}{2} = \frac{8}{2} = 4$

115. $3\frac{1}{3} \times 4\frac{2}{3} = \frac{10}{3} \times \frac{14}{3} = \frac{140}{9} = 15\frac{5}{9}$ sq. yd.

Section 2.4 – Addition and Subtraction of Fractions

Problems

1. a. $\frac{2}{11} + \frac{3}{11} = \frac{2+3}{11} = \frac{5}{11}$

  b. $\frac{1}{8} + \frac{3}{8} = \frac{1+3}{8} = \frac{4}{8} = \frac{1}{2}$

2. Multiples of 8: 8, 16, 24, 32, …. Since 6 goes into 24, The LCD is 24.
$\frac{3}{8} = \frac{3 \times 3}{8 \times 3} = \frac{9}{24}$ and $\frac{1}{6} = \frac{1 \times 4}{6 \times 4} = \frac{4}{24}$. Thus
$\frac{3}{8} + \frac{1}{6} = \frac{9}{24} + \frac{4}{24} = \frac{13}{24}$

3. Multiples of 9: 9, 18, 27, 36, 45, ….
Since 4 goes into 36, The LCD is 36.
$\frac{3}{4} = \frac{3 \times 9}{4 \times 9} = \frac{27}{36}$ and $\frac{5}{9} = \frac{5 \times 4}{9 \times 4} = \frac{20}{36}$. Thus
$\frac{3}{4} + \frac{5}{9} = \frac{27}{36} + \frac{20}{36} = \frac{47}{36}$

4. $\frac{1}{40}$ and $\frac{1}{12}$: $40 = 2^3 \cdot 5$
$\phantom{\frac{1}{40} \text{ and } \frac{1}{12}: \quad} 12 = 2^2 \cdot 3$
The LCD is $2^3 \cdot 3 \cdot 5 = 120$.
- OR -

$\begin{array}{r|ll} 2 & 40 & 12 \\ 2 & 20 & 6 \\ \hline & 10 & 3 \end{array}$ Thus, the LCD is $2 \cdot 2 \cdot 10 \cdot 3 = 120$.

5. $\frac{1}{40} = \frac{1 \times 3}{40 \times 3} = \frac{3}{120}$ and $\frac{5}{12} + \frac{5 \times 10}{12 \times 10} = \frac{50}{120}$
Thus, $\frac{1}{40} + \frac{5}{12} = \frac{3}{120} + \frac{50}{120} = \frac{53}{120}$.

6. Multiples of 12: 12, 24, 36, 48, 60, 72, 84, …. Since 8 and 9 goes into 72, The LCD is 72. $\frac{1}{8} = \frac{1 \times 9}{8 \times 9} = \frac{9}{72}$, $\frac{1}{12} = \frac{1 \times 6}{12 \times 6} = \frac{6}{72}$, and $\frac{1}{9} = \frac{1 \times 8}{9 \times 8} = \frac{8}{72}$.
Thus $\frac{1}{8} + \frac{1}{12} + \frac{1}{9} = \frac{9}{72} + \frac{6}{72} + \frac{8}{72} = \frac{23}{72}$.

7. $12 = 2^2 \cdot 3$
$10 = 2 \cdot 5$
The LCD is $2^2 \cdot 3 \cdot 5 = 60$.
- OR -

$\underline{2|12 \quad 10}$

$\quad 6 \quad 5$ Thus, the LCD is $2 \cdot 6 \cdot 5 = 60$.

$\dfrac{7}{12} = \dfrac{7 \times 5}{12 \times 5} = \dfrac{35}{60}$ and $\dfrac{1}{10} + \dfrac{1 \times 6}{10 \times 6} = \dfrac{6}{60}$

Thus $\dfrac{7}{12} - \dfrac{1}{10} = \dfrac{35}{60} - \dfrac{6}{60} = \dfrac{35 - 6}{60} = \dfrac{29}{60}$.

8. Multiples of 9: 9, 18, 27, 36, 45, 54, 63, 72,.... Since 8, 6 and 9 goes into 72, The LCD is 72.

$\dfrac{3}{8} = \dfrac{3 \times 9}{8 \times 9} = \dfrac{27}{72}, \dfrac{1}{6} = \dfrac{1 \times 12}{6 \times 12} = \dfrac{12}{72},$

and $\dfrac{2}{9} = \dfrac{2 \times 8}{9 \times 8} = \dfrac{16}{72}$. Thus

$\dfrac{3}{8} + \dfrac{1}{6} - \dfrac{2}{9} = \dfrac{27}{72} + \dfrac{12}{72} - \dfrac{16}{72}$

$\qquad = \dfrac{27 + 12 - 16}{72}$

$\qquad = \dfrac{23}{72}$.

9. a. $\dfrac{37}{100} + \dfrac{1}{10} = \dfrac{37}{100} + \dfrac{10}{100} = \dfrac{47}{100}$

The fraction of the students having brown or gray eyes is $\dfrac{47}{100}$.

b. $\dfrac{1}{5} + \dfrac{33}{100} = \dfrac{20}{100} + \dfrac{33}{100} = \dfrac{53}{100}$

The fraction of the students having green or blue eyes is $\dfrac{53}{100}$.

Exercises 2.4

1. $\dfrac{1}{3} + \dfrac{1}{3} = \dfrac{1 + 1}{3} = \dfrac{2}{3}$

3. $\dfrac{1}{7} + \dfrac{4}{7} = \dfrac{1 + 4}{7} = \dfrac{5}{7}$

5. $\dfrac{2}{9} + \dfrac{4}{9} = \dfrac{2 + 4}{9} = \dfrac{6}{9} = \dfrac{2}{3}$

7. $\dfrac{1}{6} + \dfrac{5}{6} = \dfrac{1 + 5}{6} = \dfrac{6}{6} = 1$

9. $\dfrac{3}{4} + \dfrac{5}{4} = \dfrac{3 + 5}{4} = \dfrac{8}{4} = 2$

11. The LCD of 3 and 5 is 15.

$\dfrac{1}{3} = \dfrac{1 \times 5}{3 \times 5} = \dfrac{5}{15}$ and $\dfrac{1}{5} = \dfrac{1 \times 3}{5 \times 3} = \dfrac{3}{15}$

Thus $\dfrac{1}{3} + \dfrac{1}{5} = \dfrac{5}{15} + \dfrac{3}{15} = \dfrac{8}{15}$.

13. The LCD of 2 and 6 is 6. $\dfrac{1}{2} = \dfrac{1 \times 3}{2 \times 3} = \dfrac{3}{6}$

Thus $\dfrac{1}{2} + \dfrac{1}{6} = \dfrac{3}{6} + \dfrac{1}{6} = \dfrac{4}{6} = \dfrac{2}{3}$.

15. The LCD of 2 and 5 is 10.

$\dfrac{1}{2} = \dfrac{1 \times 5}{2 \times 5} = \dfrac{5}{10}$ and $\dfrac{2}{5} = \dfrac{2 \times 2}{5 \times 2} = \dfrac{4}{10}$

Thus $\dfrac{1}{2} + \dfrac{2}{5} = \dfrac{5}{10} + \dfrac{4}{10} = \dfrac{9}{10}$.

17. The LCD of 7 and 14 is 14. $\dfrac{4}{7} = \dfrac{4 \times 2}{7 \times 2} = \dfrac{8}{14}$

Thus $\dfrac{4}{7} + \dfrac{3}{14} = \dfrac{8}{14} + \dfrac{3}{14} = \dfrac{11}{14}$.

19. The LCD of 2 and 8 is 8. $\dfrac{1}{2} = \dfrac{1 \times 4}{2 \times 4} = \dfrac{4}{8}$

Thus $\dfrac{1}{2} + \dfrac{3}{8} = \dfrac{4}{8} + \dfrac{3}{8} = \dfrac{7}{8}$.

21. $\underline{2|40 \quad 18}$

$\quad 20 \quad 9$ The LCD is $2 \cdot 20 \cdot 9 = 360$.

- OR -

$40 = 2^3 \cdot 5$

$18 = 2 \cdot 3^2$ so the LCD is $2^3 \cdot 3^2 \cdot 5 = 360$.

$\dfrac{1}{40} = \dfrac{1 \times 9}{40 \times 9} = \dfrac{9}{360}; \dfrac{1}{18} = \dfrac{1 \times 20}{18 \times 20} = \dfrac{20}{360}$.

Thus $\dfrac{1}{40} + \dfrac{1}{18} = \dfrac{9}{360} + \dfrac{20}{360} = \dfrac{29}{360}$.

23. $65 = 5 \cdot 13$

$26 = 2 \cdot 13$; The LCD is $2 \cdot 5 \cdot 13 = 130$.

- OR -

$\underline{13|65 \quad 26}$

$\quad 5 \quad 2$ so the LCD is $13 \cdot 5 \cdot 2 = 130$.

$\dfrac{2}{65} = \dfrac{2 \times 2}{65 \times 2} = \dfrac{4}{130}; \dfrac{3}{26} = \dfrac{3 \times 5}{26 \times 5} = \dfrac{15}{130}$

Thus $\dfrac{2}{65} + \dfrac{3}{26} = \dfrac{4}{130} + \dfrac{15}{130} = \dfrac{19}{130}$.

25.

$$\begin{array}{r|rr} 2 & 120 & 180 \\ 2 & 60 & 90 \\ 3 & 30 & 45 \\ 5 & 10 & 15 \\ \hline & 2 & 3 \end{array}$$

The LCD is $2^3 \cdot 3^2 \cdot 5 = 360$.

- OR -

$120 = 2^3 \cdot 3 \cdot 5$

$180 = 2^2 \cdot 3^2 \cdot 5$

so the LCD is $2^3 \cdot 3^2 \cdot 5 = 360$.

$$\frac{7}{120} = \frac{7 \times 3}{120 \times 3} = \frac{21}{360}; \quad \frac{1}{180} = \frac{1 \times 2}{180 \times 2} = \frac{2}{360}$$

Thus $\dfrac{7}{120} + \dfrac{1}{180} = \dfrac{21}{360} + \dfrac{2}{360} = \dfrac{23}{360}$.

27. Multiples of 60: 60, 120, …. Since 10 and 20 goes into 60, the LCD is 60.

$$\frac{3}{10} + \frac{7}{20} + \frac{11}{60} = \frac{3 \cdot 6}{10 \cdot 6} + \frac{7 \cdot 3}{20 \cdot 3} + \frac{11}{60}$$
$$= \frac{18}{60} + \frac{21}{60} + \frac{11}{60}$$
$$= \frac{50}{60}$$
$$= \frac{5}{6}$$

29.

$$\begin{array}{r|rrr} 2 & 14 & 6 & 9 \\ 3 & 7 & 3 & 9 \\ \hline & 7 & 1 & 3 \end{array}$$

so the LCD is $2 \cdot 3 \cdot 7 \cdot 3 = 126$.

$$\frac{11}{14} + \frac{5}{6} + \frac{8}{9} = \frac{11 \cdot 9}{14 \cdot 9} + \frac{5 \cdot 21}{6 \cdot 21} + \frac{8 \cdot 14}{9 \cdot 14}$$
$$= \frac{99}{126} + \frac{105}{126} + \frac{112}{126}$$
$$= \frac{316}{126}$$
$$= \frac{158}{63} \text{ or } 2\frac{32}{63}$$

31. $\dfrac{3}{7} - \dfrac{1}{7} = \dfrac{3-1}{7} = \dfrac{2}{7}$

33. $\dfrac{5}{6} - \dfrac{1}{6} = \dfrac{5-1}{6} = \dfrac{4}{6} = \dfrac{2}{3}$

35. The LCD of 12 and 4 is 12.

$$\frac{5}{12} - \frac{1}{4} = \frac{5}{12} - \frac{1 \cdot 3}{4 \cdot 3} = \frac{5}{12} - \frac{3}{12} = \frac{2}{12} = \frac{1}{6}$$

37. The LCD of 2 and 5 is 10.

$$\frac{1}{2} - \frac{1}{5} = \frac{1 \cdot 5}{2 \cdot 5} - \frac{1 \cdot 2}{5 \cdot 2} = \frac{5}{10} - \frac{2}{10} = \frac{3}{10}$$

39. The LCD of 20 and 40 is 40.

$$\frac{5}{20} - \frac{7}{40} = \frac{5 \cdot 2}{20 \cdot 2} - \frac{7}{40} = \frac{10}{40} - \frac{7}{40} = \frac{3}{40}$$

41. The LCD of 8 and 12 is 24.

$$\frac{7}{8} - \frac{5}{12} = \frac{7 \cdot 3}{8 \cdot 3} - \frac{5 \cdot 2}{12 \cdot 2} = \frac{21}{24} - \frac{10}{24} = \frac{11}{24}$$

43.

$$\begin{array}{r|rr} 2 & 60 & 48 \\ 2 & 30 & 24 \\ 3 & 15 & 12 \\ \hline & 5 & 4 \end{array}$$

so the LCD is $2 \cdot 2 \cdot 3 \cdot 5 \cdot 4 = 120$.

$$\frac{13}{60} - \frac{1}{48} = \frac{13 \cdot 4}{60 \cdot 4} - \frac{1 \cdot 5}{48 \cdot 5}$$
$$= \frac{52}{240} - \frac{5}{240}$$
$$= \frac{47}{240}$$

45. $\dfrac{8}{9} - \dfrac{2}{9} - \dfrac{1}{9} = \dfrac{8-2-1}{9} = \dfrac{5}{9}$

47. The LCD of 4, 12, and 6 is 12.

$$\frac{3}{4} + \frac{1}{12} - \frac{1}{6} = \frac{3 \cdot 3}{4 \cdot 3} + \frac{1}{12} - \frac{1 \cdot 2}{6 \cdot 2}$$
$$= \frac{9}{12} + \frac{1}{12} - \frac{2}{12}$$
$$= \frac{9+1-2}{12}$$
$$= \frac{8}{12}$$
$$= \frac{2}{3}$$

49. The LCD of 2 and 3 is 6.

$$\frac{9}{2} - \frac{7}{3} = \frac{9 \cdot 3}{2 \cdot 3} - \frac{7 \cdot 2}{3 \cdot 2} = \frac{27}{6} - \frac{14}{6} = \frac{13}{6} \text{ or } 2\frac{1}{6}$$

51.

$$\frac{3}{4} + \frac{3}{8} + \frac{1}{32} = \frac{3 \cdot 8}{4 \cdot 8} + \frac{3 \cdot 4}{8 \cdot 4} + \frac{1}{32}$$
$$= \frac{24}{32} + \frac{12}{32} + \frac{1}{32}$$
$$= \frac{37}{32} \text{ or } 1\frac{5}{32}$$

The result is $\dfrac{37}{32}$ $\left(\text{or } 1\dfrac{5}{32}\right)$ in. thick.

53. $\dfrac{1}{4}+\dfrac{1}{2}+\dfrac{1}{8}=\dfrac{2}{8}+\dfrac{4}{8}+\dfrac{1}{8}=\dfrac{7}{8}$ of his estate was left to his daughter, wife, and son together. Thus $1-\dfrac{7}{8}=\dfrac{8}{8}-\dfrac{7}{8}=\dfrac{1}{8}$ of his estate remains.

55. The LCD of 1, 4, and 20 is 20.
$1-\dfrac{1}{4}-\dfrac{3}{10}=\dfrac{20}{20}-\dfrac{5}{20}-\dfrac{6}{20}=\dfrac{9}{20}$

57. $\dfrac{2}{12}=\dfrac{1}{6}$. The fraction of the time spent watching TV is $\dfrac{1}{6}$.

59. $\dfrac{3}{10}+\dfrac{9}{20}=\dfrac{6}{20}+\dfrac{9}{20}=\dfrac{15}{20}=\dfrac{3}{4}$
The fraction of the expenses for benefits or salary is $\dfrac{3}{4}$.

61. $\dfrac{1}{10}+\dfrac{1}{10}=\dfrac{2}{10}=\dfrac{1}{5}$
The fraction of the days rainy or snowy is $\dfrac{1}{5}$.

63. $\dfrac{3}{10}+\dfrac{1}{5}=\dfrac{3}{10}+\dfrac{2}{10}=\dfrac{5}{10}=\dfrac{1}{2}$
The fraction of the people who walk or use a car is $\dfrac{1}{2}$.

65. $1-\dfrac{3}{10}=\dfrac{10}{10}-\dfrac{3}{10}=\dfrac{7}{10}$
The fraction of the people who do not walk is $\dfrac{7}{10}$.

67. $6\dfrac{7}{8}=\dfrac{8\cdot 6+7}{8}=\dfrac{48+7}{8}=\dfrac{55}{8}$

69. $\dfrac{10}{6}=1\dfrac{4}{6}=1\dfrac{2}{3}$

71. Multiples of 10 (hot dogs):
      10, 20, 30, 40, 50, ….
Since 8 goes into 40, the LCD is 40.
10×4 packs $=40$ hot dogs and
8×5 packs $=40$ buns . Thus, you must buy at least 4 packages of hot dogs and 5 packages of buns.

73. Yes; answers may vary.

75. Answers may vary.

Mastery Test 2.4

77. $\dfrac{1}{30}$ and $\dfrac{1}{18}$.
$30=2\cdot 3\cdot 5$
$18=2\cdot 3^2$ so the LCD $=2\cdot 3^2\cdot 5=90$.

79. $\dfrac{1}{8}+\dfrac{1}{6}=\dfrac{3}{24}+\dfrac{4}{24}=\dfrac{7}{24}$

81. $\dfrac{1}{10}+\dfrac{1}{12}+\dfrac{3}{8}=\dfrac{12}{120}+\dfrac{10}{120}+\dfrac{45}{120}$
$=\dfrac{12+10+45}{120}$
$=\dfrac{67}{120}$

83. $\dfrac{3}{10}+\dfrac{1}{12}-\dfrac{1}{8}=\dfrac{36}{120}+\dfrac{10}{120}-\dfrac{15}{120}$
$=\dfrac{36+10-15}{120}$
$=\dfrac{31}{120}$

Section 2.5 – Addition and Subtraction of Mixed Numbers

Problems

1. Method 1:

$$3\frac{1}{9}+2\frac{4}{9}=\frac{28}{9}+\frac{22}{9}=\frac{50}{9}=5\frac{5}{9}$$

Method 2:
$$\begin{array}{r}3\frac{1}{9}\\+2\frac{4}{9}\\\hline 5\frac{5}{9}\end{array}$$

2. Multiples of 6: 6, 12, 18, …. Since 4 goes into 12, the LCD of 4 and 6 is 12.

$$1\frac{3}{4}=\frac{7}{4}=\frac{7\cdot 3}{4\cdot 3}=\frac{21}{12}\text{ and }\frac{1}{6}=\frac{1\cdot 2}{6\cdot 2}=\frac{2}{12}\text{ so}$$

$$1\frac{3}{4}+\frac{1}{6}=\frac{21}{12}+\frac{2}{12}=\frac{23}{12}=1\frac{11}{12}$$

3. The LCD of 4 and 6 is 12.

$$5\frac{1}{4}=\frac{21}{4}=\frac{21\cdot 3}{4\cdot 3}=\frac{63}{12}\text{ and}$$
$$1\frac{5}{6}=\frac{11}{6}=\frac{11\cdot 2}{6\cdot 2}=\frac{22}{12}\text{ so}$$

$$5\frac{1}{4}+1\frac{5}{6}=\frac{21}{4}+\frac{11}{6}=\frac{63}{12}+\frac{22}{12}=\frac{85}{12}=7\frac{1}{12}$$

4. The LCD of 6 and 9 is 18.

$$4\frac{1}{6}-3\frac{2}{9}=\frac{25}{6}-\frac{29}{9}$$
$$=\frac{25\cdot 3}{6\cdot 3}-\frac{29\cdot 2}{9\cdot 2}$$
$$=\frac{75}{18}-\frac{58}{18}$$
$$=\frac{17}{18}$$

5.
$$\begin{array}{l}2\,|\,8\ 10\ 12\\2\,|\,4\ \ 5\ \ 6\\\ \ \ \ 2\ \ 5\ \ 3\end{array}\text{ so the LCD is }2^3\cdot 5\cdot 3=120.$$

$$1\frac{3}{8}+2\frac{3}{10}-2\frac{1}{12}=\frac{11}{8}+\frac{23}{10}-\frac{25}{12}$$
$$=\frac{11\cdot 15}{8\cdot 15}+\frac{23\cdot 12}{10\cdot 12}-\frac{25\cdot 10}{12\cdot 10}$$
$$=\frac{165}{120}+\frac{276}{120}-\frac{250}{120}$$
$$=\frac{165+276-250}{120}$$
$$=\frac{191}{120}=1\frac{71}{120}$$

6. $$20\frac{8}{12}\text{ ft}+15\text{ ft}+20\frac{8}{12}\text{ ft}+15\text{ ft}$$
$$=20\frac{2}{3}\text{ ft}+20\frac{2}{3}\text{ ft}+15\text{ ft}+15\text{ ft}$$
$$=40\frac{4}{3}\text{ ft}+30\text{ ft}$$
$$=\left(40+1\frac{1}{3}\right)\text{ ft}+30\text{ ft}$$
$$=41\frac{1}{3}\text{ ft}+30\text{ ft}$$
$$=71\frac{1}{3}\text{ ft}$$

You need $71\frac{1}{3}$ feet of molding.

Exercises 2.5

1. $3\frac{1}{7}+1\frac{3}{7}=4\frac{4}{7}$

3. $2\frac{1}{7}+\frac{3}{7}=2\frac{4}{7}$

5. $\frac{3}{8}+5\frac{1}{8}=5\frac{4}{8}=5\frac{1}{2}$

7. $1\frac{3}{5}+2\frac{4}{5}=3\frac{7}{5}=3+1\frac{2}{5}=4\frac{2}{5}$

9. $2+3\frac{1}{7}=5\frac{1}{7}$

11. The LCD of 4 and 15 is 60.

$$\frac{3}{4} = \frac{3 \cdot 15}{4 \cdot 15} = \frac{45}{60} \text{ and } \frac{2}{15} = \frac{2 \cdot 4}{15 \cdot 4} = \frac{8}{60}$$

Thus $2\frac{3}{4} + \frac{2}{15} = 2\frac{45}{60} + \frac{8}{60} = 2\frac{53}{60}$.

13. The LCD of 10 and 12 is 60.

$$\frac{3}{10} = \frac{3 \cdot 6}{10 \cdot 6} = \frac{18}{60} \text{ and } \frac{11}{12} = \frac{11 \cdot 5}{12 \cdot 5} = \frac{55}{60} \text{ so}$$

$$1\frac{3}{10} + 2\frac{11}{12} = 1\frac{18}{60} + 2\frac{55}{60}$$
$$= 3\frac{73}{60} = 3 + 1\frac{13}{60} = 4\frac{13}{60}$$

15. The LCD of 4 and 6 is 12.

$$1\frac{3}{4} + 2\frac{5}{6} = 1\frac{3 \cdot 3}{4 \cdot 3} + 2\frac{5 \cdot 2}{6 \cdot 2}$$
$$= 1\frac{9}{12} + 2\frac{10}{12}$$
$$= 3\frac{19}{12}$$
$$= 3 + 1\frac{7}{12}$$
$$= 4\frac{7}{12}$$

17. The LCD of 7 and 9 is 63.

$$8\frac{1}{7} + 3\frac{1}{9} = 8\frac{1 \cdot 9}{7 \cdot 9} + 3\frac{1 \cdot 7}{9 \cdot 7}$$
$$= 8\frac{9}{63} + 3\frac{7}{63}$$
$$= 11\frac{16}{63}$$

19. The LCD of 7 and 9 is 63.

$$9\frac{1}{11} + 3\frac{1}{10} = 9\frac{1 \cdot 10}{11 \cdot 10} + 3\frac{1 \cdot 11}{10 \cdot 11}$$
$$= 9\frac{10}{110} + 3\frac{11}{110}$$
$$= 12\frac{21}{110}$$

21. $3\frac{3}{7} - 1\frac{1}{7} = 2\frac{2}{7}$

23. $4\frac{5}{6} - 3\frac{1}{6} = 1\frac{4}{6} = 1\frac{2}{3}$

25. The LCD of 12 and 4 is 12.

$$3\frac{1}{12} - 1\frac{1}{4} = \frac{37}{12} - \frac{5 \cdot 3}{4 \cdot 3}$$
$$= \frac{37}{12} - \frac{15}{12}$$
$$= \frac{22}{12}$$
$$= \frac{11}{6}$$
$$= 1\frac{5}{6}$$

27. The LCD of 2 and 5 is 12.

$$3\frac{1}{2} - 2\frac{4}{5} = \frac{7 \cdot 5}{2 \cdot 5} - \frac{14 \cdot 2}{5 \cdot 2} = \frac{35}{10} - \frac{28}{10} = \frac{7}{10}$$

29. The LCD of 20 and 40 is 40.

$$4\frac{1}{20} - 3\frac{3}{40} = \frac{81 \cdot 2}{20 \cdot 2} - \frac{123}{40} = \frac{162}{40} - \frac{123}{40} = \frac{39}{40}$$

31. The LCD of 8 and 12 is 24.

$$3\frac{7}{8} - 1\frac{5}{12} = \frac{31 \cdot 3}{8 \cdot 3} - \frac{17 \cdot 2}{12 \cdot 2}$$
$$= \frac{93}{24} - \frac{34}{24} = \frac{59}{24} = 2\frac{11}{24}$$

33.

```
2 | 60  48
2 | 30  24
3 | 15  12
    5   4
```

so the LCD is $2 \cdot 2 \cdot 3 \cdot 5 \cdot 4 = 240$.

$$3\frac{13}{60} - 3\frac{1}{48} = 3\frac{13 \cdot 4}{60 \cdot 4} - 3\frac{1 \cdot 5}{48 \cdot 5}$$
$$= 3\frac{52}{240} - 3\frac{5}{240}$$
$$= \frac{47}{240}$$

35. Since the denominators are the same, we have $3\frac{8}{9} + 1\frac{2}{9} - 1\frac{1}{9} = 3 + 1 - 1 + \frac{8}{9} + \frac{2}{9} - \frac{1}{9}$

$$= 3 + \frac{9}{9}$$
$$= 3 + 1$$
$$= 4$$

37. The LCD of 4, 12, and 6 is 12.

$$3\frac{3}{4}+1\frac{1}{12}-1\frac{1}{6}=\frac{15\cdot3}{4\cdot3}+\frac{13}{12}-\frac{7\cdot2}{6\cdot2}$$
$$=\frac{45}{12}+\frac{13}{12}-\frac{14}{12}$$
$$=\frac{45+13-14}{12}$$
$$=\frac{44}{12}$$
$$=\frac{11}{3}=3\frac{2}{3}$$

39. The LCD of 2, 3, and 4 is 12.

$$4\frac{1}{2}-2\frac{1}{3}+3\frac{1}{4}=\frac{9\cdot6}{2\cdot6}-\frac{7\cdot4}{3\cdot4}+\frac{13\cdot3}{4\cdot3}$$
$$=\frac{54}{12}-\frac{28}{12}+\frac{39}{12}$$
$$=\frac{54-28+39}{12}$$
$$=\frac{65}{12}$$
$$=5\frac{5}{12}$$

41. The LCD of 65 and 26 is 130.

$$3\frac{1}{65}+10\frac{1}{26}-1\frac{2}{65}$$
$$=3\frac{1\cdot2}{65\cdot2}+10\frac{1\cdot5}{26\cdot5}-1\frac{2\cdot2}{65\cdot2}$$
$$=3\frac{2}{130}+10\frac{5}{130}-1\frac{4}{130}$$
$$=3+10-1+\frac{2}{130}+\frac{5}{130}-\frac{4}{130}$$
$$=12+\frac{3}{130}$$
$$=12\frac{3}{130}$$

43. The LCD of 45 and 60 is 180.

$$\begin{array}{ccc}14\frac{11}{45} & 14\frac{11\cdot4}{45\cdot4} & 14\frac{44}{180}\\[6pt]+7\frac{7}{60}=&+7\frac{7\cdot3}{60\cdot3}=&+7\frac{21}{180}\\[6pt]-3\frac{8}{45} & -3\frac{8\cdot4}{45\cdot4} & -3\frac{32}{180}\\[6pt]\hline & & =18\frac{33}{180}=18\frac{11}{60}\end{array}$$

45. $101\frac{6}{10}-98\frac{6}{10}=3$

That is 3 degrees above normal.

47. $4\frac{7}{16}-3\frac{1}{8}=4\frac{7}{8}-3\frac{1\cdot2}{8\cdot2}=4\frac{7}{8}-3\frac{2}{16}=1\frac{5}{16}$

This weight is $1\frac{5}{16}$ lb above the average.

49. $2\frac{1}{2}+\frac{3}{4}=\frac{5\cdot2}{2\cdot2}+\frac{3}{4}=\frac{10}{4}+\frac{3}{4}=\frac{13}{4}=3\frac{1}{4}$

These ingredients total $3\frac{1}{4}$ cups.

51. $\frac{1}{4}+2\frac{1}{2}+3=\frac{1}{4}+2\frac{1\cdot2}{2\cdot2}+3$
$$=\frac{1}{4}+2\frac{2}{4}+3$$
$$=2\frac{3}{4}+3$$
$$=5\frac{3}{4}$$

The total weight was $5\frac{3}{4}$ lb.

53. $1-\frac{1}{4}-\frac{9}{20}=\frac{20}{20}-\frac{5}{20}-\frac{9}{20}=\frac{6}{20}=\frac{3}{10}$

This fraction is $\frac{3}{10}$.

55. $46\frac{3}{5}-38\frac{9}{10}=46\frac{3\cdot2}{5\cdot2}-38\frac{9}{10}$
$$=46\frac{6}{10}-38\frac{9}{10}$$
$$=\frac{466}{10}-\frac{389}{10}=\frac{77}{10}=7\frac{7}{10}$$

Americans work $7\frac{7}{10}$ more hours per week than Canadians.

57. $15\frac{1}{4}+9\frac{2}{5}=15\frac{5}{20}+9\frac{8}{20}=24\frac{13}{20}$

Pedro worked a total of $24\frac{13}{20}$ hours.

59. $11\frac{4}{12}+14\frac{8}{12}+11\frac{4}{12}+14\frac{8}{12}=50\frac{24}{12}$
$$=50+2=52$$

You need 52 feet of baseboard molding.

61. $\dfrac{\overset{1}{\cancel{5}}}{\underset{2}{\cancel{18}}}\cdot\dfrac{\overset{1}{\cancel{9}}}{\underset{2}{\cancel{10}}}=\dfrac{1}{4}$

63. $\dfrac{4}{5} \div \dfrac{15}{32} = \dfrac{4}{5} \cdot \dfrac{32}{15} = \dfrac{128}{75}$

65. $\$3\dfrac{1}{4} - \$\dfrac{1}{8} = \$3\dfrac{2}{8} - \$\dfrac{1}{8} = \$3\dfrac{1}{8}$ per share

67. $\$62\dfrac{3}{8} + \$\dfrac{1}{4} = \$62\dfrac{3}{8} + \$\dfrac{2}{8} = \$62\dfrac{5}{8}$ per share

69. No; answers may vary. For example,
$\dfrac{1}{2} + \dfrac{3}{4} = \dfrac{5}{4}$.

71. Answers may vary.

73. $2\dfrac{3}{4} + \dfrac{1}{15} = 2\dfrac{45}{60} + \dfrac{4}{60} = 2\dfrac{49}{60}$

75. $2\dfrac{3}{4} + 3\dfrac{5}{6} = 2\dfrac{9}{12} + 3\dfrac{10}{12} = 5\dfrac{19}{12} = 6\dfrac{7}{12}$

77. $2\dfrac{5}{9} + 3\dfrac{7}{10} - 4\dfrac{1}{12}$
$= 2\dfrac{5 \cdot 20}{9 \cdot 20} + 3\dfrac{7 \cdot 18}{10 \cdot 18} - 4\dfrac{1 \cdot 15}{12 \cdot 15}$
$= 2\dfrac{100}{180} + 3\dfrac{126}{180} - 4\dfrac{15}{180}$
$= 1\dfrac{211}{180} = 1 + 1\dfrac{31}{180} = 2\dfrac{31}{180}$

Section 2.6 – Order of Operations and Grouping Symbols

Problems

1. a. $\dfrac{1}{3} \cdot \left(\dfrac{3}{2}\right)^2 - \dfrac{1}{12} = \dfrac{1}{3} \cdot \dfrac{9}{4} - \dfrac{1}{12}$
$= \dfrac{9}{12} - \dfrac{1}{12} = \dfrac{8}{12} = \dfrac{2}{3}$

b. $\left(\dfrac{1}{3}\right)^3 + \dfrac{2}{3} \cdot \dfrac{1}{9} = \dfrac{1}{27} + \dfrac{2}{3} \cdot \dfrac{1}{9}$
$= \dfrac{1}{27} + \dfrac{2}{27} = \dfrac{3}{27} = \dfrac{1}{9}$

2. a. $\dfrac{3}{4} \div \dfrac{5}{6} - \left(\dfrac{1}{3} + \dfrac{1}{5}\right) = \dfrac{3}{4} \div \dfrac{5}{6} - \left(\dfrac{8}{15}\right)$
$= \dfrac{3}{4} \cdot \dfrac{6}{5} - \left(\dfrac{8}{15}\right)$
$= \dfrac{9}{10} - \left(\dfrac{8}{15}\right)$
$= \dfrac{27}{30} - \dfrac{16}{30} = \dfrac{11}{30}$

b. $27 \div \dfrac{1}{3} \cdot \dfrac{1}{3} \cdot \dfrac{1}{3} + \dfrac{1}{2} - 1 = 27 \cdot \dfrac{3}{1} \cdot \dfrac{1}{3} \cdot \dfrac{1}{3} + \dfrac{1}{2} - 1$
$= 81 \cdot \dfrac{1}{3} \cdot \dfrac{1}{3} + \dfrac{1}{2} - 1$
$= 27 \cdot \dfrac{1}{3} + \dfrac{1}{2} - 1$
$= 9 + \dfrac{1}{2} - 1$
$= 9\dfrac{1}{2} - 1 = 8\dfrac{1}{2}$

3. $\left(\dfrac{1}{2}\right)^3 \div \dfrac{1}{8} \cdot \dfrac{1}{2} + \dfrac{1}{3}\left(\dfrac{3}{2} - \dfrac{1}{2}\right) - \dfrac{1}{3} \cdot \dfrac{1}{2}$
$= \left(\dfrac{1}{2}\right)^3 \div \dfrac{1}{8} \cdot \dfrac{1}{2} + \dfrac{1}{3}(1) - \dfrac{1}{3} \cdot \dfrac{1}{2}$
$= \dfrac{1}{8} \div \dfrac{1}{8} \cdot \dfrac{1}{2} + \dfrac{1}{3}(1) - \dfrac{1}{3} \cdot \dfrac{1}{2}$
$= 1 \cdot \dfrac{1}{2} + \dfrac{1}{3}(1) - \dfrac{1}{3} \cdot \dfrac{1}{2}$
$= \dfrac{1}{2} + \dfrac{1}{3} - \dfrac{1}{6}$
$= \dfrac{5}{6} - \dfrac{1}{6}$
$= \dfrac{4}{6}$
$= \dfrac{2}{3}$

4. $\dfrac{1}{6} \div 1\dfrac{1}{6} + \left\{ 27 \cdot \left(\dfrac{1}{3}\right)^2 - \left[\dfrac{1}{3} + \left(2\dfrac{1}{3} - \dfrac{1}{3}\right)\right]\right\}$

$= \dfrac{1}{6} \div 1\dfrac{1}{6} + \left\{ 27 \cdot \left(\dfrac{1}{3}\right)^2 - \left[\dfrac{1}{3} + (2)\right]\right\}$

$= \dfrac{1}{6} \div 1\dfrac{1}{6} + \left\{ 27 \cdot \left(\dfrac{1}{3}\right)^2 - 2\dfrac{1}{3}\right\}$

$= \dfrac{1}{6} \div 1\dfrac{1}{6} + \left\{ 27 \cdot \dfrac{1}{9} - 2\dfrac{1}{3}\right\}$

$= \dfrac{1}{6} \div 1\dfrac{1}{6} + \left\{ 3 - 2\dfrac{1}{3}\right\}$

$= \dfrac{1}{6} \div 1\dfrac{1}{6} + \dfrac{2}{3}$

$= \dfrac{1}{6} \cdot \dfrac{6}{7} + \dfrac{2}{3}$

$= \dfrac{1}{7} + \dfrac{2}{3}$

$= \dfrac{17}{21}$

5. $\dfrac{5\dfrac{1}{4} + 6\dfrac{1}{2} + 4\dfrac{1}{4} + 3\dfrac{1}{2}}{4} = \dfrac{18 + 1\dfrac{1}{2}}{4}$

$= \dfrac{19\dfrac{1}{2}}{4}$

$= \dfrac{\dfrac{39}{2}}{4}$

$= \dfrac{39}{2} \cdot \dfrac{1}{4}$

$= \dfrac{39}{8} = 4\dfrac{7}{8}$ lb

Exercises 2.6

1. $\left(\dfrac{1}{2}\right)^2 \cdot \dfrac{1}{5} + \dfrac{1}{6} = \dfrac{1}{4} \cdot \dfrac{1}{5} + \dfrac{1}{6}$

$= \dfrac{1}{20} + \dfrac{1}{6}$

$= \dfrac{3}{60} + \dfrac{10}{60} = \dfrac{13}{60}$

3. $\dfrac{1}{7} + \dfrac{1}{3} \cdot \left(\dfrac{1}{2}\right)^2 = \dfrac{1}{7} + \dfrac{1}{3} \cdot \dfrac{1}{4}$

$= \dfrac{1}{7} + \dfrac{1}{12}$

$= \dfrac{12}{84} + \dfrac{7}{84} = \dfrac{19}{84}$

5. $\dfrac{1}{7} \cdot \left(\dfrac{1}{2}\right)^3 - \dfrac{1}{56} = \dfrac{1}{7} \cdot \dfrac{1}{8} - \dfrac{1}{56} = \dfrac{1}{56} - \dfrac{1}{56} = 0$

7. $\dfrac{1}{2} - \dfrac{1}{3} \cdot \dfrac{1}{5} = \dfrac{1}{2} - \dfrac{1}{15} = \dfrac{15}{30} - \dfrac{2}{30} = \dfrac{13}{30}$

9. $12 \div 6 - \left(\dfrac{1}{3} + \dfrac{1}{2}\right) = 12 \div 6 - \left(\dfrac{5}{6}\right)$

$= 2 - \dfrac{5}{6}$

$= \dfrac{12}{6} - \dfrac{5}{6} = \dfrac{7}{6}$

11. $\dfrac{1}{3} \cdot \dfrac{1}{4} \div \dfrac{1}{2} + \left(\dfrac{5}{6} - \dfrac{1}{2}\right) = \dfrac{1}{3} \cdot \dfrac{1}{4} \div \dfrac{1}{2} + \left(\dfrac{1}{3}\right)$

$= \dfrac{1}{12} \div \dfrac{1}{2} + \left(\dfrac{1}{3}\right)$

$= \dfrac{1}{12} \cdot \dfrac{2}{1} + \left(\dfrac{1}{3}\right)$

$= \dfrac{1}{6} + \dfrac{1}{3}$

$= \dfrac{1}{6} + \dfrac{2}{6} = \dfrac{3}{6} = \dfrac{1}{2}$

13. $\dfrac{1}{6} \div \dfrac{1}{3} \cdot \dfrac{1}{3} \cdot \dfrac{1}{3} + \left(\dfrac{1}{4} - \dfrac{1}{9}\right) = \dfrac{1}{6} \div \dfrac{1}{3} \cdot \dfrac{1}{3} \cdot \dfrac{1}{3} + \left(\dfrac{5}{36}\right)$

$= \dfrac{1}{6} \cdot \dfrac{3}{1} \cdot \dfrac{1}{3} \cdot \dfrac{1}{3} + \left(\dfrac{5}{36}\right)$

$= \dfrac{1}{2} \cdot \dfrac{1}{3} \cdot \dfrac{1}{3} + \left(\dfrac{5}{36}\right)$

$= \dfrac{1}{6} \cdot \dfrac{1}{3} + \left(\dfrac{5}{36}\right)$

$= \dfrac{1}{18} + \dfrac{5}{36}$

$= \dfrac{2}{36} + \dfrac{5}{36} = \dfrac{7}{36}$

15. $8 \div \dfrac{1}{2} \cdot \dfrac{1}{2} \cdot \dfrac{1}{2} - \left(\dfrac{1}{3} + \dfrac{1}{5}\right) = 8 \div \dfrac{1}{2} \cdot \dfrac{1}{2} \cdot \dfrac{1}{2} - \left(\dfrac{8}{15}\right)$

$= 8 \cdot \dfrac{2}{1} \cdot \dfrac{1}{2} \cdot \dfrac{1}{2} - \left(\dfrac{8}{15}\right)$

$= 16 \cdot \dfrac{1}{2} \cdot \dfrac{1}{2} - \left(\dfrac{8}{15}\right)$

$= 8 \cdot \dfrac{1}{2} - \left(\dfrac{8}{15}\right)$

$= 4 - \dfrac{8}{15}$

$= \dfrac{60}{15} - \dfrac{8}{15}$

$= \dfrac{52}{15} = 3\dfrac{7}{15}$

17. $\dfrac{1}{10} \div \dfrac{1}{5} \cdot \dfrac{1}{2} + \dfrac{1}{8}\left(\dfrac{4}{5} - \dfrac{1}{2}\right) + \left(\dfrac{1}{8} \div \dfrac{1}{4}\right)$

$= \dfrac{1}{10} \div \dfrac{1}{5} \cdot \dfrac{1}{2} + \dfrac{1}{8}\left(\dfrac{3}{10}\right) + \left(\dfrac{1}{8} \cdot \dfrac{4}{1}\right)$

$= \dfrac{1}{10} \div \dfrac{1}{5} \cdot \dfrac{1}{2} + \dfrac{1}{8}\left(\dfrac{3}{10}\right) + \left(\dfrac{1}{2}\right)$

$= \dfrac{1}{10} \cdot \dfrac{5}{1} \cdot \dfrac{1}{2} + \dfrac{1}{8}\left(\dfrac{3}{10}\right) + \left(\dfrac{1}{2}\right)$

$= \dfrac{1}{2} \cdot \dfrac{1}{2} + \dfrac{1}{8}\left(\dfrac{3}{10}\right) + \left(\dfrac{1}{2}\right)$

$= \dfrac{1}{4} + \dfrac{3}{80} + \left(\dfrac{1}{2}\right)$

$= \dfrac{20}{80} + \dfrac{3}{80} + \dfrac{40}{80}$

$= \dfrac{23}{80} + \dfrac{40}{80}$

$= \dfrac{63}{80}$

19. $\dfrac{1}{5} \div \dfrac{1}{3} \cdot \dfrac{1}{3} + \dfrac{1}{2}\left(\dfrac{1}{2} - \dfrac{1}{5}\right) + \left(\dfrac{1}{8} \div \dfrac{1}{4}\right)$

$= \dfrac{1}{5} \div \dfrac{1}{3} \cdot \dfrac{1}{3} + \dfrac{1}{2}\left(\dfrac{3}{10}\right) + \left(\dfrac{1}{8} \cdot \dfrac{4}{1}\right)$

$= \dfrac{1}{5} \div \dfrac{1}{3} \cdot \dfrac{1}{3} + \dfrac{1}{2}\left(\dfrac{3}{10}\right) + \left(\dfrac{1}{2}\right)$

$= \dfrac{1}{5} \cdot \dfrac{3}{1} \cdot \dfrac{1}{3} + \dfrac{1}{2}\left(\dfrac{3}{10}\right) + \left(\dfrac{1}{2}\right)$

$= \dfrac{3}{5} \cdot \dfrac{1}{3} + \dfrac{1}{2}\left(\dfrac{3}{10}\right) + \left(\dfrac{1}{2}\right)$

$= \dfrac{1}{5} + \dfrac{1}{2}\left(\dfrac{3}{10}\right) + \left(\dfrac{1}{2}\right)$

$= \dfrac{1}{5} + \dfrac{3}{20} + \left(\dfrac{1}{2}\right)$

$= \dfrac{7}{20} + \dfrac{1}{2}$

$= \dfrac{17}{20}$

21. $\dfrac{1}{20} \div \dfrac{1}{5} + \left\{\dfrac{1}{3} \div \dfrac{1}{4} - \left[\dfrac{1}{4} + \left(\dfrac{1}{3} - \dfrac{1}{5}\right)\right]\right\}$

$= \dfrac{1}{20} \div \dfrac{1}{5} + \left\{\dfrac{1}{3} \div \dfrac{1}{4} - \left[\dfrac{1}{4} + \left(\dfrac{2}{15}\right)\right]\right\}$

$= \dfrac{1}{20} \div \dfrac{1}{5} + \left\{\dfrac{1}{3} \div \dfrac{1}{4} - \left[\dfrac{23}{60}\right]\right\}$

$= \dfrac{1}{20} \div \dfrac{1}{5} + \left\{\dfrac{1}{3} \cdot \dfrac{4}{1} - \left[\dfrac{23}{60}\right]\right\}$

$= \dfrac{1}{20} \div \dfrac{1}{5} + \left\{\dfrac{4}{3} - \left[\dfrac{23}{60}\right]\right\}$

$= \dfrac{1}{20} \div \dfrac{1}{5} + \left\{\dfrac{19}{20}\right\}$

$= \dfrac{1}{4} + \dfrac{19}{20} = \dfrac{6}{5} = 1\dfrac{1}{5}$

23. $\dfrac{7}{30} \div \dfrac{1}{15} \cdot \left\{\dfrac{1}{10} \div \dfrac{1}{20} - \left[\dfrac{1}{2} \cdot \dfrac{1}{2} + \dfrac{1}{2}\right]\right\}$

$= \dfrac{7}{30} \div \dfrac{1}{15} \cdot \left\{\dfrac{1}{10} \div \dfrac{1}{20} - \left[\dfrac{1}{4} + \dfrac{1}{2}\right]\right\}$

$= \dfrac{7}{30} \div \dfrac{1}{15} \cdot \left\{\dfrac{1}{10} \div \dfrac{1}{20} - \left[\dfrac{3}{4}\right]\right\}$

$= \dfrac{7}{30} \div \dfrac{1}{15} \cdot \left\{2 - \left[\dfrac{3}{4}\right]\right\}$

$= \dfrac{7}{30} \div \dfrac{1}{15} \cdot \left\{\dfrac{5}{4}\right\} = \dfrac{7}{2} \cdot \dfrac{5}{4} = \dfrac{35}{8} = 4\dfrac{3}{8}$

25. $\left\{\dfrac{1}{4} \div \dfrac{1}{12} \cdot \dfrac{1}{6} + \left[\dfrac{1}{5}\left(\dfrac{1}{3} + \dfrac{1}{2}\right) - \dfrac{1}{6}\right] - \left(\dfrac{1}{3} + \dfrac{1}{2} \cdot \dfrac{1}{3}\right)\right\}$

$= \left\{\dfrac{1}{4} \div \dfrac{1}{12} \cdot \dfrac{1}{6} + \left[\dfrac{1}{5}\left(\dfrac{5}{6}\right) - \dfrac{1}{6}\right] - \left(\dfrac{1}{3} + \dfrac{1}{6}\right)\right\}$

$= \left\{\dfrac{1}{4} \div \dfrac{1}{12} \cdot \dfrac{1}{6} + \left[\dfrac{1}{6} - \dfrac{1}{6}\right] - \left(\dfrac{1}{2}\right)\right\}$

$= \left\{\dfrac{1}{4} \div \dfrac{1}{12} \cdot \dfrac{1}{6} + [0] - \left(\dfrac{1}{2}\right)\right\}$

$= \left\{\dfrac{1}{4} \cdot \dfrac{12}{1} \cdot \dfrac{1}{6} - \left(\dfrac{1}{2}\right)\right\}$

$= \left\{3 \cdot \dfrac{1}{6} - \left(\dfrac{1}{2}\right)\right\} = \left\{\dfrac{1}{2} - \dfrac{1}{2}\right\} = 0$

27. $\dfrac{2\frac{9}{10}+2\frac{4}{5}}{2}=\dfrac{\frac{29}{10}+\frac{14}{5}}{2}$

$=\dfrac{\frac{29}{10}+\frac{28}{10}}{2}$

$=\dfrac{\frac{57}{10}}{2}=\dfrac{57}{10}\cdot\dfrac{1}{2}=\dfrac{57}{20}=2\dfrac{17}{20}$ in.

29. $\dfrac{29+32\frac{1}{2}+32\frac{3}{5}}{3}=\dfrac{29+\frac{65}{2}+\frac{163}{5}}{3}$

$=\dfrac{\frac{290}{10}+\frac{325}{10}+\frac{326}{10}}{3}$

$=\dfrac{\frac{941}{10}}{3}$

$=\dfrac{941}{10}\cdot\dfrac{1}{3}=\dfrac{931}{30}=31\dfrac{11}{30}$ lb

31. a. $\dfrac{150\frac{2}{5}+148\frac{1}{2}+148\frac{1}{5}}{3}$

$=\dfrac{150\frac{4}{10}+148\frac{5}{10}+148\frac{2}{10}}{3}$

$=\dfrac{446\frac{11}{10}}{3}$

$=\dfrac{\frac{4471}{10}}{3}$

$=\dfrac{4471}{10}\cdot\dfrac{1}{3}=\dfrac{4471}{30}=\$149\dfrac{1}{30}$ million

b. $\$498\dfrac{17}{30}-\$149\dfrac{1}{30}=\$349\dfrac{16}{30}$

$=\$349\dfrac{8}{15}$ million

33. $\dfrac{6\frac{4}{5}+8\frac{1}{10}+7\frac{7}{10}}{3}=\dfrac{6\frac{4}{5}+15\frac{8}{10}}{3}$

$=\dfrac{6\frac{4}{5}+15\frac{4}{5}}{3}$

$=\dfrac{21\frac{8}{5}}{3}$

$=\dfrac{\frac{113}{5}}{3}$

$=\dfrac{113}{5}\cdot\dfrac{1}{3}=8\dfrac{8}{15}$ nights

35. $\dfrac{13\frac{4}{5}+14\frac{2}{5}+14\frac{3}{10}}{3}$

$=\dfrac{27\frac{6}{5}+14\frac{3}{10}}{3}$

$=\dfrac{27\frac{12}{10}+14\frac{3}{10}}{3}$

$=\dfrac{41\frac{15}{10}}{3}=\dfrac{41\frac{3}{2}}{3}$

$=\dfrac{85}{5}\cdot\dfrac{1}{3}=20\dfrac{1}{2}$ hr per week

37. $\dfrac{19\frac{4}{5}+21\frac{1}{10}+20\frac{1}{5}}{3}=\dfrac{39\frac{5}{5}+21\frac{1}{10}}{3}$

$=\dfrac{40+21\frac{1}{10}}{3}$

$=\dfrac{61\frac{1}{10}}{3}$

$=\dfrac{\frac{611}{10}}{3}$

$=\dfrac{611}{10}\cdot\dfrac{1}{3}$

$=20\dfrac{11}{30}$ hr per week

39. Women 18 and over

41.
$$x + 7 = 13$$
$$x + 7 - 7 = 13 - 7$$
$$x = 6$$

43.
$$15 = 5x$$
$$15 \div 5 = 5x \div 5$$
$$3 = x$$

45. a. Divisors of 8: 8, 4, 2, 1

b. $A_8 = \dfrac{8+4+2+1}{4} = \dfrac{15}{4} = 3\dfrac{3}{4}$

c. $H_8 = \dfrac{4}{\dfrac{1}{8}+\dfrac{1}{4}+\dfrac{1}{2}+\dfrac{1}{1}}$

$= \dfrac{4}{\dfrac{1}{8}+\dfrac{2}{8}+\dfrac{4}{8}+\dfrac{8}{8}}$

$= \dfrac{4}{\dfrac{15}{8}} = \dfrac{4}{1}\cdot\dfrac{8}{15} = \dfrac{32}{15} = 2\dfrac{2}{15}$

d. $A_8 \cdot H_8 = 3\dfrac{3}{4}\cdot 2\dfrac{2}{15} = \dfrac{15}{4}\cdot\dfrac{32}{15} = 8$; yes

47. Answers may vary.

Mastery Test 2.6

49. $\dfrac{3}{8}\div\dfrac{1}{12}-\left(\dfrac{1}{4}+\dfrac{1}{10}\right) = \dfrac{3}{8}\div\dfrac{1}{12}-\left(\dfrac{5}{20}+\dfrac{2}{20}\right)$

$= \dfrac{3}{8}\div\dfrac{1}{12}-\left(\dfrac{7}{20}\right)$

$= \dfrac{3}{8}\cdot\dfrac{12}{1}-\left(\dfrac{7}{20}\right)$

$= \dfrac{9}{2}-\dfrac{7}{20}$

$= \dfrac{83}{20} = 4\dfrac{3}{20}$

51. $\dfrac{1}{3}\cdot\left(\dfrac{3}{2}\right)^2 - \dfrac{1}{18} = \dfrac{1}{3}\cdot\dfrac{9}{4}-\dfrac{1}{18}$

$= \dfrac{3}{4}-\dfrac{1}{18} = \dfrac{27}{36}-\dfrac{2}{36} = \dfrac{25}{36}$

53. $\left(\dfrac{1}{3}\right)^3 \div \dfrac{1}{3}\cdot\dfrac{1}{9}+\dfrac{1}{2}\left(\dfrac{5}{3}-\dfrac{1}{3}\right)-\dfrac{1}{3}\cdot\dfrac{1}{2}$

$= \left(\dfrac{1}{3}\right)^3 \div \dfrac{1}{3}\cdot\dfrac{1}{9}+\dfrac{1}{2}\left(\dfrac{4}{3}\right)-\dfrac{1}{3}\cdot\dfrac{1}{2}$

$= \dfrac{1}{27}\div\dfrac{1}{3}\cdot\dfrac{1}{9}+\dfrac{1}{2}\left(\dfrac{4}{3}\right)-\dfrac{1}{3}\cdot\dfrac{1}{2}$

$= \dfrac{1}{27}\cdot\dfrac{3}{1}\cdot\dfrac{1}{9}+\dfrac{1}{2}\left(\dfrac{4}{3}\right)-\dfrac{1}{3}\cdot\dfrac{1}{2}$

$= \dfrac{1}{9}\cdot\dfrac{1}{9}+\dfrac{1}{2}\left(\dfrac{4}{3}\right)-\dfrac{1}{3}\cdot\dfrac{1}{2}$

$= \dfrac{1}{81}+\dfrac{2}{3}-\dfrac{1}{6}$

$= \dfrac{1}{81}+\dfrac{54}{81}-\dfrac{1}{6}$

$= \dfrac{55}{81}-\dfrac{1}{6}$

$= \dfrac{110}{162}-\dfrac{27}{162} = \dfrac{83}{162}$

55. $\dfrac{2\dfrac{1}{2}+5\dfrac{1}{4}+3\dfrac{1}{2}+4\dfrac{1}{4}}{4} = \dfrac{6+9\dfrac{1}{2}}{4}$

$= \dfrac{15\dfrac{1}{2}}{4}$

$= \dfrac{\dfrac{31}{2}}{4}$

$= \dfrac{31}{2}\cdot\dfrac{1}{4} = 3\dfrac{7}{8}$

Section 2.7 – Equations and Problem Solving

Problems

1. a. $n+5=6$
  b. $n-8=2$
  c. $3\cdot n=12$

2. a.　$n + \dfrac{1}{4} = \dfrac{3}{5}$

$n + \dfrac{1}{4} - \dfrac{1}{4} = \dfrac{3}{5} - \dfrac{1}{4}$

$n = \dfrac{12}{20} - \dfrac{5}{20} = \dfrac{7}{20}$

b.　$m - \dfrac{1}{3} = \dfrac{3}{5}$

$m - \dfrac{1}{3} + \dfrac{1}{3} = \dfrac{3}{5} + \dfrac{1}{3}$

$m = \dfrac{9}{15} + \dfrac{5}{15}$

$m = \dfrac{14}{15}$

c.　$\dfrac{q}{7} = \dfrac{3}{5}$

$\cancel{7} \cdot \dfrac{q}{\cancel{7}} = \dfrac{3}{5} \cdot 7$

$q = \dfrac{21}{5} = 4\dfrac{1}{5}$

3.　$\dfrac{5}{6} \cdot n = 15$

$\dfrac{\cancel{\dfrac{5}{6}} \cdot n}{\cancel{\dfrac{5}{6}}} = \dfrac{15}{\dfrac{5}{6}}$

$n = {}^{3}\cancel{15} \cdot \dfrac{6}{\cancel{5}_{1}}$

$n = 18$

4.　$n \cdot 2\dfrac{1}{2} = 3\dfrac{1}{4}$

$n \cdot \dfrac{5}{2} = \dfrac{13}{4}$

$\dfrac{n \cdot \cancel{\dfrac{5}{2}}}{\cancel{\dfrac{5}{2}}} = \dfrac{\dfrac{13}{4}}{\dfrac{5}{2}}$

$n = \dfrac{13}{{}_{2}\cancel{4}} \cdot \dfrac{\cancel{2}^{1}}{5} = \dfrac{13}{2} \cdot \dfrac{1}{5} = \dfrac{13}{10}$

5.　$\dfrac{2}{7} \cdot n = 1\dfrac{1}{2}$

$\dfrac{2}{7} \cdot n = \dfrac{3}{2}$

$\dfrac{\cancel{\dfrac{2}{7}} \cdot n}{\cancel{\dfrac{2}{7}}} = \dfrac{\dfrac{3}{2}}{\dfrac{2}{7}}$

$n = \dfrac{3}{2} \cdot \dfrac{7}{2} = \dfrac{21}{4}$

6.　$2\dfrac{1}{4} \cdot 1\dfrac{1}{3} = n$

$\dfrac{{}^{3}\cancel{9}}{\cancel{4}} \cdot \dfrac{\cancel{4}}{\cancel{3}_{1}} = n$

$3 = n$

7.　$1\dfrac{1}{4} = \dfrac{5}{4}$ pints cost 60¢

1 pint cost $60 \div \dfrac{5}{4} = {}^{12}\cancel{60} \cdot \dfrac{4}{\cancel{5}_{1}} = 48$ ¢

Thus, 2 pints will cost $48 \cdot 2 = 96$ ¢.

8.　$6\dfrac{1}{3} = \dfrac{19}{3}$ minutes to go $1\dfrac{1}{2} = \dfrac{3}{2}$ miles

In 1 minute, the runner goes

$\dfrac{3}{2} \div \dfrac{19}{3} = \dfrac{3}{2} \cdot \dfrac{3}{19} = \dfrac{9}{38}$ mile

In $12\dfrac{2}{3} = \dfrac{38}{3}$ minutes, the runner goes

$\dfrac{{}^{3}\cancel{9}}{\cancel{38}} \cdot \dfrac{\cancel{38}}{\cancel{3}_{1}} = 3$ miles.

9.　$\dfrac{3}{25} + \dfrac{9}{100} + \dfrac{1}{25} + c = 1$

$\dfrac{4}{25} + \dfrac{9}{100} + c = 1$

$\dfrac{16}{100} + \dfrac{9}{100} + c = 1$

$\dfrac{25}{100} + c = \dfrac{100}{100}$

$c = \dfrac{100}{100} - \dfrac{25}{100} = \dfrac{75}{100} = \dfrac{3}{4}$

The fraction of coin that would be copper is $\dfrac{3}{4}$.

Exercises 2.7

1. $=$

3. $\times$

5. $+$

7. $-$

9. $2 \cdot n$

11. $5 + n$

13. $n - 7$

15. $\dfrac{3}{4} \div n = 5$

17. $\dfrac{1}{2} \cdot 3 \cdot n = 2$

19. $\dfrac{1}{2} \cdot n - 4 = \dfrac{3}{2}$

21. $\quad m + \dfrac{1}{8} = \dfrac{3}{7}$

$m + \dfrac{1}{8} - \dfrac{1}{8} = \dfrac{3}{7} - \dfrac{1}{8}$

$m = \dfrac{24}{56} - \dfrac{7}{56} = \dfrac{17}{56}$

23. $\quad p + \dfrac{2}{5} = 1\dfrac{3}{4}$

$p + \dfrac{2}{5} = \dfrac{7}{4}$

$p + \dfrac{2}{5} - \dfrac{2}{5} = \dfrac{7}{4} - \dfrac{2}{5}$

$p = \dfrac{35}{20} - \dfrac{8}{20} = \dfrac{27}{20}$

25. $\quad y - \dfrac{3}{4} = \dfrac{4}{5}$

$y - \dfrac{3}{4} + \dfrac{3}{4} = \dfrac{4}{5} + \dfrac{3}{4}$

$y = \dfrac{16}{20} + \dfrac{15}{20} = \dfrac{31}{20} = 1\dfrac{11}{20}$

27. $\quad \dfrac{u}{6} = 3\dfrac{1}{2}$

$\dfrac{u}{6} = \dfrac{7}{2}$

$\cancel{6} \cdot \dfrac{u}{\cancel{6}} = \dfrac{7}{\cancel{2}_1} \cdot \cancel{6}^{\,3}$

$u = 7 \cdot 3 = 21$

29. $\quad 3 \cdot t = 2\dfrac{1}{5}$

$3 \cdot t = \dfrac{11}{5}$

$\dfrac{\cancel{3} \cdot t}{\cancel{3}} = \dfrac{\dfrac{11}{5}}{3}$

$t = \dfrac{11}{5} \cdot \dfrac{1}{3} = \dfrac{11}{15}$

31. $\quad 1\dfrac{1}{2} \cdot n = 7\dfrac{1}{2}$

$\dfrac{3}{2} \cdot n = \dfrac{15}{2}$

$\dfrac{\dfrac{\cancel{3}}{\cancel{2}} \cdot n}{\dfrac{\cancel{3}}{\cancel{2}}} = \dfrac{\dfrac{15}{2}}{\dfrac{3}{2}}$

$n = \dfrac{\overset{5}{\cancel{15}}}{\cancel{2}} \cdot \dfrac{\cancel{2}}{\cancel{3}_1} = 5$

33. $\quad n \cdot 1\dfrac{2}{3} = 4$

$n \cdot \dfrac{5}{3} = 4$

$\dfrac{n \cdot \dfrac{\cancel{5}}{\cancel{3}}}{\dfrac{\cancel{5}}{\cancel{3}}} = \dfrac{4}{\dfrac{5}{3}}$

$n = 4 \cdot \dfrac{3}{5} = \dfrac{12}{5}$

35. $n \cdot 2\frac{1}{2} = 6\frac{1}{4}$

$n \cdot \frac{5}{2} = \frac{25}{4}$

$\dfrac{n \cdot \frac{5}{2}}{\frac{5}{2}} = \dfrac{\frac{25}{4}}{\frac{5}{2}}$

$n = \dfrac{{}^5\cancel{25}}{{}_2\cancel{4}} \cdot \dfrac{\cancel{2}^{1}}{\cancel{5}_1} = \dfrac{5}{2} = 2\frac{1}{2}$

37. $1\frac{1}{3} \cdot n = 4\frac{2}{3}$

$\frac{4}{3} \cdot n = \frac{14}{3}$

$\dfrac{\frac{4}{3} \cdot n}{\frac{4}{3}} = \dfrac{\frac{14}{3}}{\frac{4}{3}}$

$n = \dfrac{{}^7\cancel{14}}{\cancel{3}} \cdot \dfrac{\cancel{3}}{\cancel{4}_2} = \dfrac{7}{2} = 3\frac{1}{2}$

39. $1\frac{1}{8} \cdot 2\frac{1}{2} = n$

$\frac{9}{8} \cdot \frac{5}{2} = n$

$\frac{45}{16} = n$

$n = \frac{45}{16} = 2\frac{13}{16}$

41. Let ℓ = length of the loaf

$\ell \div 16 = 22\frac{1}{2} \div 12$

$\dfrac{\ell}{16} = \dfrac{{}^{15}\cancel{45}}{2} \cdot \dfrac{1}{\cancel{12}_4}$

$\cancel{16} \cdot \dfrac{\ell}{\cancel{16}} = \dfrac{15}{{}_1\cancel{8}} \cdot \cancel{16}^{2}$

$\ell = 30$

The loaf would be 30 inches long.

43. $2\frac{1}{2} = \frac{5}{2}$ dozen cookies calls for $1\frac{7}{8} = \frac{15}{8}$ cups of flour. Therefore 1 dozen cookies calls for $\dfrac{15}{8} \div \dfrac{5}{2} = \dfrac{{}^3\cancel{15}}{{}_4\cancel{8}} \cdot \dfrac{\cancel{2}^{1}}{\cancel{5}_1} = \dfrac{3}{4}$ cup of

flour. Thus $1\frac{1}{3} = \frac{4}{3}$ dozen cookies calls for $\frac{4}{3} \cdot \frac{3}{4} = 1$ cup of milk.

45. 20 people ate 15 lb of ham so one person would have eaten $\frac{15}{20} = \frac{3}{4}$ lb of ham. Thus, 32 people would have needed ${}^8\cancel{32} \cdot \dfrac{3}{\cancel{4}_1} = 24$ pounds of ham.

47. $4\frac{1}{2} = \frac{9}{2}$ minutes to go $1\frac{1}{2} = \frac{3}{2}$ km; in 1 minute the runner goes

$\dfrac{3}{2} \div \dfrac{9}{2} = \dfrac{{}^1\cancel{3}}{\cancel{2}} \cdot \dfrac{\cancel{2}}{\cancel{9}_3} = \dfrac{1}{3}$ km. Thus in

$7\frac{1}{2} = \frac{15}{2}$ minutes the runner can go

$\dfrac{{}^5\cancel{15}}{2} \cdot \dfrac{1}{\cancel{3}_1} = \dfrac{5}{2} = 2\frac{1}{2}$ km.

49. $3\frac{1}{4} = \frac{13}{4}$ lb costs 91¢ so 1 lb cost

$91 \div \dfrac{13}{4} = {}^7\cancel{91} \cdot \dfrac{4}{\cancel{13}_1} = 28$ ¢. Thus $2\frac{1}{2} = \frac{5}{2}$

lb would cost $\dfrac{5}{{}_1\cancel{2}} \cdot \cancel{28}^{14} = 70$ ¢.

51. $1 - \dfrac{12}{25} = \dfrac{25}{25} - \dfrac{12}{25} = \dfrac{13}{25}$

The fraction of the patents that were foreign is $\frac{13}{25}$.

53. $F = 1\frac{4}{5}C + 32$

$= \dfrac{9}{\cancel{5}} (\cancel{25})^{5} + 32 = 45 + 32 = 77$

The temperature is 77 degrees Fahrenheit.

55. $F = \dfrac{c}{4} + 39$

$= \dfrac{120}{4} + 39 = 30 + 39 = 69$

The temperature is 69 degrees Fahrenheit.

57. a. $\dfrac{1}{6}$

    b. $\dfrac{1}{3}$

    c. $1 - \dfrac{1}{6} - \dfrac{1}{3} = \dfrac{6}{6} - \dfrac{1}{6} - \dfrac{2}{6} = \dfrac{3}{6} = \dfrac{1}{2}$

    d. $50 \cdot \dfrac{1}{2} = 25$ lb

59. a. $6 + \dfrac{1}{4} + \dfrac{1}{2} + \dfrac{1}{3} = 6 + \dfrac{3}{12} + \dfrac{6}{12} + \dfrac{4}{12}$

$$= 6 + \dfrac{13}{12}$$

$$= 6 + 1\dfrac{1}{12}$$

$$= 7\dfrac{1}{12} \text{ cups}$$

    b. $7\dfrac{1}{12} \cdot 8 = \dfrac{85}{{}_3\cancel{12}} \cdot \cancel{8}^{\,2} = \dfrac{170}{3} = 56\dfrac{2}{3}$ oz

    c. $56\dfrac{2}{3} \div 10 = \dfrac{17\cancel{0}}{3} \cdot \dfrac{1}{1\cancel{0}} = \dfrac{17}{3} = 5\dfrac{2}{3}$ oz per

    serving

61. $1\underline{8}5 \rightarrow 190$

63. $8 \div 4 \cdot \dfrac{11}{2} - \left[3\left(\dfrac{5}{3} - \dfrac{1}{3} \right) + 1 \right]$

$$= 8 \div 4 \cdot \dfrac{11}{2} - \left[3\left(\dfrac{4}{3} \right) + 1 \right]$$

$$= 8 \div 4 \cdot \dfrac{11}{2} - [4 + 1]$$

$$= 8 \div 4 \cdot \dfrac{11}{2} - [5]$$

$$= 2 \cdot \dfrac{11}{2} - [5] = 11 - 5 = 6$$

65. $45 \div 4\dfrac{1}{2} = 45 \div \dfrac{9}{2} = {}^5\cancel{45} \cdot \dfrac{2}{\cancel{9}_1} = 10$ ¢/oz;

$66 \div 5\dfrac{1}{2} = 66 \div \dfrac{11}{2} = {}^6\cancel{66} \cdot \dfrac{2}{\cancel{11}_1} = 12$ ¢/oz.

The $4\dfrac{1}{2}$ oz item is the better buy.

67. $70 \div 3\dfrac{1}{2} = 70 \div \dfrac{7}{2} = {}^{10}\cancel{70} \cdot \dfrac{2}{\cancel{7}_1} = 20$ ¢/oz;

$98 \div 4 = {}^{49}\cancel{98} \cdot \dfrac{1}{\cancel{4}_2} = \dfrac{49}{2} = 24.5$ ¢/oz. The

$3\dfrac{1}{2}$ oz item is the better buy.

69. $33 \div 5\dfrac{1}{2} = 33 \div \dfrac{11}{2} = {}^3\cancel{33} \cdot \dfrac{2}{\cancel{11}_1} = 6$ ¢/oz;

$29 \div 5 = 29 \cdot \dfrac{1}{5} = \dfrac{29}{5} = 5\dfrac{4}{5} = 5.8$ ¢/oz. The

5 oz item is the better buy.

71. Answers may vary.

73. Answers may vary.

Mastery Test 2.7

75. $n \cdot 1\dfrac{1}{2} = 3\dfrac{3}{4}$

$$\dfrac{n \cdot \frac{\cancel{3}}{\cancel{2}}}{\frac{\cancel{3}}{\cancel{2}}} = \dfrac{\frac{15}{4}}{\frac{3}{2}}$$

$$n = \dfrac{{}^5\cancel{15}}{{}_2\cancel{4}} \cdot \dfrac{\cancel{2}^{\,1}}{\cancel{3}_1} = \dfrac{5}{2} = 2\dfrac{1}{2}$$

77. a. $3 \cdot n = 9$

    b. $n - 5 = 2$

    c. $n + 8 = 7$

79. $2\dfrac{1}{2} \cdot 3\dfrac{5}{7} = n$

$$\dfrac{5}{{}_1\cancel{2}} \cdot \dfrac{\cancel{26}^{\,13}}{7} = n$$

$$\dfrac{65}{7} = n$$

$$9\dfrac{2}{7} = n$$

81. $1\frac{1}{2}=\frac{3}{2}$ miles in $8\frac{1}{2}=\frac{17}{2}$ minutes means

she ran $\frac{3}{2}\div\frac{17}{2}=\frac{3}{2}\cdot\frac{\cancel{2}}{17}=\frac{3}{17}$ mile in 1

minute. Thus in 17 minutes she can run

$\cancel{17}\cdot\frac{3}{\cancel{17}}=3$ miles.

Review Exercises – Chapter 2

1. a. Proper
  b. Proper
  c. Improper
  d. Proper
  e. Improper

2. a. $\frac{22}{7}=3$ with remainder 1 so $\frac{22}{7}=3\frac{1}{7}$

  b. $\frac{18}{7}=2$ with remainder 4 so $\frac{18}{7}=2\frac{4}{7}$

  c. $\frac{29}{3}=9$ with remainder 2 so $\frac{29}{3}=9\frac{2}{3}$

  d. $\frac{14}{4}=3$ with remainder 2; $\frac{14}{4}=3\frac{2}{4}=3\frac{1}{2}$

  e. $\frac{19}{11}=1$ with remainder 8 so $\frac{19}{11}=1\frac{8}{11}$

3. a. $4\frac{1}{2}=\frac{2\times4+1}{2}=\frac{9}{2}$

  b. $3\frac{1}{9}=\frac{3\times9+1}{9}=\frac{28}{9}$

  c. $4\frac{2}{5}=\frac{5\times4+2}{5}=\frac{22}{5}$

  d. $8\frac{3}{14}=\frac{14\times8+3}{14}=\frac{115}{14}$

  e. $7\frac{7}{8}=\frac{8\times7+7}{8}=\frac{63}{8}$

4. a. P/E ratio $=\frac{80}{10}=8$

  b. P/E ratio $=\frac{80}{8}=10$

  c. P/E ratio $=\frac{80}{20}=4$

  d. P/E ratio $=\frac{80}{40}=2$

  e. P/E ratio $=\frac{80}{16}=5$

5. a. $\frac{4}{3}=\frac{?}{6}$; $\frac{4}{3}=\frac{4\cdot2}{3\cdot2}=\frac{8}{6}$. The missing number is 8.

  b. $\frac{3}{5}=\frac{?}{25}$; $\frac{3}{5}=\frac{3\cdot5}{5\cdot5}=\frac{15}{25}$. The missing number is 15.

  c. $\frac{8}{9}=\frac{?}{27}$; $\frac{8}{9}=\frac{8\cdot3}{9\cdot3}=\frac{24}{27}$. The missing number is 24.

  d. $\frac{14}{21}=\frac{?}{42}$; $\frac{14}{21}=\frac{14\cdot2}{21\cdot2}=\frac{28}{42}$. The missing number is 28.

  e. $\frac{3}{9}=\frac{?}{54}$; $\frac{3}{9}=\frac{3\cdot6}{9\cdot6}=\frac{18}{54}$. The missing number is 18.

6. a. $\frac{6}{21}=\frac{2}{?}$; $\frac{6}{21}=\frac{6\div3}{21\div3}=\frac{2}{7}$. The missing number is 7.

  b. $\frac{8}{10}=\frac{4}{?}$; $\frac{8}{10}=\frac{8\div2}{10\div2}=\frac{4}{5}$. The missing number is 5.

  c. $\frac{18}{24}=\frac{6}{?}$; $\frac{18}{24}=\frac{18\div3}{24\div3}=\frac{6}{8}$. The missing number is 8.

  d. $\frac{24}{48}=\frac{4}{?}$; $\frac{24}{48}=\frac{24\div6}{48\div6}=\frac{4}{8}$. The missing number is 8.

  e. $\frac{18}{30}=\frac{6}{?}$; $\frac{18}{30}=\frac{18\div3}{30\div3}=\frac{6}{10}$. The missing number is 10.

7. a. $\frac{4}{8}=\frac{\overset{1}{\cancel{2}}\cdot\overset{1}{\cancel{2}}}{\underset{1}{\cancel{2}}\cdot\underset{1}{\cancel{2}}\cdot2}=\frac{1}{2}$

  b. $\frac{6}{9}=\frac{2\cdot\cancel{3}}{3\cdot\cancel{3}}=\frac{2}{3}$

  c. $\frac{14}{35}=\frac{2\cdot\cancel{7}}{5\cdot\cancel{7}}=\frac{2}{5}$

  d. $\frac{8}{28}=\frac{\cancel{2}\cdot\cancel{2}\cdot2}{\cancel{2}\cdot\cancel{2}\cdot7}=\frac{2}{7}$

  e. $\frac{10}{95}=\frac{2\cdot\cancel{5}}{\cancel{5}\cdot19}=\frac{2}{19}$

8. a. $\frac{1}{3}$ and $\frac{3}{10}$; $1 \times 10 = 10 > 3 \times 3 = 9$ so

$\frac{1}{3} > \frac{3}{10}$

b. $\frac{2}{3}$ and $\frac{3}{7}$; $2 \times 7 = 14 > 3 \times 3 = 9$ so

$\frac{2}{3} > \frac{3}{7}$

c. $\frac{4}{5}$ and $\frac{5}{7}$; $4 \times 7 = 28 > 5 \times 5 = 25$ so

$\frac{4}{5} > \frac{5}{7}$

d. $\frac{2}{9}$ and $\frac{3}{7}$; $2 \times 7 = 14 < 9 \times 3 = 27$ so

$\frac{2}{9} < \frac{3}{7}$

e. $\frac{3}{8}$ and $\frac{5}{32}$; $3 \times 32 = 96 > 8 \times 5 = 40$ so

$\frac{3}{8} > \frac{5}{32}$

9. a. $\frac{1}{3} \cdot \frac{2}{7} = \frac{2}{21}$

b. $\frac{2}{\cancel{8}} \cdot \frac{\cancel{8}}{9} = \frac{2}{9}$

c. $\frac{{}^1\cancel{3}}{\cancel{7}} \cdot \frac{\cancel{7}}{\cancel{9}_3} = \frac{1}{3}$

d. $\frac{{}^1\cancel{4}}{{}_1\cancel{6}} \cdot \frac{\cancel{15}^{\,3}}{\cancel{8}_2} = \frac{3}{2} = 1\frac{1}{2}$

e. $\frac{\cancel{7}}{\cancel{8}} \cdot \frac{\cancel{8}}{\cancel{7}} = 1$

10. a. $\frac{4}{7} \cdot 3\frac{1}{6} = \frac{4}{7} \cdot \frac{19}{3} = \frac{38}{21} = 1\frac{17}{21}$

b. $\frac{3}{5} \cdot 3\frac{1}{3} = \frac{\cancel{3}}{{}_1\cancel{5}} \cdot \frac{\cancel{10}^{\,2}}{\cancel{3}} = \frac{2}{1} = 2$

c. $\frac{6}{7} \cdot 1\frac{3}{4} = \frac{{}^3\cancel{6}}{\cancel{7}} \cdot \frac{\cancel{7}}{\cancel{4}_2} = \frac{3}{2} = 1\frac{1}{2}$

d. $\frac{9}{10} \cdot 2\frac{1}{4} = \frac{9}{10} \cdot \frac{9}{4} = \frac{81}{40} = 2\frac{1}{40}$

e. $\frac{6}{7} \cdot 4\frac{2}{3} = \frac{{}^2\cancel{6}}{{}_1\cancel{7}} \cdot \frac{\cancel{14}^{\,2}}{\cancel{3}_1} = \frac{4}{1} = 4$

11. a. $\left(\frac{2}{5}\right)^2 \cdot \frac{5}{6} = \frac{{}^2\cancel{4}}{{}_5\cancel{25}} \cdot \frac{\cancel{5}^{\,1}}{\cancel{6}_3} = \frac{2}{15}$

b. $\left(\frac{3}{2}\right)^2 \cdot \frac{4}{9} = \frac{\cancel{9}}{\cancel{4}} \cdot \frac{\cancel{4}}{\cancel{9}} = 1$

c. $\left(\frac{3}{2}\right)^2 \cdot \frac{8}{27} = \frac{{}^1\cancel{9}}{{}_1\cancel{4}} \cdot \frac{\cancel{8}^{\,2}}{\cancel{27}_3} = \frac{2}{3}$

d. $\left(\frac{3}{2}\right)^2 \cdot \frac{14}{27} = \frac{{}^1\cancel{9}}{{}_2\cancel{4}} \cdot \frac{\cancel{14}^{\,7}}{\cancel{27}_3} = \frac{7}{6} = 1\frac{1}{6}$

e. $\left(\frac{3}{2}\right)^2 \cdot \frac{8}{9} = \frac{\cancel{9}}{{}_1\cancel{4}} \cdot \frac{\cancel{8}^{\,2}}{\cancel{9}} = \frac{2}{1} = 2$

12. a. $\frac{3}{4} \div \frac{6}{7} = \frac{{}^1\cancel{3}}{4} \cdot \frac{7}{\cancel{6}_2} = \frac{7}{8}$

b. $\frac{3}{8} \div \frac{6}{7} = \frac{{}^1\cancel{3}}{8} \cdot \frac{7}{\cancel{6}_2} = \frac{7}{16}$

c. $\frac{4}{5} \div \frac{5}{9} = \frac{4}{5} \cdot \frac{9}{5} = \frac{36}{25} = 1\frac{11}{25}$

d. $\frac{5}{3} \div \frac{7}{9} = \frac{5}{{}_1\cancel{3}} \cdot \frac{\cancel{9}^{\,3}}{7} = \frac{15}{7} = 2\frac{1}{7}$

e. $\frac{6}{7} \div \frac{12}{7} = \frac{{}^1\cancel{6}}{\cancel{7}} \cdot \frac{\cancel{7}}{\cancel{12}_2} = \frac{1}{2}$

13. a. $2\frac{1}{4} \div \frac{4}{5} = \frac{9}{4} \div \frac{4}{5} = \frac{9}{4} \cdot \frac{5}{4} = \frac{45}{16} = 2\frac{13}{16}$

b. $3\frac{1}{7} \div \frac{7}{8} = \frac{22}{7} \div \frac{7}{8} = \frac{22}{7} \cdot \frac{8}{7} = \frac{176}{49} = 3\frac{29}{49}$

c. $6\frac{1}{2} \div \frac{4}{13} = \frac{13}{2} \div \frac{4}{13} = \frac{13}{2} \cdot \frac{13}{4} = \frac{169}{8} = 21\frac{1}{8}$

d. $1\frac{1}{9} \div \frac{20}{27} = \frac{{}^1\cancel{10}}{{}_1\cancel{9}} \cdot \frac{\cancel{27}^{\,3}}{\cancel{20}_2} = \frac{3}{3} = 1\frac{1}{2}$

e. $4\frac{1}{7} \div \frac{14}{15} = \frac{29}{7} \cdot \frac{15}{14} = \frac{435}{98} = 4\frac{43}{98}$

14. a. $\dfrac{3}{5} \div 1\dfrac{1}{5} = \dfrac{3}{5} \div \dfrac{6}{5} = \dfrac{\cancel{3}^{1}}{\cancel{5}} \cdot \dfrac{\cancel{5}}{\cancel{6}_{2}} = \dfrac{1}{2}$

b. $\dfrac{4}{7} \div 2\dfrac{3}{7} = \dfrac{4}{7} \div \dfrac{17}{7} = \dfrac{4}{\cancel{7}} \cdot \dfrac{\cancel{7}}{17} = \dfrac{4}{17}$

c. $\dfrac{3}{5} \div 3\dfrac{1}{5} = \dfrac{3}{5} \div \dfrac{16}{5} = \dfrac{3}{\cancel{5}} \cdot \dfrac{\cancel{5}}{16} = \dfrac{3}{16}$

d. $\dfrac{1}{7} \div 2\dfrac{1}{2} = \dfrac{1}{7} \div \dfrac{5}{2} = \dfrac{1}{7} \cdot \dfrac{2}{5} = \dfrac{2}{35}$

e. $\dfrac{2}{9} \div 3\dfrac{1}{8} = \dfrac{2}{9} \div \dfrac{25}{8} = \dfrac{2}{9} \cdot \dfrac{8}{25} = \dfrac{16}{225}$

15. a. $3\dfrac{1}{3} \cdot 4\dfrac{2}{3} = \dfrac{10}{3} \cdot \dfrac{14}{3} = \dfrac{140}{9} = 15\dfrac{5}{9}$ sq. yd.

b. $3\dfrac{1}{2} \cdot 4\dfrac{1}{2} = \dfrac{7}{2} \cdot \dfrac{9}{2} = \dfrac{63}{4} = 15\dfrac{3}{4}$ sq. yd.

c. $3\dfrac{1}{3} \cdot 4\dfrac{1}{2} = \dfrac{\cancel{10}^{5}}{\cancel{3}_{1}} \cdot \dfrac{\cancel{9}^{3}}{\cancel{2}_{1}} = 15$ sq. yd.

d. $3\dfrac{1}{2} \cdot 4\dfrac{1}{3} = \dfrac{7}{2} \cdot \dfrac{13}{3} = \dfrac{91}{6} = 15\dfrac{1}{6}$ sq. yd.

e. $4\dfrac{1}{2} \cdot 5\dfrac{1}{2} = \dfrac{9}{2} \cdot \dfrac{11}{2} = \dfrac{99}{4} = 24\dfrac{3}{4}$ sq. yd.

16. a. $\dfrac{1}{5} + \dfrac{2}{5} = \dfrac{3}{5}$

b. $\dfrac{2}{3} + \dfrac{1}{3} = \dfrac{3}{3} = 1$

c. $\dfrac{3}{7} + \dfrac{1}{7} = \dfrac{4}{7}$

d. $\dfrac{2}{9} + \dfrac{1}{9} = \dfrac{3}{9} = \dfrac{1}{3}$

e. $\dfrac{7}{2} + \dfrac{9}{2} = \dfrac{16}{2} = 8$

17. a. $\dfrac{1}{3} + \dfrac{5}{6} = \dfrac{1 \cdot 2}{3 \cdot 2} + \dfrac{5}{6} = \dfrac{2}{6} + \dfrac{5}{6} = \dfrac{7}{6} = 1\dfrac{1}{6}$

b. $\dfrac{1}{5} + \dfrac{1}{9} = \dfrac{1 \cdot 9}{5 \cdot 9} + \dfrac{1 \cdot 5}{9 \cdot 5} = \dfrac{9}{45} + \dfrac{5}{14} = \dfrac{14}{45}$

c. $\dfrac{3}{7} + \dfrac{5}{6} = \dfrac{3 \cdot 6}{7 \cdot 6} + \dfrac{5 \cdot 7}{6 \cdot 7}$
$= \dfrac{18}{42} + \dfrac{35}{42} = \dfrac{53}{42} = 1\dfrac{11}{42}$

d. $\dfrac{1}{6} + \dfrac{9}{20} = \dfrac{1 \cdot 10}{6 \cdot 10} + \dfrac{9 \cdot 3}{20 \cdot 3} = \dfrac{10}{60} + \dfrac{27}{60} = \dfrac{37}{60}$

e. $\dfrac{2}{7} + \dfrac{3}{15} = \dfrac{2 \cdot 15}{7 \cdot 15} + \dfrac{3 \cdot 7}{15 \cdot 7}$
$= \dfrac{30}{105} + \dfrac{21}{105} = \dfrac{51}{105} = \dfrac{17}{35}$

18. a. $\dfrac{15}{4} + \dfrac{16}{3} = \dfrac{15 \cdot 3}{4 \cdot 3} + \dfrac{16 \cdot 4}{3 \cdot 4}$
$= \dfrac{45}{12} + \dfrac{64}{12} = \dfrac{109}{12} = 9\dfrac{1}{12}$

b. $\dfrac{7}{2} + \dfrac{5}{3} = \dfrac{7 \cdot 3}{2 \cdot 3} + \dfrac{5 \cdot 2}{3 \cdot 2}$
$= \dfrac{21}{6} + \dfrac{10}{6} = \dfrac{31}{6} = 5\dfrac{1}{6}$

c. $\dfrac{17}{4} + \dfrac{33}{16} = \dfrac{17 \cdot 4}{4 \cdot 4} + \dfrac{33}{16}$
$= \dfrac{68}{16} + \dfrac{33}{16} = \dfrac{101}{16} = 6\dfrac{5}{16}$

d. $\dfrac{19}{9} + \dfrac{13}{3} = \dfrac{19}{9} + \dfrac{13 \cdot 3}{3 \cdot 3}$
$= \dfrac{19}{9} + \dfrac{39}{9} = \dfrac{58}{9} = 6\dfrac{4}{9}$

e. $\dfrac{9}{8} + \dfrac{19}{9} = \dfrac{9 \cdot 9}{8 \cdot 9} + \dfrac{19 \cdot 8}{9 \cdot 8}$
$= \dfrac{81}{72} + \dfrac{152}{72} = \dfrac{233}{72} = 3\dfrac{17}{72}$

19. a. $\begin{array}{r|ccc} 2 & 7 & 6 & 12 \\ 3 & 7 & 3 & 6 \\ \hline & 7 & 1 & 2 \end{array}$ so the LCD is $2 \cdot 3 \cdot 7 \cdot 2 = 84$;
$\dfrac{5}{7} + \dfrac{1}{6} + \dfrac{1}{12} = \dfrac{60}{84} + \dfrac{14}{84} + \dfrac{7}{84} = \dfrac{81}{84} = \dfrac{27}{28}$

b. Multiples of 12: 12, 24, 36, …. Since 4 and 8 go into 24, the LCD is 24.
$\dfrac{3}{4} + \dfrac{1}{8} + \dfrac{1}{12} = \dfrac{18}{24} + \dfrac{3}{24} + \dfrac{2}{24} = \dfrac{23}{24}$

c. Since 8 and 4 go into 16, the LCD is 16.
$\dfrac{5}{8} + \dfrac{3}{4} + \dfrac{1}{16} = \dfrac{10}{16} + \dfrac{12}{16} + \dfrac{1}{16} = \dfrac{23}{16} = 1\dfrac{7}{16}$

d. Multiples of 15: 15, 30, 45, …. Since 5 and 6 go into 30, the LCD is 30.
$\dfrac{3}{5} + \dfrac{2}{6} + \dfrac{1}{15} = \dfrac{18}{30} + \dfrac{10}{30} + \dfrac{2}{30} = \dfrac{30}{30} = 1$

e. Multiples of 12: 12, 24, 36, 48, ….
Since 9 and 4 go into 36, the LCD is 36.
$\dfrac{6}{9} + \dfrac{2}{4} + \dfrac{1}{12} = \dfrac{24}{36} + \dfrac{18}{36} + \dfrac{3}{36}$
$= \dfrac{45}{36} = 1\dfrac{9}{36} = 1\dfrac{1}{4}$

20. a. $\frac{1}{2}+\frac{1}{4}=\frac{2}{4}+\frac{1}{4}=\frac{3}{4}$

   b. $\frac{13}{100}+\frac{1}{4}=\frac{13}{100}+\frac{25}{100}=\frac{38}{100}=\frac{19}{50}$

   c. $\frac{1}{10}+\frac{13}{100}=\frac{10}{100}+\frac{13}{100}=\frac{23}{100}$

   d. $\frac{1}{50}+\frac{1}{2}=\frac{1}{50}+\frac{25}{50}=\frac{26}{50}=\frac{13}{25}$

   e. $\frac{1}{10}+\frac{1}{50}=\frac{5}{50}+\frac{1}{50}=\frac{6}{50}=\frac{3}{25}$

21. a. $4\frac{1}{5}+3\frac{1}{6}=\frac{21}{5}+\frac{19}{6}$

   $=\frac{126}{30}+\frac{95}{30}=\frac{221}{30}=7\frac{11}{30}$

   b. $2\frac{1}{3}+3\frac{1}{12}=\frac{7}{3}+\frac{37}{12}$

   $=\frac{28}{12}+\frac{37}{12}=\frac{65}{12}=5\frac{5}{12}$

   c. $4\frac{4}{7}+3\frac{2}{8}=4\frac{4}{7}+3\frac{1}{4}$

   $=\frac{32}{7}+\frac{13}{4}$

   $=\frac{128}{28}+\frac{91}{28}=\frac{219}{28}=7\frac{23}{28}$

   d. $5\frac{1}{3}+2\frac{1}{9}=\frac{16}{3}+\frac{19}{9}=\frac{48}{9}+\frac{19}{9}=\frac{67}{9}=7\frac{4}{9}$

   e. $3\frac{5}{8}+5\frac{3}{12}=\frac{29}{8}+\frac{63}{12}$

   $=\frac{87}{24}+\frac{126}{24}=\frac{213}{24}=8\frac{7}{8}$

22. a. $2\frac{7}{8}-2\frac{2}{3}=\frac{23}{8}-\frac{8}{3}=\frac{69}{24}-\frac{64}{24}=\frac{5}{24}$

   b. $3\frac{1}{3}-1\frac{3}{5}=\frac{10}{3}-\frac{8}{5}=\frac{50}{15}-\frac{24}{15}=\frac{26}{15}=1\frac{11}{15}$

   c. $3\frac{1}{5}-2\frac{1}{3}=\frac{16}{5}-\frac{7}{3}=\frac{48}{15}-\frac{35}{15}=\frac{13}{15}$

   d. $4\frac{3}{5}-3\frac{5}{8}=\frac{23}{5}-\frac{29}{8}=\frac{184}{40}-\frac{145}{40}=\frac{39}{40}$

   e. $1\frac{7}{8}-1\frac{5}{9}=\frac{15}{8}-\frac{14}{9}=\frac{135}{72}-\frac{112}{72}=\frac{23}{72}$

23. a. $2\frac{5}{9}+3\frac{1}{8}-2\frac{1}{10}=\frac{23}{9}+\frac{25}{8}-\frac{21}{10}$

   $=\frac{920}{360}+\frac{1125}{360}-\frac{756}{360}$

   $=\frac{1289}{360}=3\frac{209}{360}$

   b. $3\frac{5}{9}+3\frac{1}{6}-2\frac{1}{10}=\frac{32}{9}+\frac{19}{6}-\frac{21}{10}$

   $=\frac{320}{90}+\frac{285}{90}-\frac{189}{90}$

   $=\frac{416}{90}$

   $=\frac{208}{45}=4\frac{28}{45}$

   c. $4\frac{5}{9}+3\frac{1}{12}-2\frac{1}{8}=\frac{41}{9}+\frac{37}{12}-\frac{17}{8}$

   $=\frac{328}{72}+\frac{222}{72}-\frac{153}{72}$

   $=\frac{397}{72}=5\frac{37}{72}$

   d. $5\frac{5}{9}+3\frac{1}{12}-2\frac{1}{6}=\frac{50}{9}+\frac{37}{12}-\frac{13}{6}$

   $=\frac{200}{36}+\frac{111}{36}-\frac{78}{36}$

   $=\frac{233}{36}=6\frac{17}{36}$

   e. $6\frac{5}{9}+3\frac{1}{8}-2\frac{1}{6}=\frac{59}{9}+\frac{25}{8}-\frac{13}{6}$

   $=\frac{472}{72}+\frac{225}{72}-\frac{156}{72}$

   $=\frac{541}{72}=7\frac{37}{72}$

24. a. $4\frac{1}{4}+5\frac{1}{2}+4\frac{1}{4}+5\frac{1}{2}=8\frac{2}{4}+10\frac{2}{2}$

   $=8\frac{1}{2}+10+1$

   $=8\frac{1}{2}+11=19\frac{1}{2}$ yd

   b. $3\frac{1}{2}+4\frac{1}{3}+3\frac{1}{2}+4\frac{1}{3}=6\frac{2}{2}+8\frac{2}{3}$

   $=6+1+8\frac{2}{3}$

   $=7+8\frac{2}{3}=15\frac{2}{3}$ yd

   c. $4\frac{1}{3}+5\frac{1}{2}+4\frac{1}{3}+5\frac{1}{2}=8\frac{2}{3}+10\frac{2}{2}$

   $=8\frac{2}{3}+10+1$

   $=8\frac{2}{3}+11=19\frac{2}{3}$ yd

d. $3\frac{1}{2}+5\frac{1}{3}+3\frac{1}{2}+5\frac{1}{3}=6\frac{2}{2}+10\frac{2}{3}$

$\qquad = 6+1+10\frac{2}{3}$

$\qquad = 7+10\frac{2}{3}=17\frac{2}{3}$ yd

e. $3\frac{1}{6}+2\frac{5}{6}+3\frac{1}{6}+2\frac{5}{6}=10\frac{12}{6}$

$\qquad = 10+2=12$ yd

25. a. $\frac{1}{2}\cdot\left(\frac{2}{3}\right)^2-\frac{1}{9}=\frac{1}{\underset{1}{\cancel{2}}}\cdot\frac{\cancel{4}^2}{9}-\frac{1}{9}=\frac{2}{9}-\frac{1}{9}=\frac{1}{9}$

b. $\frac{1}{3}\cdot\left(\frac{3}{4}\right)^2-\frac{1}{16}=\frac{1}{\underset{1}{\cancel{3}}}\cdot\frac{\cancel{9}^3}{16}-\frac{1}{16}$

$\qquad = \frac{3}{16}-\frac{1}{16}=\frac{2}{16}=\frac{1}{8}$

c. $\frac{1}{5}\cdot\left(\frac{5}{6}\right)^2-\frac{1}{36}=\frac{1}{\underset{1}{\cancel{5}}}\cdot\frac{\cancel{25}^5}{36}-\frac{1}{36}$

$\qquad = \frac{5}{36}-\frac{1}{36}=\frac{4}{36}=\frac{1}{9}$

d. $\frac{1}{6}\cdot\left(\frac{6}{7}\right)^2-\frac{1}{49}=\frac{1}{\underset{1}{\cancel{6}}}\cdot\frac{\cancel{36}^6}{49}-\frac{1}{49}$

$\qquad = \frac{6}{49}-\frac{1}{49}=\frac{5}{49}$

e. $\frac{1}{7}\cdot\left(\frac{7}{8}\right)^2-\frac{1}{64}=\frac{1}{\underset{1}{\cancel{7}}}\cdot\frac{\cancel{49}^7}{64}-\frac{1}{64}$

$\qquad = \frac{7}{64}-\frac{1}{64}=\frac{6}{64}=\frac{3}{32}$

26. a. $4\div\frac{1}{2}\cdot\frac{1}{2}\cdot\frac{1}{2}+\frac{1}{3}-2=4\cdot\frac{2}{1}\cdot\frac{1}{2}\cdot\frac{1}{2}+\frac{1}{3}-2$

$\qquad = 8\cdot\frac{1}{2}\cdot\frac{1}{2}+\frac{1}{3}-2$

$\qquad = 4\cdot\frac{1}{2}+\frac{1}{3}-2$

$\qquad = 2+\frac{1}{3}-2$

$\qquad = 2\frac{1}{3}-2=\frac{1}{3}$

b. $6\div\frac{1}{2}\cdot\frac{1}{2}\cdot\frac{1}{2}+\frac{1}{3}-3=6\cdot\frac{2}{1}\cdot\frac{1}{2}\cdot\frac{1}{2}+\frac{1}{3}-3$

$\qquad = 12\cdot\frac{1}{2}\cdot\frac{1}{2}+\frac{1}{3}-3$

$\qquad = 6\cdot\frac{1}{2}+\frac{1}{3}-3$

$\qquad = 3+\frac{1}{3}-3$

$\qquad = 3\frac{1}{3}-3=\frac{1}{3}$

c. $8\div\frac{1}{2}\cdot\frac{1}{2}\cdot\frac{1}{2}+\frac{1}{3}-4=8\cdot\frac{2}{1}\cdot\frac{1}{2}\cdot\frac{1}{2}+\frac{1}{3}-4$

$\qquad = 16\cdot\frac{1}{2}\cdot\frac{1}{2}+\frac{1}{3}-4$

$\qquad = 8\cdot\frac{1}{2}+\frac{1}{3}-4$

$\qquad = 4+\frac{1}{3}-4$

$\qquad = 4\frac{1}{3}-4=\frac{1}{3}$

d. $10\div\frac{1}{2}\cdot\frac{1}{2}\cdot\frac{1}{2}+\frac{1}{3}-5=10\cdot\frac{2}{1}\cdot\frac{1}{2}\cdot\frac{1}{2}+\frac{1}{3}-5$

$\qquad = 20\cdot\frac{1}{2}\cdot\frac{1}{2}+\frac{1}{3}-5$

$\qquad = 10\cdot\frac{1}{2}+\frac{1}{3}-5$

$\qquad = 5+\frac{1}{3}-5$

$\qquad = 5\frac{1}{3}-5=\frac{1}{3}$

e. $12\div\frac{1}{2}\cdot\frac{1}{2}\cdot\frac{1}{2}+\frac{1}{3}-6=12\cdot\frac{2}{1}\cdot\frac{1}{2}\cdot\frac{1}{2}+\frac{1}{3}-6$

$\qquad = 24\cdot\frac{1}{2}\cdot\frac{1}{2}+\frac{1}{3}-6$

$\qquad = 12\cdot\frac{1}{2}+\frac{1}{3}-6$

$\qquad = 6+\frac{1}{3}-6$

$\qquad = 6\frac{1}{3}-6=\frac{1}{3}$

27. a. $\left(\dfrac{1}{2}\right)^3 \div \dfrac{1}{3} \cdot \dfrac{1}{4} + \dfrac{1}{3}\left(\dfrac{7}{2} - \dfrac{1}{2}\right) - \dfrac{1}{3} \cdot \dfrac{1}{2}$

$= \left(\dfrac{1}{2}\right)^3 \div \dfrac{1}{3} \cdot \dfrac{1}{4} + \dfrac{1}{3}\left(\dfrac{6}{2}\right) - \dfrac{1}{3} \cdot \dfrac{1}{2}$

$= \dfrac{1}{8} \div \dfrac{1}{3} \cdot \dfrac{1}{4} + \dfrac{1}{3}(3) - \dfrac{1}{3} \cdot \dfrac{1}{2}$

$= \dfrac{1}{8} \cdot \dfrac{3}{1} \cdot \dfrac{1}{4} + \dfrac{1}{3}(3) - \dfrac{1}{3} \cdot \dfrac{1}{2}$

$= \dfrac{3}{8} \cdot \dfrac{1}{4} + \dfrac{1}{3}(3) - \dfrac{1}{3} \cdot \dfrac{1}{2}$

$= \dfrac{3}{32} + 1 - \dfrac{1}{6} = \dfrac{35}{32} - \dfrac{1}{6} = \dfrac{89}{96}$

b. $\left(\dfrac{1}{2}\right)^3 \div \dfrac{1}{3} \cdot \dfrac{1}{4} + \dfrac{1}{3}\left(\dfrac{9}{2} - \dfrac{1}{2}\right) - \dfrac{1}{3} \cdot \dfrac{1}{2}$

$= \left(\dfrac{1}{2}\right)^3 \div \dfrac{1}{3} \cdot \dfrac{1}{4} + \dfrac{1}{3}\left(\dfrac{8}{2}\right) - \dfrac{1}{3} \cdot \dfrac{1}{2}$

$= \dfrac{1}{8} \div \dfrac{1}{3} \cdot \dfrac{1}{4} + \dfrac{1}{3}(4) - \dfrac{1}{3} \cdot \dfrac{1}{2}$

$= \dfrac{1}{8} \cdot \dfrac{3}{1} \cdot \dfrac{1}{4} + \dfrac{1}{3}(4) - \dfrac{1}{3} \cdot \dfrac{1}{2}$

$= \dfrac{3}{8} \cdot \dfrac{1}{4} + \dfrac{1}{3}(4) - \dfrac{1}{3} \cdot \dfrac{1}{2}$

$= \dfrac{3}{32} + \dfrac{4}{3} - \dfrac{1}{6} = \dfrac{137}{96} - \dfrac{1}{6} = \dfrac{121}{96} = 1\dfrac{25}{96}$

c. $\left(\dfrac{1}{2}\right)^3 \div \dfrac{1}{3} \cdot \dfrac{1}{4} + \dfrac{1}{3}\left(\dfrac{5}{2} - \dfrac{1}{2}\right) - \dfrac{1}{3} \cdot \dfrac{1}{2}$

$= \left(\dfrac{1}{2}\right)^3 \div \dfrac{1}{3} \cdot \dfrac{1}{4} + \dfrac{1}{3}\left(\dfrac{4}{2}\right) - \dfrac{1}{3} \cdot \dfrac{1}{2}$

$= \dfrac{1}{8} \div \dfrac{1}{3} \cdot \dfrac{1}{4} + \dfrac{1}{3}(2) - \dfrac{1}{3} \cdot \dfrac{1}{2}$

$= \dfrac{1}{8} \cdot \dfrac{3}{1} \cdot \dfrac{1}{4} + \dfrac{1}{3}(2) - \dfrac{1}{3} \cdot \dfrac{1}{2}$

$= \dfrac{3}{8} \cdot \dfrac{1}{4} + \dfrac{1}{3}(2) - \dfrac{1}{3} \cdot \dfrac{1}{2}$

$= \dfrac{3}{32} + \dfrac{2}{3} - \dfrac{1}{6} = \dfrac{73}{96} - \dfrac{1}{6} = \dfrac{19}{32}$

d. $\left(\dfrac{1}{2}\right)^3 \div \dfrac{1}{8} \cdot \dfrac{1}{2} + \dfrac{1}{3}\left(\dfrac{3}{2} - \dfrac{1}{2}\right) - \dfrac{1}{3} \cdot \dfrac{1}{2}$

$= \left(\dfrac{1}{2}\right)^3 \div \dfrac{1}{8} \cdot \dfrac{1}{2} + \dfrac{1}{3}(1) - \dfrac{1}{3} \cdot \dfrac{1}{2}$

$= \dfrac{1}{8} \div \dfrac{1}{8} \cdot \dfrac{1}{2} + \dfrac{1}{3}(1) - \dfrac{1}{3} \cdot \dfrac{1}{2}$

$= \dfrac{1}{8} \cdot \dfrac{8}{1} \cdot \dfrac{1}{2} + \dfrac{1}{3}(1) - \dfrac{1}{3} \cdot \dfrac{1}{2}$

$= 1 \cdot \dfrac{1}{2} + \dfrac{1}{3}(1) - \dfrac{1}{3} \cdot \dfrac{1}{2}$

$= \dfrac{1}{2} + \dfrac{1}{3} - \dfrac{1}{6} = \dfrac{5}{6} - \dfrac{1}{6} = \dfrac{4}{6} = \dfrac{2}{3}$

e. $\left(\dfrac{1}{2}\right)^3 \div \dfrac{1}{4} \cdot \dfrac{1}{2} + \dfrac{1}{3}\left(\dfrac{11}{2} - \dfrac{1}{2}\right) - \dfrac{1}{3} \cdot \dfrac{1}{2}$

$= \left(\dfrac{1}{2}\right)^3 \div \dfrac{1}{4} \cdot \dfrac{1}{2} + \dfrac{1}{3}\left(\dfrac{10}{2}\right) - \dfrac{1}{3} \cdot \dfrac{1}{2}$

$= \dfrac{1}{8} \div \dfrac{1}{4} \cdot \dfrac{1}{2} + \dfrac{1}{3}(5) - \dfrac{1}{3} \cdot \dfrac{1}{2}$

$= \dfrac{1}{8} \cdot \dfrac{4}{1} \cdot \dfrac{1}{2} + \dfrac{1}{3}(5) - \dfrac{1}{3} \cdot \dfrac{1}{2}$

$= \dfrac{1}{2} \cdot \dfrac{1}{2} + \dfrac{1}{3}(5) - \dfrac{1}{3} \cdot \dfrac{1}{2}$

$= \dfrac{1}{4} + \dfrac{5}{3} - \dfrac{1}{6} = \dfrac{23}{12} - \dfrac{1}{6} = \dfrac{7}{4} = 1\dfrac{3}{4}$

28. a. $\dfrac{1}{6} \div 1\dfrac{1}{6} + \left\{ 16 \cdot \left(\dfrac{1}{2}\right)^2 - \left[\dfrac{1}{3} + \left(3\dfrac{1}{2} - \dfrac{1}{2}\right) \right] \right\}$

$= \dfrac{1}{6} \div 1\dfrac{1}{6} + \left\{ 16 \cdot \left(\dfrac{1}{2}\right)^2 - \left[\dfrac{1}{3} + (3) \right] \right\}$

$= \dfrac{1}{6} \div 1\dfrac{1}{6} + \left\{ 16 \cdot \left(\dfrac{1}{2}\right)^2 - \left[3\dfrac{1}{3} \right] \right\}$

$= \dfrac{1}{6} \div 1\dfrac{1}{6} + \left\{ 16 \cdot \dfrac{1}{4} - \left[3\dfrac{1}{3} \right] \right\}$

$= \dfrac{1}{6} \div \dfrac{7}{6} + \left\{ 4 - \left[\dfrac{10}{3} \right] \right\}$

$= \dfrac{1}{6} \div \dfrac{7}{6} + \left\{ \dfrac{2}{3} \right\}$

$= \dfrac{1}{\cancel{6}} \cdot \dfrac{\cancel{6}}{7} + \dfrac{2}{3} = \dfrac{1}{7} + \dfrac{2}{3} = \dfrac{17}{21}$

b. $\dfrac{1}{5} \div 1\dfrac{1}{5} + \left\{ 20 \cdot \left(\dfrac{1}{2}\right)^2 - \left[\dfrac{1}{3} + \left(4\dfrac{1}{2} - \dfrac{1}{2}\right) \right] \right\}$

$= \dfrac{1}{5} \div 1\dfrac{1}{5} + \left\{ 20 \cdot \left(\dfrac{1}{2}\right)^2 - \left[\dfrac{1}{3} + (4) \right] \right\}$

$= \dfrac{1}{5} \div 1\dfrac{1}{5} + \left\{ 20 \cdot \left(\dfrac{1}{2}\right)^2 - \left[4\dfrac{1}{3} \right] \right\}$

$= \dfrac{1}{5} \div 1\dfrac{1}{5} + \left\{ 20 \cdot \dfrac{1}{4} - \left[\dfrac{13}{3} \right] \right\}$

$= \dfrac{1}{5} \div \dfrac{6}{5} + \left\{ 5 - \left[\dfrac{13}{3} \right] \right\}$

$= \dfrac{1}{5} \div \dfrac{6}{5} + \left\{ \dfrac{2}{3} \right\}$

$= \dfrac{1}{\cancel{5}} \cdot \dfrac{\cancel{5}}{6} + \dfrac{2}{3} = \dfrac{1}{6} + \dfrac{2}{3} = \dfrac{5}{6}$

c. $\dfrac{1}{4} \div 1\dfrac{1}{4} + \left\{24 \cdot \left(\dfrac{1}{2}\right)^2 - \left[\dfrac{1}{3} + \left(5\dfrac{1}{2} - \dfrac{1}{2}\right)\right]\right\}$

$= \dfrac{1}{4} \div 1\dfrac{1}{4} + \left\{24 \cdot \left(\dfrac{1}{2}\right)^2 - \left[\dfrac{1}{3} + (5)\right]\right\}$

$= \dfrac{1}{4} \div 1\dfrac{1}{4} + \left\{24 \cdot \left(\dfrac{1}{2}\right)^2 - \left[5\dfrac{1}{3}\right]\right\}$

$= \dfrac{1}{4} \div 1\dfrac{1}{4} + \left\{24 \cdot \dfrac{1}{4} - \left[\dfrac{16}{3}\right]\right\}$

$= \dfrac{1}{4} \div \dfrac{5}{4} + \left\{6 - \left[\dfrac{16}{3}\right]\right\}$

$= \dfrac{1}{4} \div \dfrac{5}{4} + \left\{\dfrac{2}{3}\right\}$

$= \dfrac{1}{\cancel{4}} \cdot \dfrac{\cancel{4}}{5} + \dfrac{2}{3} = \dfrac{1}{5} + \dfrac{2}{3} = \dfrac{13}{15}$

d. $\dfrac{1}{3} \div 1\dfrac{1}{3} + \left\{28 \cdot \left(\dfrac{1}{2}\right)^2 - \left[\dfrac{1}{3} + \left(6\dfrac{1}{2} - \dfrac{1}{2}\right)\right]\right\}$

$= \dfrac{1}{3} \div 1\dfrac{1}{3} + \left\{28 \cdot \left(\dfrac{1}{2}\right)^2 - \left[\dfrac{1}{3} + (6)\right]\right\}$

$= \dfrac{1}{3} \div 1\dfrac{1}{3} + \left\{28 \cdot \left(\dfrac{1}{2}\right)^2 - \left[6\dfrac{1}{3}\right]\right\}$

$= \dfrac{1}{3} \div 1\dfrac{1}{3} + \left\{28 \cdot \dfrac{1}{4} - \left[\dfrac{19}{3}\right]\right\}$

$= \dfrac{1}{3} \div \dfrac{4}{3} + \left\{7 - \left[\dfrac{19}{3}\right]\right\}$

$= \dfrac{1}{3} \div \dfrac{4}{3} + \left\{\dfrac{2}{3}\right\}$

$= \dfrac{1}{\cancel{3}} \cdot \dfrac{\cancel{3}}{4} + \dfrac{2}{3} = \dfrac{1}{4} + \dfrac{2}{3} = \dfrac{11}{12}$

e. $\dfrac{1}{2} \div 1\dfrac{1}{2} + \left\{32 \cdot \left(\dfrac{1}{2}\right)^2 - \left[\dfrac{1}{3} + \left(7\dfrac{1}{2} - \dfrac{1}{2}\right)\right]\right\}$

$= \dfrac{1}{2} \div 1\dfrac{1}{2} + \left\{32 \cdot \left(\dfrac{1}{2}\right)^2 - \left[\dfrac{1}{3} + (7)\right]\right\}$

$= \dfrac{1}{2} \div 1\dfrac{1}{2} + \left\{32 \cdot \left(\dfrac{1}{2}\right)^2 - \left[7\dfrac{1}{3}\right]\right\}$

$= \dfrac{1}{2} \div 1\dfrac{1}{2} + \left\{32 \cdot \dfrac{1}{4} - \left[\dfrac{22}{3}\right]\right\}$

$= \dfrac{1}{2} \div \dfrac{3}{2} + \left\{8 - \left[\dfrac{22}{3}\right]\right\}$

$= \dfrac{1}{2} \div \dfrac{3}{2} + \left\{\dfrac{2}{3}\right\}$

$= \dfrac{1}{\cancel{2}} \cdot \dfrac{\cancel{2}}{3} + \dfrac{2}{3} = \dfrac{1}{3} + \dfrac{2}{3} = 1$

29. a. $\dfrac{3\dfrac{1}{2} + 4\dfrac{1}{4} + 2\dfrac{1}{2} + 7\dfrac{1}{4}}{4}$

$= \dfrac{5\dfrac{2}{2} + 11\dfrac{2}{4}}{4}$

$= \dfrac{5 + 1 + 11\dfrac{1}{2}}{4}$

$= \dfrac{17\dfrac{1}{2}}{4} = \dfrac{\dfrac{35}{2}}{4} = \dfrac{35}{2} \cdot \dfrac{1}{4} = \dfrac{35}{8} = 4\dfrac{3}{8}$ lb

b. $\dfrac{4\dfrac{1}{2} + 5\dfrac{1}{4} + 3\dfrac{1}{2} + 8\dfrac{1}{4}}{4}$

$= \dfrac{7\dfrac{2}{2} + 13\dfrac{2}{4}}{4}$

$= \dfrac{7 + 1 + 13\dfrac{1}{2}}{4}$

$= \dfrac{21\dfrac{1}{2}}{4} = \dfrac{\dfrac{43}{2}}{4} = \dfrac{43}{2} \cdot \dfrac{1}{4} = \dfrac{43}{8} = 5\dfrac{3}{8}$ lb

c. $\dfrac{5\dfrac{1}{2} + 6\dfrac{1}{4} + 4\dfrac{1}{2} + 9\dfrac{1}{4}}{4}$

$= \dfrac{9\dfrac{2}{2} + 15\dfrac{2}{4}}{4}$

$= \dfrac{9 + 1 + 15\dfrac{1}{2}}{4}$

$= \dfrac{25\dfrac{1}{2}}{4} = \dfrac{\dfrac{51}{2}}{4} = \dfrac{51}{2} \cdot \dfrac{1}{4} = \dfrac{51}{8} = 6\dfrac{3}{8}$ lb

d. $\dfrac{6\dfrac{1}{2} + 7\dfrac{1}{4} + 5\dfrac{1}{2} + 10\dfrac{1}{4}}{4}$

$= \dfrac{11\dfrac{2}{2} + 17\dfrac{2}{4}}{4}$

$= \dfrac{11 + 1 + 17\dfrac{1}{2}}{4}$

$= \dfrac{29\dfrac{1}{2}}{4} = \dfrac{\dfrac{59}{2}}{4} = \dfrac{59}{2} \cdot \dfrac{1}{4} = \dfrac{59}{8} = 7\dfrac{3}{8}$ lb

e. $\dfrac{7\frac{1}{2}+8\frac{1}{4}+6\frac{1}{2}+11\frac{1}{4}}{4}$

$=\dfrac{13\frac{2}{2}+19\frac{2}{4}}{4}$

$=\dfrac{13+1+19\frac{1}{2}}{4}$

$=\dfrac{33\frac{1}{2}}{4}=\dfrac{\frac{67}{2}}{4}=\dfrac{67}{2}\cdot\dfrac{1}{4}=\dfrac{67}{8}=8\dfrac{3}{8}$ lb

30. a. $n+8=10$
   b. $n-5=1$
   c. $2n=12$
   d. $\dfrac{n}{2}=8$
   e. $\dfrac{n}{7}=3$

31. a. $p+\dfrac{1}{6}=\dfrac{1}{3}$

      $p+\dfrac{1}{6}-\dfrac{1}{6}=\dfrac{1}{3}-\dfrac{1}{6}$

              $p=\dfrac{1}{6}$

  b. $q+\dfrac{1}{5}=\dfrac{1}{4}$

      $q+\dfrac{1}{5}-\dfrac{1}{5}=\dfrac{1}{4}-\dfrac{1}{5}$

              $q=\dfrac{1}{20}$

  c. $r+\dfrac{1}{4}=\dfrac{2}{5}$

      $r+\dfrac{1}{4}-\dfrac{1}{4}=\dfrac{2}{5}-\dfrac{1}{4}$

              $r=\dfrac{3}{20}$

  d. $s+\dfrac{1}{3}=\dfrac{5}{6}$

      $s+\dfrac{1}{3}-\dfrac{1}{3}=\dfrac{5}{6}-\dfrac{1}{3}$

              $s=\dfrac{1}{2}$

  e. $t+\dfrac{1}{2}=\dfrac{6}{7}$

      $t+\dfrac{1}{2}-\dfrac{1}{2}=\dfrac{6}{7}-\dfrac{1}{2}$

              $t=\dfrac{5}{14}$

32. a. $r-\dfrac{1}{6}=\dfrac{2}{7}$

      $r-\dfrac{1}{6}+\dfrac{1}{6}=\dfrac{2}{7}+\dfrac{1}{6}$

              $r=\dfrac{19}{42}$

  b. $s-\dfrac{1}{5}=\dfrac{3}{7}$

      $s-\dfrac{1}{5}+\dfrac{1}{5}=\dfrac{3}{7}+\dfrac{1}{5}$

              $s=\dfrac{22}{35}$

  c. $t-\dfrac{1}{4}=\dfrac{4}{7}$

      $t-\dfrac{1}{4}+\dfrac{1}{4}=\dfrac{4}{7}+\dfrac{1}{4}$

              $t=\dfrac{23}{28}$

  d. $u-\dfrac{1}{3}=\dfrac{5}{7}$

      $u-\dfrac{1}{3}+\dfrac{1}{3}=\dfrac{5}{7}+\dfrac{1}{3}$

           $u=\dfrac{22}{21}=1\dfrac{1}{21}$

  e. $v-\dfrac{1}{2}=\dfrac{6}{7}$

      $v-\dfrac{1}{2}+\dfrac{1}{2}=\dfrac{6}{7}+\dfrac{1}{2}$

           $v=\dfrac{19}{14}=1\dfrac{5}{14}$

33. a. $\dfrac{v}{3}=\dfrac{2}{7}$

      $\cancel{3}\cdot\dfrac{v}{\cancel{3}}=\dfrac{2}{7}\cdot3$

          $v=\dfrac{6}{7}$

  b. $\dfrac{v}{4}=\dfrac{3}{7}$

      $\cancel{4}\cdot\dfrac{v}{\cancel{4}}=\dfrac{3}{7}\cdot4$

         $v=\dfrac{12}{7}$ or $1\dfrac{5}{7}$

  c. $\dfrac{v}{5}=\dfrac{4}{7}$

      $\cancel{5}\cdot\dfrac{v}{\cancel{5}}=\dfrac{4}{7}\cdot5$

         $v=\dfrac{20}{7}$ or $2\dfrac{6}{7}$

d. $\dfrac{v}{6} = \dfrac{5}{7}$

$\cancel{6} \cdot \dfrac{v}{\cancel{6}} = \dfrac{5}{7} \cdot 6$

$v = \dfrac{30}{7}$ or $4\dfrac{2}{7}$

e. $\dfrac{v}{7} = \dfrac{6}{7}$

$\cancel{7} \cdot \dfrac{v}{\cancel{7}} = \dfrac{6}{7} \cdot \cancel{7}$

$v = 6$

34. a. $\dfrac{1}{2} \cdot n = 8$

$\cancel{2} \cdot \dfrac{1}{\cancel{2}} \cdot n = 8 \cdot 2$

$n = 16$

b. $\dfrac{2}{3} \cdot n = 4$

$\dfrac{\cancel{\frac{2}{3}} \cdot n}{\cancel{\frac{2}{3}}} = \dfrac{4}{\frac{2}{3}}$

$n = {}^2\cancel{4} \cdot \dfrac{3}{\cancel{2}_1} = 6$

c. $\dfrac{3}{5} \cdot n = 27$

$\dfrac{\cancel{\frac{3}{5}} \cdot n}{\cancel{\frac{3}{5}}} = \dfrac{27}{\frac{3}{5}}$

$n = {}^9\cancel{27} \cdot \dfrac{5}{\cancel{3}_1} = 45$

d. $\dfrac{2}{7} \cdot n - 14$

$\dfrac{\cancel{\frac{2}{7}} \cdot n}{\cancel{\frac{2}{7}}} = \dfrac{14}{\frac{2}{7}}$

$n = {}^7\cancel{14} \cdot \dfrac{7}{\cancel{2}_1} = 49$

e. $\dfrac{6}{5} \cdot n = 12$

$\dfrac{\cancel{\frac{6}{5}} \cdot n}{\cancel{\frac{6}{5}}} = \dfrac{12}{\frac{6}{5}}$

$n = {}^2\cancel{12} \cdot \dfrac{5}{\cancel{6}_1} = 10$

35. a. $1 - \dfrac{3}{25} - \dfrac{7}{100} - \dfrac{1}{25} = 1 - \dfrac{12}{100} - \dfrac{7}{100} - \dfrac{4}{100}$

$= 1 - \dfrac{23}{100}$

$= \dfrac{100}{100} - \dfrac{23}{100}$

$= \dfrac{77}{100}$

b. $1 - \dfrac{6}{25} - \dfrac{7}{100} - \dfrac{2}{25} = 1 - \dfrac{24}{100} - \dfrac{7}{100} - \dfrac{8}{100}$

$= 1 - \dfrac{39}{100}$

$= \dfrac{100}{100} - \dfrac{39}{100}$

$= \dfrac{61}{100}$

c. $1 - \dfrac{7}{25} - \dfrac{7}{100} - \dfrac{3}{25} = 1 - \dfrac{28}{100} - \dfrac{7}{100} - \dfrac{12}{100}$

$= 1 - \dfrac{47}{100}$

$= \dfrac{100}{100} - \dfrac{47}{100}$

$= \dfrac{53}{100}$

d. $1 - \dfrac{8}{25} - \dfrac{7}{100} - \dfrac{4}{25} = 1 - \dfrac{32}{100} - \dfrac{7}{100} - \dfrac{16}{100}$

$- 1 - \dfrac{55}{100}$

$= \dfrac{100}{100} - \dfrac{55}{100}$

$= \dfrac{45}{100}$

$= \dfrac{9}{20}$

e. $1 - \dfrac{9}{25} - \dfrac{7}{100} - \dfrac{6}{25} = 1 - \dfrac{36}{100} - \dfrac{7}{100} - \dfrac{24}{100}$

$= 1 - \dfrac{67}{100}$

$= \dfrac{100}{100} - \dfrac{67}{100}$

$= \dfrac{33}{100}$

Cumulative Review Chapters 1–2

1. $438 = 400 + 30 + 8$

2. $900 + 80 + 4 = 984$

3. 74,008: Seventy-four thousand, eight

4. 6710

5. $8\underline{6}49 \rightarrow 8600$

6.
$$
\begin{array}{r}
\overset{1}{2776} \\
+\ 903 \\
\hline
3679
\end{array}
$$

7.
$$
\begin{array}{r}
\overset{4\ 12}{6\cancel{8}2} \\
-498 \\
\hline
4
\end{array}
\rightarrow
\begin{array}{r}
\overset{5\ 14}{\cancel{6}\cancel{8}2} \\
-4\ 98 \\
\hline
\boxed{1\ 5\ 4}
\end{array}
$$

8.
$$
\begin{array}{r}
\overset{12}{\ }\overset{3\ 6}{\ }\\
137 \\
\times\ 319 \\
\hline
1233 \\
137\ \ \\
411\ \ \ \\
\hline
43,703
\end{array}
$$

9.
$$
\begin{array}{r}
310 \\
\times\ \ 12 \\
\hline
620 \\
310\ \ \\
\hline
3720
\end{array}
$$

The total amount of money paid is \$3720.

10.
$$
\begin{array}{r}
34 \\
26\overline{)889} \\
\underline{78}\ \ \\
109 \\
\underline{104} \\
5
\end{array}
$$
Answer: 34 r 5

11.
$$
\begin{array}{l}
2\underline{|24} \\
2\underline{|12} \\
2\underline{|\ 6} \\
\ \ \ 3
\end{array}
$$
Thus, the prime factors of 24 are 2 and 3.

12. $180 = 18 \times 10$
$$
\begin{aligned}
&= 2 \times 9 \times 2 \times 5 \\
&= 2 \times 3 \times 3 \times 2 \times 5 \\
&= 2^2 \times 3^2 \times 5
\end{aligned}
$$

13. $2^3 \times 4 \times 7^0 = 8 \times 4 \times 1 = 32 \times 1 = 32$

14. $36 \div 6 \cdot 6 + 8 - 4 = 6 \cdot 6 + 8 - 4$
$$
\begin{aligned}
&= 36 + 8 - 4 \\
&= 44 - 4 \\
&= 40
\end{aligned}
$$

15. $\quad 26 = m + 3$
$$
\begin{aligned}
26 - 3 &= m + 3 - 6 \\
23 &= m
\end{aligned}
$$

16. $21 = 7x$
$$
\begin{aligned}
\frac{21}{7} &= \frac{\cancel{7}x}{\cancel{7}} \\
3 &= x
\end{aligned}
$$

17. $\frac{2}{3}$ is proper, since the numerator is less than the denominator.

18. $\frac{11}{2} = 5\frac{1}{2}$

19. $2\frac{1}{4} = \frac{4 \times 2 + 1}{4} = \frac{8 + 1}{4} = \frac{9}{4}$

21. $\frac{2}{3} = \frac{18}{?}$; $\frac{2}{3} = \frac{2 \cdot 9}{3 \cdot 9} = \frac{18}{27}$. The missing number is 27.

22. $\frac{10}{12} = \frac{\cancel{2} \times 5}{\cancel{2} \times 6} = \frac{5}{6}$

23. $\frac{3}{4}$ and $\frac{5}{6}$; $3 \times 6 = 18 < 4 \times 5 = 20$ so

$$\frac{3}{4} < \frac{5}{6}$$

24. $\frac{1}{2} \cdot 6\frac{1}{3} = \frac{1}{2} \cdot \frac{19}{3} = \frac{19}{6} = 3\frac{1}{6}$

25. $\left(\frac{7}{6}\right)^2 \cdot \frac{1}{49} = \left(\frac{7}{6} \cdot \frac{7}{6}\right) \cdot \frac{1}{49} = \frac{\cancel{49}}{36} \cdot \frac{1}{\cancel{49}} = \frac{1}{36}$

26. $\frac{6}{7} \div 1\frac{1}{3} = \frac{6}{7} \div \frac{4}{3} = \frac{\cancel{6}^{3}}{7} \cdot \frac{3}{\cancel{4}_{2}} = \frac{9}{14}$

27. $7\frac{1}{3} + 9\frac{3}{10} = \frac{22}{3} + \frac{93}{10}$

$$= \frac{22 \cdot 10}{3 \cdot 10} + \frac{93 \cdot 3}{10 \cdot 3}$$

$$= \frac{220}{30} + \frac{279}{30}$$

$$= \frac{220 + 279}{30} = \frac{499}{30} = 16\frac{19}{30}$$

28. $8\frac{1}{7} - 1\frac{8}{9} = \frac{57}{7} - \frac{17}{9}$

$$= \frac{57 \cdot 9}{7 \cdot 9} - \frac{17 \cdot 7}{9 \cdot 7}$$

$$= \frac{513}{63} - \frac{119}{63}$$

$$= \frac{513 - 119}{63} = \frac{394}{63} = 6\frac{16}{63}$$

29. $z - \frac{6}{7} = \frac{4}{9}$

$$z - \frac{6}{7} + \frac{6}{7} = \frac{4}{9} + \frac{6}{7}$$

$$z = \frac{28}{63} + \frac{54}{63}$$

$$z = \frac{28 + 54}{63} = \frac{82}{63}$$

30. Let n = the number.

$$\frac{9}{10}n = 5\frac{1}{5}$$

$$\frac{9}{10}n = \frac{26}{5}$$

$$\frac{\frac{\cancel{9}}{\cancel{10}}n}{\frac{\cancel{9}}{\cancel{10}}} = \frac{\frac{26}{5}}{\frac{9}{10}}$$

$$n = \frac{26}{\cancel{5}} \cdot \frac{\cancel{10}^{2}}{9} = \frac{52}{9} = 5\frac{7}{9}$$

The number is $5\frac{7}{9}$.

31. $3\frac{1}{2} = \frac{7}{2}$ lb cost 49¢ so 1 lb would cost

$49 \div \frac{7}{2} = {}^{7}\cancel{49} \cdot \frac{2}{\cancel{7}_{1}} = 14$ ¢. Thus 8 lb

would cost $8 \cdot 14 = 112$ ¢ or $1.12.

32. Perimeter $= 4\frac{1}{3} + 6\frac{2}{3} + 4\frac{1}{3} + 6\frac{2}{3}$

$$= 20\frac{1+2+1+2}{3}$$

$$= 20\frac{6}{3} = 20 + 2 = 22 \text{ yards}$$

33. Area $= \left(4\frac{1}{3}\right) \cdot \left(6\frac{2}{3}\right)$

$$= \frac{13}{3} \cdot \frac{20}{3} = \frac{260}{9} = 28\frac{8}{9} \text{ sq. yd.}$$

Chapter 3

Decimals

Section 3.1 – Addition and Subtraction of Decimals

Problems

1. One hundred forty-seven and seventeen hundredths

2. $47.321 = 40 + 7 + \dfrac{3}{10} + \dfrac{2}{100} + \dfrac{1}{1000}$

3.
$$\begin{array}{r} 4.0 \\ +12.4 \\ \hline 16.4 \end{array}$$

4.
$$\begin{array}{r} {}^{1\ 1\ 1} \\ 49.280 \\ +\ \ 7.921 \\ \hline 57.201 \end{array}$$

5.
$$\begin{array}{r} {}^{1\ 11 2\ 1} \\ 11{,}022.77 \\ 1\,521.54 \\ 674.00 \\ +\ \ 1\,618.73 \\ \hline 14{,}837.04 \end{array}$$

6.
$$\begin{array}{r} {}^{3\ 11\,13} \\ 7\,\cancel{4}\,\cancel{2}.3\,2 \\ -\ \ \ 1\,3.6\,0 \\ \hline 7\,2\,8.7\,2 \end{array}$$

7.
$$\begin{array}{r} {}^{3\ 12\ \ 59\ 10} \\ \cancel{4}\,2\,9.\cancel{6}\cancel{0}\,0 \\ -2\,3\,3.3\,8\,1 \\ \hline 1\,9\,6.2\,1\,9 \end{array}$$

8. **a.** La Quinta Inn is 3.5 miles to the left.

   b.
   $$\begin{array}{r} 4.6 \\ -0.5 \\ \hline 4.1 \end{array}$$
   It is 4.1 miles from the Holiday Inn to Shoney's Inn.

c.
$$\begin{array}{r} 3.7 \\ +0.8 \\ \hline 4.5 \end{array}$$
It is 4.5 miles from the Happy Traveler RV Part to the Wingate Inn.

Exercises 3.1

1. 3.8: Three and eight tenths

3. 13.12: Thirteen and twelve hundredths

5. 132.34: One hundred thirty-two and thirty-four hundredths

7. 5.183: Five and one hundred eighty-three thousandths

9. 0.2172: Two thousand, one hundred seventy-two ten-thousandths

11. $3.21 = 3 + \dfrac{2}{10} + \dfrac{1}{100}$

13. $41.38 = 40 + 1 + \dfrac{3}{10} + \dfrac{8}{100}$

15. $89.123 = 80 + 9 + \dfrac{1}{10} + \dfrac{2}{100} + \dfrac{3}{1000}$

17. $238.392 = 200 + 30 + 8 + \dfrac{3}{10} + \dfrac{9}{100} + \dfrac{2}{1000}$

19. $301.5879 =$
$300 + 1 + \dfrac{5}{10} + \dfrac{8}{100} + \dfrac{7}{1000} + \dfrac{9}{1000}$

21.
$$\begin{array}{r} 0.4 \\ +0.1 \\ \hline 0.5 \end{array}$$

23.
$$
\begin{array}{r}
\overset{1}{0.6} \\
+\,0.9 \\
\hline
1.5
\end{array}
$$

25.
$$
\begin{array}{r}
0.3 \\
-\,0.1 \\
\hline
0.2
\end{array}
$$

27.
$$
\begin{array}{r}
8.3 \\
-\,5.2 \\
\hline
3.1
\end{array}
$$

29.
$$
\begin{array}{r}
\overset{4\ \ 10}{\cancel{5}.0} \\
-\,3.2 \\
\hline
1.8
\end{array}
$$

31.
$$
\begin{array}{r}
\overset{8\ \ 10}{\cancel{9}.0} \\
-\,4.1 \\
\hline
4.9
\end{array}
$$

33.
$$
\begin{array}{r}
\overset{2\ \ 18}{\cancel{3}.8} \\
-\,1.9 \\
\hline
1.9
\end{array}
$$

35.
$$
\begin{array}{r}
\overset{0\ \ 11}{\cancel{1}.1} \\
-\,0.8 \\
\hline
0.3
\end{array}
$$

37.
$$
\begin{array}{r}
\overset{1}{12.23} \\
+\ \ 9.00 \\
\hline
21.23
\end{array}
$$

39.
$$
\begin{array}{r}
\overset{1\ 1}{4.60} \\
+\,18.73 \\
\hline
23.33
\end{array}
$$

41.
$$
\begin{array}{r}
\overset{16\ \ 13}{\cancel{17}.35} \\
-\ \ 8.40 \\
\hline
8.95
\end{array}
$$

43.
$$
\begin{array}{r}
\$648.01 \\
+\$341.06 \\
\hline
\$989.07
\end{array}
$$

45.
$$
\begin{array}{r}
\overset{1}{72.030} \\
+\,847.124 \\
\hline
919.154
\end{array}
$$

47.
$$
\begin{array}{r}
\overset{1}{104.000} \\
+\ \ 78.103 \\
\hline
182.103
\end{array}
$$

49.
$$
\begin{array}{r}
\overset{1}{0.350} \\
3.600 \\
+\,0.127 \\
\hline
4.077
\end{array}
$$

51.
$$
\begin{array}{r}
\overset{6\ \ 11\ 10}{2\cancel{7}.\cancel{2}0} \\
-\ \ 0.3\,5 \\
\hline
2\,6.8\,5
\end{array}
$$

53.
$$
\begin{array}{r}
\overset{8\ \ 9\ 10}{1\cancel{9}.\cancel{0}0} \\
-\ \ 16.6\,2 \\
\hline
2.3\,8
\end{array}
$$

55.
$$
\begin{array}{r}
\overset{2\ \ 10}{9.4\cancel{3}0} \\
-\ \ 6.4\,0\,6 \\
\hline
3.0\,2\,4
\end{array}
$$

57.
$$
\begin{array}{r}
\overset{7\ \ 11\ 9\ 10}{\cancel{8}.\cancel{2}\cancel{0}0} \\
-\ \ 1.3\,5\,6 \\
\hline
6.8\,4\,4
\end{array}
$$

59.
$$
\begin{array}{r}
6.0900 \\
+\,3.0046 \\
\hline
9.0946
\end{array}
$$

61.
$$
\begin{array}{r}
\overset{6\ \ 11}{18,\cancel{7}19.7} \\
-\,18,327.2 \\
\hline
392.5
\end{array}
$$

The trip was 392.5 miles.

63.
$$\begin{array}{r} \overset{1\ 1\ 1}{18.47} \\ 23.48 \\ +12.63 \\ \hline 54.58 \end{array}$$
She spent $54.58.

65.
$$\begin{array}{r} \overset{3\ 14}{75\cancel{4}.41} \\ -730.80 \\ \hline 23.61 \end{array}$$
He scored 23.61 fewer points in 1988.

67.
$$\begin{array}{r} \overset{2}{26.0\%} \\ 43.0\% \\ 7.0\% \\ +17.5\% \\ \hline 93.5\% \end{array}$$

69. From exercise #67, we have 93.5% from skin, muscle, blood, and bone. Get the total by adding the remaining values to this percent.
$$\begin{array}{r} \overset{1\ 3}{93.5\%} \\ 2.7\% \\ 2.2\% \\ 2.2\% \\ 1.5\% \\ 0.5\% \\ 0.5\% \\ 0.2\% \\ +\ 0.1\% \\ \hline 103.4\% \end{array}$$

71.
$$\begin{array}{r} 23 \\ \times 10 \\ \hline 230 \end{array}$$

73.
$$\begin{array}{r} 240 \\ \times\ 1000 \\ \hline 240,000 \end{array}$$

75. $3\underline{4}87 \rightarrow 3500$

77.
$$\begin{array}{r} \$27.50 \\ 50.00 \\ +\ 10.00 \\ \hline \$87.50 \end{array}$$
The sum of these checks is $87.50.

79.
$$\begin{array}{r} \overset{2}{568.40} \\ 90.00 \\ +\ 60.30 \\ \hline 718.70 \end{array}$$
The total is $718.70.

81. 3.7 miles

83. $4.6 - 3.5 = 1.1$ miles

85. 2.3 miles

87. $2.5 - 2.3 = 0.2$ mile

89. "And" indicates a decimal point, but this is a whole number.

91. Answers may vary.
$4.8 - 3 = 1.8$ and $6.6 - 4.8 = 1.8$.

93.
$$\begin{array}{r} 3.4y^2 + 0.5y \\ +\ 7.2y^2 + 0.7y \\ \hline 10.6y^2 + 1.2y \end{array}$$

95.
$$\begin{array}{r} \overset{6\ 10}{\cancel{7}.0y^2 + 0.9y} \\ -\ 6.2y^2 + 0.4y \\ \hline 0.8y^2 + 0.5y \end{array}$$

Mastery Test 3.1

97.
$$\begin{array}{r} \overset{1\ 1\ 1}{38.773} \\ +\ 3.690 \\ \hline 42.463 \end{array}$$

99. $41.208 = 10 + 1 + \dfrac{2}{10} + \dfrac{0}{100} + \dfrac{8}{1000}$
$$= 10 + 1 + \dfrac{2}{10} + \dfrac{8}{1000}$$

101.
$$
\begin{array}{r}
{\scriptstyle 6\ 12\ 14} \\
4\,7\,3.4\,3 \\
-\ \ \ 1\,8.6\,0 \\
\hline
4\,5\,4.8\,3
\end{array}
$$

103.
$$
\begin{array}{r}
{\scriptstyle 1\ 15} \\
2.5 \\
-\,0.9 \\
\hline
1.6 \text{ miles}
\end{array}
$$

Section 3.2 – Multiplication and Divison of Decimals

Problems

1. a.
$$
\begin{array}{r}
3.12 \\
\times\ 5.3 \\
\hline
936 \\
1560 \\
\hline
16.536
\end{array}
$$

b.
$$
\begin{array}{r}
12.172 \\
\times\ \ \ 5.1 \\
\hline
12172 \\
60860 \\
\hline
62.0772
\end{array}
$$

2. a.
$$
\begin{array}{r}
3.201 \\
\times\,31.02 \\
\hline
6402 \\
3201 \\
9603 \\
\hline
99.29502
\end{array}
$$

b.
$$
\begin{array}{r}
4.132 \\
\times\ 0.0021 \\
\hline
4132 \\
8264 \\
\hline
.0086772
\end{array}
$$

3. a. $58.12 \times 100 = 5812$
b. $43.1 \times 1000 = 43,100$
c. $10.296 \times 10 = 102.96$

4. $(.99 \times 10)$ cents $= 9.9$ cents

5. a. 692 cents $= 692 \times \$0.01 = \$6.92$
   b. $\$7.92 = 7.92 \times 100$ cents $= 792$ cents

6. $0.035\overline{)1.4} \rightarrow 35\overline{)1400.}$ so $\dfrac{1.4}{0.035} = 40$
$$
\begin{array}{r}
40. \\
35\overline{)1400.} \\
140 \\
\hline
00
\end{array}
$$

7. $13\overline{)0.0065}$ so $\dfrac{0.0065}{13} = 0.0005$
$$
\begin{array}{r}
.0005 \\
13\overline{)0.0065} \\
65 \\
\hline
0
\end{array}
$$

8. a. $\underline{2}7.752 \rightarrow 30$
   b. $27.\underline{7}52 \rightarrow 27.8$
   c. $27.7\underline{5}2 \rightarrow 27.75$

9. $0.12\overline{)56} \rightarrow 12\overline{)5600.000}$
$$
\begin{array}{r}
466.666 \\
12\overline{)5600.000} \\
48 \\
\hline
80 \\
72 \\
\hline
80 \\
72 \\
\hline
80 \\
72 \\
\hline
80 \\
72 \\
\hline
8
\end{array}
$$
Thus, $56 \div 0.12 = 466.67$.

10. a. $352.9 \div 100 = 3.529$
   b. $3.27 \div 1000 = 0.00327$
   c. $9.35 \div 10 = 0.935$

11. $22\overline{)27.00}$ so $T = \dfrac{27}{22} = 1.2$ years
$$
\begin{array}{r}
1.22 \\
22\overline{)27.00} \\
22 \\
\hline
50 \\
44 \\
\hline
60 \\
44 \\
\hline
16
\end{array}
$$

Exercises 3.2

1. $0.5 \cdot 0.7 = 0.35$

3. $0.8 \cdot 0.8 = 0.64$

5. $0.005 \cdot 0.07 = 0.00035$

7.
$$
\begin{array}{r}
0.613 \\
\times\ \ 9.2 \\
\hline
1226 \\
5517\ \ \\
\hline
5.6396
\end{array}
$$

9.
$$
\begin{array}{r}
8.7 \\
\times\ 11 \\
\hline
87 \\
87\ \ \\
\hline
95.7
\end{array}
$$

11.
$$
\begin{array}{r}
7.03 \\
\times\ 0.0035 \\
\hline
3515 \\
2109\ \ \\
\hline
0.024605
\end{array}
$$

13.
$$
\begin{array}{r}
3.0012 \\
\times\ \ \ \ 4.3 \\
\hline
90036 \\
120048\ \ \\
\hline
12.90516
\end{array}
$$

15.
$$
\begin{array}{r}
0.0031 \\
\times\ \ \ 0.82 \\
\hline
62 \\
248\ \ \\
\hline
0.002542
\end{array}
$$

17. $42.33 \cdot 10 = 423.3$ (since multiplying by 10, move decimal 1 place to the right)

19. $19.5 \cdot 100 = 1950$ (since multiplying by 100, move the decimal 2 places to the right)

21. $32.89 \cdot 1000 = 32,890$ (since multiplying by 1000, move the decimal 3 places to the right)

23. $0.48 \cdot 10 = 4.8$ (since multiplying by 10, move decimal 1 place to the right)

25. $0.039 \cdot 100 = 3.9$ (since multiplying by 100, move decimal 2 places to the right)

27.
$$
\begin{array}{r}
0.6\ \ \\
15\overline{)9.0} \\
9\,0\ \\
\hline
0
\end{array}
$$
Answer: 0.6

29.
$$
\begin{array}{r}
6.4\ \ \\
5\overline{)32.0} \\
30\ \ \ \\
\hline
20 \\
20 \\
\hline
0
\end{array}
$$
Answer: 6.4

31. $0.005\overline{)8.5} \rightarrow$
$$
\begin{array}{r}
1700 \\
5\overline{)8500} \\
5\ \ \ \ \\
\hline
35\ \ \\
35\ \ \\
\hline
00
\end{array}
$$
Thus $8.5 \div 0.005 = 1700$.

33. $0.05\overline{)4} \rightarrow$
$$
\begin{array}{r}
80 \\
5\overline{)400} \\
40\ \\
\hline
00
\end{array}
$$
Thus $4 \div 0.05 = 80$.

35.
$$
\begin{array}{r}
0.046 \\
60\overline{)2.760} \\
2\,40\ \ \\
\hline
360 \\
360 \\
\hline
0
\end{array}
$$
Thus $2.76 \div 60 = 0.046$.

37. $3\underline{4}.8 \rightarrow 30$

39. $96.87 = \underline{0}96.87 \rightarrow 100$

41. $3.\underline{1}5 \rightarrow 3.2$

43. $7.81 = \underline{0}7.81 \rightarrow 10$

45. $338.1\underline{2}3 \rightarrow 338.12$

47. $7.8 \div 100 = 0.078$ (since dividing by 100, move the decimal 2 places to the left)

49. $0.05 \div 100 = 0.0005$ (since dividing by 100, move the decimal 2 places to the left)

51.
$$
\begin{array}{r}
0.333 \\
3\overline{)1.000} \\
\underline{9} \\
10 \\
\underline{9} \\
10 \\
\underline{9} \\
1
\end{array}
$$
Thus $1 \div 3 = 0.33$

53. $0.70\overline{)0.06} \rightarrow 7\overline{)0.600}$
$$
\begin{array}{r}
0.085 \\
\underline{56} \\
40 \\
\underline{35} \\
5
\end{array}
$$
Thus $0.06 \div 0.70 = 0.09$.

55. $12.243\overline{)2.8} \rightarrow 12,243\overline{)2800.000}$
$$
\begin{array}{r}
0.228 \\
2448\,6 \\
351\,40 \\
244\,86 \\
106540 \\
\underline{97944} \\
8596
\end{array}
$$
Thus $12.243\overline{)2.8} = 0.23$.

57. $8.156 \div 1000 = 0.00\underline{8}156 = 0.01$

59.
$$
\begin{array}{r}
0.027 \\
20\overline{)0.545} \\
\underline{40} \\
140 \\
\underline{140} \\
0
\end{array}
$$
Thus $20\overline{)0.545} = 0.03$

61.
$$
\begin{array}{r}
13.5 \\
\times 1.61 \\
\hline
135 \\
810 \\
135 \\
\hline
21.735
\end{array}
$$
The cost is $21.74.

63. a.
$$
\begin{array}{r}
24 \\
\times 0.67 \\
\hline
168 \\
144 \\
\hline
16.08
\end{array}
$$
It would cost $16.08.

b.
$$
\begin{array}{r}
16.08 \\
\times 30 \\
\hline
482.40
\end{array}
$$
It would cost $482.40.

65.
$$
\begin{array}{r}
1.8 \\
\times 30 \\
\hline
54.0
\end{array}
$$
The total cost is $54.00.

67. $48 - 15 = 33$
$$
\begin{array}{r}
1.09 \\
\times 15 \\
\hline
545 \\
109 \\
\hline
16.35
\end{array}
\quad \text{and} \quad
\begin{array}{r}
1.27 \\
\times 23 \\
\hline
381 \\
254 \\
\hline
29.21
\end{array}
$$
$16.35 + 41.91 = 58.26$
The total gas bill is $58.26.

69. Mileage cost:
$$
\begin{array}{r}
348 \\
\times .49 \\
\hline
3132 \\
1392 \\
\hline
170.52
\end{array}
$$

Rental cost:
$$
\begin{array}{r}
69.99 \\
+ 3 \\
\hline
209.97
\end{array}
$$
$170.52 + 209.97 = 380.49$
The total rental cost is $380.49.

70. Mileage cost: $257 \times \$0.49 = \$125.93$
Rental cost: $\$79.99 \times 3 = \$239.97$
Total cost $= \$125.93 + \$239.97 = \$365.90$

71.
$$
\begin{array}{r}
0.915 \\
2\overline{)1.830} \\
\underline{1\,8} \\
3 \\
\underline{2} \\
10 \\
\underline{10} \\
0
\end{array}
$$

The download time for text using a 9600 bps modem is 0.915 sec.

73.
$$
\begin{array}{r}
0.305 \\
2\overline{)0.610} \\
\underline{6} \\
10 \\
\underline{10} \\
0
\end{array}
$$

The download time for text using a 28,800 bps modem is 0.305 sec.

75. $33.3 - 22.2 = 11.1$ min
$$
\begin{array}{r}
5.55 \\
2\overline{)11.10} \\
\underline{10} \\
11 \\
\underline{10} \\
10 \\
\underline{10} \\
0
\end{array}
$$

The time difference is 5.55 min.

77.
$$
\begin{array}{r}
2.83 \\
\times\ \ \ 3 \\
\hline
8.\underline{49} \to 8.50
\end{array}
$$

The price for 3 ounces is $8.50.

79.
$$
\begin{array}{r}
752 \\
\times 0.0068 \\
\hline
6016 \\
4512 \\
\hline
5.1136
\end{array}
$$

The paper is 5.1136 cm thick.

81.
$$
\begin{array}{r}
376.59 \\
\times\ \ \ \ \ 3 \\
\hline
1129.77
\end{array}
$$

He could go 1129.77 miles on 3 gallons.

83.
$$
\begin{array}{r}
0.29 \\
6\overline{)1.74} \\
\underline{1\,2} \\
54 \\
\underline{54} \\
0
\end{array}
$$
The cost per bottle is $0.29.

85.
$$
\begin{array}{r}
9.94 \\
101\overline{)1004.00} \\
\underline{909} \\
95\,0 \\
\underline{90\,9} \\
4\,10 \\
\underline{4\,04} \\
6
\end{array}
$$

His average was 9.9 yards per carry.

87.
$$
\begin{array}{r}
5.09 \\
155\overline{)790.00} \\
\underline{775} \\
15\,00 \\
\underline{13\,95} \\
1\,05
\end{array}
$$
$\to$ payback time = 5.1 years

89.
$$
\begin{array}{r}
2.27 \\
18\overline{)41.00} \\
\underline{36} \\
5\,0 \\
\underline{3\,6} \\
1\,40 \\
\underline{1\,26} \\
14
\end{array}
$$
$\to$ payback time = 2.3 years

91.
$$
\begin{array}{r}
3.42 \\
92\overline{)315.00} \\
\underline{276} \\
39\,0 \\
\underline{36\,8} \\
2\,20 \\
\underline{1\,84} \\
36
\end{array}
$$
$\to$ payback time = 3.4 years

93. $\dfrac{3}{10} = \dfrac{?}{100}$; $\dfrac{3}{10} = \dfrac{3\cdot 10}{10\cdot 10} = \dfrac{30}{100}$. The missing number is 30.

95. $(2.38\times 16.2) + 30.97 = 38.56 + 30.97$
$$= 69.53$$
$$= 69.5 \text{ in.}$$

97. $(3.34 \times 8.25) + 31.98 = 27.56 + 31.98$
$= 59.54$
$= 59.5$ in.

99. $(1.88 \times 18.8) + 32.01 = 35.34 + 32.01$
$= 67.35$
$= 67.4$ in.

101. $(2.89 \times 15.9) + 27.81 = 45.95 + 27.81$
$= 73.76$
$= 73.8$ in.

103. $(1.88 \times 29.9) + 32.01 = 56.21 + 32.01$
$= 88.22$
$= 88.2$ in.

The chart shows the relationship only for persons whose height is between 60 and 85 inches and therefore does not apply.

105. Answers may vary.

107. Answers may vary.

Mastery Test 3.2

109.

$$
0.14\overline{)90} \rightarrow 14\overline{)9000.000}
$$
$$
\begin{array}{r}
642.857 \\
\hline
84 \\
\hline
60 \\
56 \\
\hline
40 \\
28 \\
\hline
12\,0 \\
11\,2 \\
\hline
80 \\
70 \\
\hline
100 \\
98 \\
\hline
2
\end{array}
$$

Thus $90 \div 0.14 = 642.86$

111.
$$
12\overline{)0.0060}
$$
$$
\begin{array}{r}
0.0005 \\
\hline
60 \\
\hline
0
\end{array}
$$
Thus $\dfrac{0.0060}{12} = 0.0005$.

113. a. $32.423 \times 10 = 324.23$
b. $48.4 \times 1000 = 48,400$
c. $0.328 \times 100 = 32.8$

115. a.
$$
\begin{array}{r}
4.41 \\
\times\ 3.2 \\
\hline
882 \\
1323 \\
\hline
14.112
\end{array}
$$
b.
$$
\begin{array}{r}
14.724 \\
\times\ \ \ 5.1 \\
\hline
14724 \\
73620 \\
\hline
75.0924
\end{array}
$$

Section 3.3 – Fractions and Decimals

Problems

1. a. $\dfrac{3}{5} = 3 \div 5$ or $5\overline{)3.0}$
$$
\begin{array}{r}
0.6 \\
\hline
3\,0 \\
\hline
0
\end{array}
$$
Hence $\dfrac{3}{5} = 0.6$.

b. $\dfrac{3}{40} = 3 \div 40$ or $40\overline{)3.000}$
$$
\begin{array}{r}
0.075 \\
\hline
2\,80 \\
\hline
200 \\
200 \\
\hline
0
\end{array}
$$
Hence $\dfrac{3}{40} = 0.075$.

2. $\frac{2}{7} = 2 \div 7$ or

$$
\begin{array}{r}
0.285714 \\
7{\overline{\smash{\big)}\,2.000000}} \\
\underline{1\,4} \\
60 \\
\underline{56} \\
40 \\
\underline{35} \\
50 \\
\underline{49} \\
10 \\
\underline{7} \\
30 \\
\underline{28} \\
2
\end{array}
$$

Thus $\frac{2}{7} = 0.\overline{285714}$.

3. a. $0.050 = \dfrac{5\cancel{0}}{100\cancel{0}} = \dfrac{5}{100} = \dfrac{1}{20}$

   b. $0.0350 = \dfrac{35\cancel{0}}{10,00\cancel{0}} = \dfrac{35}{1000} = \dfrac{7}{200}$

4. a. $1.17 = 1 + \dfrac{17}{100} = 1\dfrac{17}{100} = \dfrac{117}{100}$

   b. $4.35 = 4 + \dfrac{\overset{7}{\cancel{35}}}{\underset{20}{\cancel{100}}} = 4\dfrac{7}{20} = \dfrac{87}{20}$

5. a. $0.\overline{41} = \dfrac{41}{99}$

   b. $0.\overline{105} = \dfrac{105}{999} = \dfrac{35 \cdot \cancel{3}}{333 \cdot \cancel{3}} = \dfrac{35}{333}$

6. $d \times 16 = 7$

$$
d = 7 \div 16 = 16{\overline{\smash{\big)}\,7.0000}}
$$
$$
\begin{array}{r}
0.4375 \\
\underline{64} \\
60 \\
\underline{48} \\
120 \\
\underline{112} \\
80 \\
\underline{80} \\
0
\end{array}
$$

Thus, 7 is 0.4375 of 16.

7. $d \times 800,000 = 128,000$

$$
d = \frac{128,000}{800,000} = \frac{128}{800} = \frac{4}{25}
$$

$$
\begin{array}{r}
0.16 \\
25{\overline{\smash{\big)}\,4.00}} \\
\underline{2\,5} \\
1\,50 \\
\underline{1\,50} \\
0
\end{array}
$$

Thus $0.16 of every dollar spent is for breakfast.

Exercises 3.3

1. $\frac{1}{2} = 1 \div 2$ or $2{\overline{\smash{\big)}\,1.0}}$ Hence $\frac{1}{2} = 0.5$.

$$
\begin{array}{r}
0.5 \\
\underline{1\,0} \\
0
\end{array}
$$

3. $\frac{11}{16} = 3 \div 40$ or $16{\overline{\smash{\big)}\,11.0000}}$

$$
\begin{array}{r}
0.6875 \\
\underline{9\,6} \\
1\,40 \\
\underline{1\,28} \\
120 \\
\underline{112} \\
80 \\
\underline{80} \\
0
\end{array}
$$

Hence $\frac{11}{16} = 0.6875$.

5. $\frac{9}{20} = 9 \div 20$ or $20{\overline{\smash{\big)}\,9.00}}$

$$
\begin{array}{r}
0.45 \\
\underline{8\,0} \\
1\,00 \\
\underline{1\,00} \\
0
\end{array}
$$

Hence $\frac{9}{20} = 0.45$.

7. $\frac{9}{10} = 9 \div 10 = 0.9$ (move the decimal on place to the left since dividing by 10)

9. $\dfrac{1}{4}=1\div 4$ or $4\overline{)1.00}$ 　Hence $\dfrac{1}{4}=0.25$.

$$\begin{array}{r}0.25\\4\overline{)1.00}\\\underline{8}\\20\\\underline{20}\\0\end{array}$$

11. $\dfrac{5}{6}=5\div 6=6\overline{)5.000}$ 　Thus $\dfrac{5}{6}\approx 0.83$.

$$\begin{array}{r}.83\overline{3}\\6\overline{)5.000}\\\underline{48}\\20\\\underline{18}\\2\end{array}$$

13. $\dfrac{3}{7}=3\div 7=7\overline{)3.000}$ 　Thus $\dfrac{3}{7}\approx 0.43$.

$$\begin{array}{r}0.428\\7\overline{)3.000}\\\underline{28}\\20\\\underline{14}\\60\\\underline{56}\\4\end{array}$$

15. $\dfrac{8}{3}=2\dfrac{2}{3}$; $\dfrac{2}{3}=2\div 3=3\overline{)2.000}$

$$\begin{array}{r}0.66\overline{6}\\3\overline{)2.000}\\\underline{18}\\20\\\underline{18}\\2\end{array}$$

　　Thus $\dfrac{8}{3}\approx 2.67$.

17. $\dfrac{1}{3}=1\div 3=3\overline{)1.000}$ 　Thus $\dfrac{1}{3}\approx 0.33$.

$$\begin{array}{r}.33\overline{3}\\3\overline{)1.000}\\\underline{9}\\10\\\underline{9}\\1\end{array}$$

19. $\dfrac{2}{11}=2\div 11=11\overline{)2.000}$ 　Thus $\dfrac{2}{11}\approx 0.18$.

$$\begin{array}{r}0.181\\11\overline{)2.000}\\\underline{11}\\90\\\underline{88}\\20\\\underline{11}\\9\end{array}$$

21. $0.8=\dfrac{8}{100}=\dfrac{\cancel{4}\cdot 2}{25\cdot\cancel{4}}=\dfrac{2}{25}$

23. $0.19=\dfrac{19}{100}$

25. $0.030=\dfrac{3\cancel{0}}{100\cancel{0}}=\dfrac{3}{100}$

27. $3.10=3+\dfrac{10}{100}=3+\dfrac{1}{10}=3\dfrac{1}{10}=\dfrac{31}{10}$

29. $0.\overline{5}=\dfrac{5}{9}$

31. $0.\overline{21}=\dfrac{21}{99}=\dfrac{7}{33}$

33. $0.\overline{11}=\dfrac{11}{99}=\dfrac{1}{9}$

35. $d\times 8=3$

$$d=3\div 8=8\overline{)3.000}$$
$$\begin{array}{r}0.375\\8\overline{)3.000}\\\underline{24}\\60\\\underline{56}\\40\\\underline{40}\\0\end{array}$$

　　Thus 3 is 0.375 of 8.

37. $d\times 1.5=37.5$
$$d=37.5\div 1.5$$
$$=1.5\overline{)37.5}\rightarrow 15\overline{)375}$$
$$\begin{array}{r}25\\15\overline{)375}\\\underline{30}\\75\\\underline{75}\\0\end{array}$$

　　Thus 37.5 is 25 of 1.5.

39. $0.25\times\square=1.2$
$$\square=\dfrac{1.2}{0.25}$$
$$=0.25\overline{)1.2}\rightarrow 25\overline{)120.0}$$
$$\begin{array}{r}4.8\\25\overline{)120.0}\\\underline{100}\\20\,0\\\underline{20\,0}\\0\end{array}$$

　　The number is 4.8.

41. $2.5 \times 14 =$

$$\begin{array}{r} 2.5 \\ \times\ 14 \\ \hline 100 \\ 25 \\ \hline 35.0 = 35 \end{array}$$

43. $1 \div 3 = 3\overline{)1.00}$ with quotient $0.3\overline{3}$

$$\begin{array}{r} 0.3\overline{3} \\ 3\overline{)1.00} \\ 9 \\ \hline 1 \end{array}$$

The batting average is 0.333.

45.
$$\begin{array}{r} 0.71 \\ 59\overline{)42.00} \\ 41\ 3 \\ \hline 70 \\ 59 \\ \hline 11 \end{array}$$
Hence $42 \div 59 = 0.7$.

47. $\dfrac{\$24}{11} = 24 \div 11 = 11\overline{)24.000}$

$$\begin{array}{r} 2.181 \\ 11\overline{)24.000} \\ 22 \\ \hline 2\ 0 \\ 1\ 1 \\ \hline 90 \\ 88 \\ \hline 20 \\ 11 \\ \hline 9 \end{array}$$

This is $2.18 per pound.

49. $\dfrac{5}{8} = 5 \div 8 = 8\overline{)5.000}$

$$\begin{array}{r} 0.625 \\ 8\overline{)5.000} \\ 4\ 8 \\ \hline 20 \\ 16 \\ \hline 40 \\ 40 \\ \hline 0 \end{array}$$
Thus $\dfrac{5}{8} = 0.625$.

51. $\dfrac{1}{18} = 1 \div 18 = 18\overline{)1.0000}$

$$\begin{array}{r} 0.0555\overline{} \\ 18\overline{)1.0000} \\ 90 \\ \hline 100 \\ 90 \\ \hline 100 \\ 90 \\ \hline 10 \end{array}$$

Hence $\dfrac{1}{18} = 0.05\overline{5} \approx 0.056$.

53. $\dfrac{29}{60} = 29 \div 60 = 60\overline{)29.000}$

$$\begin{array}{r} 0.483 \\ 60\overline{)29.000} \\ 24\ 0 \\ \hline 5\ 00 \\ 4\ 80 \\ \hline 200 \\ 180 \\ \hline 20 \end{array}$$

Hence $\dfrac{29}{60} = 0.48\overline{3} \approx 0.48$.

55. $\dfrac{1}{12} = 1 \div 12 = 12\overline{)1.0000}$

$$\begin{array}{r} 0.0833\overline{} \\ 12\overline{)1.0000} \\ 96 \\ \hline 40 \\ 36 \\ \hline 40 \end{array}$$

Hence $\dfrac{1}{12} = 0.083\overline{3}$.

57. $\dfrac{\$92,000}{2.3} = \dfrac{\$\cancel{92}^{\ 4}0,000}{\cancel{23}_1} = \$40,000$

59. $\dfrac{\$69,000}{2.3} = \dfrac{\$\cancel{69}^{\ 3}0,000}{\cancel{23}_1} = \$30,000$

61. $\dfrac{3}{10}$ and $\dfrac{4}{13}$; $3 \times 13 = 39 < 10 \times 4 = 40$

Thus $\dfrac{3}{10} < \dfrac{4}{13}$.

63. $\dfrac{4}{9}$ and $\dfrac{3}{7}$; $4 \times 7 = 28 > 9 \times 3 = 27$

Thus $\dfrac{4}{9} > \dfrac{3}{7}$.

65. $\dfrac{4}{70} = \dfrac{2}{35} = 35\overline{)2.000}$

$$\begin{array}{r} 0.057 \\ 35\overline{)2.000} \\ 1\ 75 \\ \hline 250 \\ 245 \\ \hline 5 \end{array}$$
so $\dfrac{4}{70} = \dfrac{2}{35} \approx 0.06$

67. $\dfrac{5}{70} = \dfrac{1}{14} = 14\overline{)1.000}^{\,0.071}$ so $\dfrac{5}{70} = \dfrac{1}{14} \approx 0.07$

$$\begin{array}{r} 0.071 \\ 14\overline{)1.000} \\ \underline{98} \\ 20 \\ \underline{14} \\ 6 \end{array}$$

69. $\dfrac{7}{70} = \dfrac{1}{10} = 0.1$

71. a. $\dfrac{5}{8} = 0.625$ **b.** $\dfrac{1}{2} = 0.5$

73. Terminating; answers may vary.

75. Answers may vary.

Mastery Test 3.3

77. $\dfrac{\text{math sales}}{\text{total sales}} = \dfrac{4800}{8000} = \dfrac{48}{80} = \dfrac{3}{5} = 0.6$

79. a. $0.41 = \dfrac{41}{100}$ **b.** $0.303 = \dfrac{303}{1000}$

81. a. $0.035 = \dfrac{35}{1000} = \dfrac{7}{200}$

     b. $0.0375 = \dfrac{375}{10,000} = \dfrac{3}{80}$

83. a. $\dfrac{3}{5} = 3 \div 5 = 5\overline{)3.0}^{\,0.6}$ Hence $\dfrac{3}{5} = 0.6$.

$$\begin{array}{r} 0.6 \\ 5\overline{)3.0} \\ \underline{30} \\ 0 \end{array}$$

     b. $\dfrac{9}{40} = 9 \div 40 = 40\overline{)9.000}^{\,0.225}$

$$\begin{array}{r} 0.225 \\ 40\overline{)9.000} \\ \underline{80} \\ 100 \\ \underline{80} \\ 200 \\ \underline{200} \\ 0 \end{array}$$

     Hence $\dfrac{9}{40} = 0.225$.

Section 3.4 – Decimals, Fractions and Order of Operations

Problems

1. Write all numbers so that they have three decimal places: 6.024, 6.020, 6.002
Thus, $6.024 > 6.02 > 6.002$.

2. $2.\overline{145} = 2.145145\cdots$
$2.1\overline{45} = 2.145454\cdots$
$2.145 = 2.145000$
Comparing the fourth decimal digit, we have $2.1\overline{45} > 2.\overline{145} > 2.145$.

3. $\dfrac{4}{7} = 7\overline{)4.000}^{\,0.571}$

$$\begin{array}{r} 0.571 \\ 7\overline{)4.000} \\ \underline{35} \\ 50 \\ \underline{49} \\ 10 \\ \underline{7} \\ 3 \end{array}$$

Since $0.572 > \dfrac{4}{7} = 0.571$, the second analysis showed more oil.

4. a. $\dfrac{4}{25} = \dfrac{4 \cdot 4}{25 \cdot 4} = \dfrac{16}{100} = 0.16$ so 0.163 is larger than $\dfrac{4}{25}$

     b. $\dfrac{7}{100} = 0.07$; 0.067 is samller than $\dfrac{7}{100}$

5. $\text{BMI} = \dfrac{180}{60^2} \times 703$

$$= \dfrac{\cancel{180}^{\,3}}{_{20}\cancel{60} \cdot \cancel{60}_{\,1}} \times 703 = \dfrac{703}{20}$$

$$\begin{array}{r} 35.1 \\ 20\overline{)703.0} \\ \underline{60} \\ 103 \\ \underline{100} \\ 30 \\ \underline{20} \\ 10 \end{array}$$
   Hence, the BMI is 35.

Exercises 3.4

1. $66.606 > 66.066 > 66.06$

3. $0.5101 > 0.51 > 0.501$

5. $9.999 > 9.909 > 9.099$

7. $7.430 > 7.403 > 7.043$

9. $3.1\overline{4} = 3.1444\cdots$
    $3.\overline{14} = 3.1414\cdots$
    $3.14 = 3.1400$
    Comparing the fourth decimal digit, we have $3.1\overline{4} > 3.\overline{14} > 3.14$.

11. $5.1 = 5.10$
    $5.\overline{1} = 5.11\cdots$
    $5.12 = 5.12$
    Comparing the second decimal digit, we have $5.12 > 5.\overline{1} > 5.1$.

13. $0.333 = 0.3330$
    $0.\overline{3} = 0.3333\cdots$
    $0.33 = 0.3300$
    Comparing the fourth decimal digit, we have $0.\overline{3} > 0.333 > 0.33$.

15. $0.88 = 0.8800$
    $0.\overline{8} = 0.888\cdots$
    $0.\overline{81} = 0.8181\cdots$
    so we have $0.\overline{8} > 0.88 > 0.\overline{81}$.

17. $\dfrac{1}{9} = 1 \div 9 = 9\overline{)\begin{array}{c} 0.\overline{1} \\ 1.00 \\ \underline{9} \\ 1 \end{array}}$ so $\dfrac{1}{9} = 0.1111\cdots$ and

    thus $\dfrac{1}{9}$ is greater than 0.111.

19. $\dfrac{1}{6} = 1 \div 6 = 6\overline{)\begin{array}{c} 0.1\overline{6} \\ 1.00 \\ \underline{6} \\ 40 \\ \underline{36} \\ 4 \end{array}}$ so $\dfrac{1}{6} = 0.16666\cdots$ and

    thus $\dfrac{1}{6}$ is greater than 0.1666.

21. $\dfrac{2}{7} = 2 \div 7 = 7\overline{)\begin{array}{c} 0.2857 \\ 2.0000 \\ \underline{14} \\ 60 \\ \underline{56} \\ 40 \\ \underline{35} \\ 50 \\ \underline{49} \\ 1 \end{array}}$ so $\dfrac{2}{7} = 0.2857\cdots$

    and thus $\dfrac{2}{7}$ is greater than 0.285.

23. $\dfrac{1}{7} = 1 \div 7 = 7\overline{)\begin{array}{c} 0.142 \\ 1.000 \\ \underline{7} \\ 30 \\ \underline{28} \\ 20 \\ \underline{14} \\ 6 \end{array}}$ so $\dfrac{1}{7} = 0.142\cdots$ and

    $0.1\overline{4} = 0.144\cdots$ so $0.1\overline{4}$ is greater than $\dfrac{1}{7}$.

25. $0.\overline{9} = 0.99\cdots$ and $\dfrac{1}{11} = 0.90\overline{90}$ so $0.\overline{9}$ is

    greater than $\dfrac{1}{11}$.

27. Nashnush $= 8 + \dfrac{1}{25} = 8 + 0.04 = 8.04$

    Monjane $= 8 + \dfrac{3}{4} = 8 + 0.75 = 8.75$

    Constantine $= 8 + 0.8 = 8.8$
    Thus Constantine > Monjane > Nashnush.

29. $\text{BMI} = \dfrac{150}{59^2} \times 703$

    $= \dfrac{150}{59 \cdot 59} \times 703$

    $= \dfrac{150}{3481} \times 703 \approx 0.043 \times 703 = 30.229$
    The BMI is 30.

31. Since $\dfrac{3}{4} = 0.75$ is a terminating decimal,

    convert this fraction to a decimal first and then multiply.

33. Since $2\frac{3}{5} = 2 + \frac{3}{5} = 2 + 0.6 = 2.6$ is a terminating decimal, convert $2\frac{3}{5}$ to a decimal first and multiply.

$$2\frac{3}{5} \cdot 4.65 = 2.6 \cdot 4.65$$

$$
\begin{array}{r}
4.65 \\
\times\ 2.6 \\
\hline
2790 \\
930 \\
\hline
\end{array}
$$

$$= 12.\overline{090} = 12.09$$

35. Since $\frac{10}{9} = 1\frac{1}{9} = 1.\overline{1}$ is not a terminating decimal, and the numerator is 10, first write 8.25 as $\frac{8.25}{1}$ and multiply.

$$\frac{10}{9} \cdot 8.25 = \frac{10}{9} \cdot \frac{8.25}{1} = \frac{82.5}{9}$$

Now divide:

$$
\begin{array}{r}
9.1\overline{6} \\
9\overline{)82.50} \\
81 \\
\hline
15 \\
9 \\
\hline
60 \\
54 \\
\hline
6 \\
\end{array}
$$

Thus $\frac{10}{9} \cdot 8.25 = 9.1\overline{6}$.

37. $4\frac{3}{4} - 1.75 = 4.75 - 1.75 = 3$

39. Note that 0.2135 is divisible by 7 ("21" and "35") and 0.248 is divisible by 4. Therefore

$$\frac{2}{7} \cdot 0.2135 + \frac{7}{4} \cdot 0.248$$

$$= \frac{2}{\underset{1}{\cancel{7}}} \cdot \frac{\overset{0.0305}{\cancel{0.2135}}}{1} + \frac{7}{\underset{1}{\cancel{4}}} \cdot \frac{\overset{0.062}{\cancel{0.248}}}{1}$$

$$= 0.061 + 0.434$$

$$= 0.495$$

41. Note that 0.92 is divisible by 4 and 0.96 is divisible by 8. Thus

$$\frac{3}{4} \cdot 0.92 - 0.96 \cdot \frac{3}{8}$$

$$= \frac{3}{\underset{1}{\cancel{4}}} \cdot \frac{\overset{0.23}{\cancel{0.92}}}{1} - \frac{\overset{0.12}{\cancel{0.96}}}{1} \cdot \frac{3}{\underset{1}{\cancel{8}}}$$

$$= 0.69 - 0.36$$

$$= 0.33$$

43. $\frac{4}{3} \cdot 128.1 - 31.5 \div \frac{7}{5}$

$$= \frac{4}{\underset{1}{\cancel{3}}} \cdot \frac{\overset{42.7}{\cancel{128.1}}}{1} - \frac{\overset{4.5}{\cancel{31.5}}}{1} \cdot \frac{5}{\underset{1}{\cancel{7}}}$$

$$= 170.8 - 22.5$$

$$= 148.3$$

45. $14.05 \div \frac{5}{8} + \frac{2}{5} \cdot 15.5$

$$= \frac{\overset{2.81}{\cancel{14.05}}}{1} \cdot \frac{8}{\underset{1}{\cancel{5}}} + \frac{2}{\underset{1}{\cancel{5}}} \cdot \frac{\overset{3.1}{\cancel{15.5}}}{1}$$

$$= 22.48 + 6.2$$

$$= 28.68$$

47. $3.4 \cdot 8.12 \rightarrow$

$$
\begin{array}{r}
8.12 \\
\times\ 3.4 \\
\hline
3248 \\
2436 \\
\hline
27.608 \\
\end{array}
$$

49. $0.6 \div 0.03 = 0.03\overline{)0.6} \rightarrow 3\overline{)60}$

$$
\begin{array}{r}
20 \\
3\overline{)60} \\
60 \\
\hline
0 \\
\end{array}
$$

$-$ OR $-$

$$0.6 \div 0.03 = \frac{0.6}{0.03} = \frac{60}{3} = 20$$

Thus $0.6 \div 0.03 = 20$

51. $3.8 + 3.2 \cdot 4 \div 2 = 3.8 + 12.8 \div 2$

$$= 3.8 + 6.4$$

$$= 10.2$$

53. $0.3195 > 0.3175 > 0.3121 > 0.3103$ so Copper > Nickel > Cadmium > Brass

55. Using long division,

$\frac{22}{7} = 22 \div 7 = 3.142857\cdots$. Compared to

3.141592 we see that $\frac{22}{7}$ is the greatest.

Mastery Test 3.4

57. $BMI = \dfrac{230}{70^2} \times 703$

$= \dfrac{23\cancel{0}}{7\cancel{0} \cdot 70} \times 703$

$= \dfrac{23}{490} \times 703 = \dfrac{16,169}{490}$

$\begin{array}{r} 32.9 \\ 490\overline{)16169.0} \\ \underline{1470} \\ 1469 \\ \underline{980} \\ 489\ 0 \\ \underline{441\ 0} \\ 48\ 0 \end{array}$ Hence, the BMI is 33.

Section 3.5 – Equations and Problem Solving

Problems

1. $y + 6.4 - 6.4 = 9.8 - 6.4$
$y = 9.8 - 6.4$
$y = 3.4$

2. $x - 5.4 = 6.7$
$x - 5.4 + 5.4 = 6.7 + 5.4$
$x = 6.7 + 5.4$
$x = 12.1$

3. $6.3 = 0.9z$
$6.3 \div 0.9 = 0.9z \div 0.9$
$6.3 \div 0.9 = z$
$7 = z$

$0.9\overline{)6.3} \rightarrow 9\overline{)63}$
$\phantom{0.9\overline{)6.3} \rightarrow 9\overline{)}}\underline{63}$
$\phantom{0.9\overline{)6.3} \rightarrow 9\overline{)6}}0$

4. $5 = \dfrac{n}{2.4}$
$2.4 \cdot 5 = \dfrac{n}{\cancel{2.4}} \cdot \cancel{2.4}$
$12 = n$

5. Let d = the DJIA the day before.
$d + 102.27 = 1840.69$
$d + 102.27 - 102.27 = 1840.69 - 102.27$
$d = 1840.69 - 102.27$
$d = 1738.42$
The DJIA was 1738.42 the day before.

59. Using long division, $\dfrac{3}{7} = 0.42\cdots > 0.4$ so class A has more women.

61. $8.015 = 8.015$
$8.01 = 8.010$
$8.005 = 8.005$
so $8.015 > 8.01 > 8.005$

Verify: $1738.42 + 102.27 = 1840.69$ is true.

6. The mile marker is the unknown.
a. The mile marker is $111 + 50 = 161$.
b. The mile marker is $111 + 75 = 186$.
c. The mile marker is $111 + m$.
d. $111 + m = 263.4$
$111 + m - 111 = 263.4 - 111$
$m = 152.4$
You have traveled 152.4 miles.
Verify: $111 + 152.4 = 263.4$ is true.

7. Let m = number of extra minutes
$35 + .45m = 62.90$
$35 + .45m - 35 = 62.90 - 35$
$0.45m = 27.90$
$\dfrac{0.45m}{0.45} = \dfrac{27.90}{0.45}$
$m = \dfrac{2790}{45}$
$m = 62$
The number of extra minutes used is 62.
Total bill $= \$35 + \$0.45 \cdot 62$
$= \$35 + \$27.90 = \$62.90$

Exercises 3.5

1.
$$x + 8.2 = 9.7$$
$$x + 8.2 - 8.2 = 9.7 - 8.2$$
$$x = 1.5$$

3.
$$y + 3.6 = 10.1$$
$$y + 3.6 - 3.6 = 10.1 - 3.6$$
$$y = 6.5$$

5.
$$z - 3.5 = 2.1$$
$$z - 3.5 + 3.5 = 2.1 + 3.5$$
$$z = 5.6$$

7.
$$z - 6.4 = 10.1$$
$$z - 6.4 + 6.4 = 10.1 + 6.4$$
$$z = 16.5$$

9.
$$4.2 = 0.7m$$
$$\frac{4.2}{0.7} = \frac{0.7m}{0.7}$$
$$\frac{42}{7} = m$$
$$6 = m$$

11.
$$0.63 = 0.9m$$
$$\frac{0.63}{0.9} = \frac{0.9m}{0.9}$$
$$\frac{63}{90} = m$$
$$m = \frac{63}{90} = \frac{\cancel{9} \cdot 7}{\cancel{9} \cdot 10} = \frac{7}{10} = 0.7$$

13.
$$7 = \frac{n}{3.4}$$
$$3.4 \cdot 7 = \frac{n}{\cancel{3.4}} \cdot \cancel{3.4}$$
$$23.8 = n$$

15.
$$3.4 = \frac{n}{4}$$
$$4 \cdot 3.1 = \frac{n}{\cancel{4}} \cdot \cancel{4}$$
$$12.4 = n$$

17. Let a = GPA before
$$a + 0.32 = 3.21$$
$$a + 0.32 - 0.32 = 3.21 - 0.32$$
$$a = 2.89$$
The student's GPA before was 2.89.
Verify: $2.89 + 0.32 = 3.21$ ✓

19. Let t = body temperature of a polar bear.
Polar bear temp + 4.7°F is same as goat
$$t + 4.7 = 103.8$$
$$t + 4.7 - 4.7 = 103.8 - 4.7$$
$$t = 99.1$$
The polar bear's body temperature is 99.1°F.
Verify: $99.1° + 4.7° = 103.8°$ ✓

21. Let g = pounds of garbage produced now.
amt. in 1960 + 2.7 = amt. now
$$2.5 + 2.7 = g$$
$$5.2 = g$$
Each person now produces 5.2 pounds of garbage.
Verify: 5.2 lb − 2.5 lb = 2.7 lb ✓

23. Let w = worth of their silver in March.
January worth − $4 billion is March worth
$$4.5 - 4 = w$$
$$0.5 = w$$
Their silver was worth $0.5 billion in March.
Verify:
$4 billion + $0.5 billion = $4.5 billion ✓

25. Let e = billions exported before.
amt. before decreased $8.061 to $27.969
$$e - 8.061 = 27.969$$
$$e - 8.061 + 8.061 = 27.969 + 8.061$$
$$e = 36.03$$
There was $36.03 billion exported to Latin America before.
Verify: $36.03 billion − $8.061 billion
      = $27.969 billion ✓

27. Let w = amount of the wages.
wages times 0.0751 is the deduction
$$w \cdot 0.0751 = 22.53$$
$$\frac{w \cdot 0.0751}{0.0751} = \frac{22.53}{0.0751}$$
$$w = \frac{225{,}300}{751}$$
$$w = 300$$

$$751 \overline{)225300} \; \begin{array}{r} 300 \\ \end{array}$$
$$\underline{2253}$$
$$00$$

The amount of the wages was $300.
Verify: $300 \cdot 0.0751 = $22.53 ✓

29. Let m = miles driven.

$$0.36m = 368.28$$
$$\frac{0.36m}{0.36} = \frac{368.28}{0.36}$$
$$m = \frac{36,828}{36}$$
$$m = 1023$$

The person drove 1023 miles.
Verify: $\$0.36 \times 1023 = \$368.28$ ✓

31. $E = \dfrac{D}{1.19} = \dfrac{50}{1.19} = \dfrac{5000}{119}$;

$$\begin{array}{r} 42.0 \\ 119\overline{)5000.0} \\ \underline{476} \\ 240 \\ \underline{238} \\ 2\,0 \end{array}$$

You have about 42 Euros.

33. $Q = 0.95L = 0.95(2) = 1.9$ quarts

35. Let m = number of miles traveled.

$$0.108m = 885.60$$
$$\frac{0.108m}{0.108} = \frac{885.60}{0.108}$$
$$m = \frac{885,600}{108}$$
$$m = 8200$$

The number of miles traveled was 8200.
Verify: $\$0.108 \cdot 8200 = \$885.60$ ✓

37. Let k = number of kilometers driven.

$$39.99 + 0.12k = 60.63$$
$$39.99 + 0.12k - 39.99 = 60.63 - 39.99$$
$$0.12k = 20.64$$
$$\frac{0.12k}{0.12} = \frac{20.64}{0.12}$$
$$k = \frac{2064}{12} = 172$$

You drove 172 kilometers.
Verify: $\$39.99 + \$0.12 \cdot 172 = \$60.63$ ✓

39.
$$6000 + 0.30C = 9000$$
$$6000 + 0.30C - 6000 = 9000 - 6000$$
$$0.30C = 3000$$
$$\frac{0.30C}{0.30} = \frac{3000}{0.30}$$
$$C = \frac{30,000}{3} = 10,000$$

The total annual cost is $10,000.
Verify: $\$6000 + 0.30 \cdot \$10,000 = \$9000$ ✓

41. $\dfrac{1}{7} + \dfrac{2}{5} = \dfrac{1 \cdot 5}{7 \cdot 5} + \dfrac{2 \cdot 7}{5 \cdot 7} = \dfrac{5}{35} + \dfrac{14}{35} = \dfrac{19}{35}$

43. $5\dfrac{3}{7} - 3\dfrac{4}{5} = \dfrac{38}{7} - \dfrac{19}{5}$

$$= \frac{38 \cdot 5}{7 \cdot 5} - \frac{19 \cdot 7}{5 \cdot 7}$$
$$= \frac{190}{35} - \frac{133}{35} = \frac{57}{35} = 1\frac{22}{35}$$

45.
$$D = R \cdot T$$
$$137.5 = R \cdot 2.5$$
$$\frac{137.5}{2.5} = \frac{R \cdot \cancel{2.5}}{\cancel{2.5}}$$
$$\frac{1375}{25} = R$$
$$55 = R$$

It was traveling 55 mph.

47.
$$D = R \cdot T$$
$$500 = 156 \cdot T$$
$$\frac{500}{156} = \frac{\cancel{156} \cdot T}{\cancel{156}}$$
$$\frac{1375}{.25} = T$$
$$3.20 \approx T$$

$$\begin{array}{r} 3.20 \\ 156\overline{)500.00} \\ \underline{468} \\ 32\,0 \\ \underline{31\,2} \\ 80 \end{array}$$

It took 3.2 hours.

49. Answers may vary.

Mastery Test 3.5

51.
$$8 = \frac{n}{2.3}$$
$$2.3 \cdot 8 = \frac{n}{\cancel{2.3}} \cdot \cancel{2.3}$$
$$18.4 = n$$

53.
$$x - 7.2 = 2.9$$
$$x - 7.2 + 7.2 = 2.9 + 7.2$$
$$x = 10.1$$

55. Let a = amount in the account originally.

$$a - 304.57 = 202.59$$
$$a - 304.57 + 304.57 = 202.59 + 304.57$$
$$a = 507.16$$

She had $507.16 in the account originally.
Verify: $\$507.16 - \$304.57 = \$202.59$ ✓

<u>Review Exercises – Chapter 3</u>

1. a. 23.389: Twenty-three and three
hundred eighty-nine thousandths
b. 22.34: Twenty-two and thirty four
hundredths
c. 24.564: Twenty-four and five hundred
sixty-four thousandths
d. 27.8: Twenty-seven and eight tenths
e. 29.67: Twenty-nine and sixty-seven
hundredths

2. a. $37.4 = 30 + 7 + \dfrac{4}{10}$

b. $59.09 = 50 + 9 + \dfrac{9}{100}$

c. $145.035 = 100 + 40 + 5 + \dfrac{3}{100} + \dfrac{5}{1000}$

d. $150.309 = 100 + 50 + \dfrac{3}{10} + \dfrac{9}{1000}$

e. $234.003 = 200 + 30 + 4 + \dfrac{3}{1000}$

3. a.
$$\begin{array}{r} 8.51 \\ +13.43 \\ \hline 21.94 \end{array}$$

b.
$$\begin{array}{r} 9.6457 \\ +15.7800 \\ \hline 25.4257 \end{array}$$

c.
$$\begin{array}{r} 5.7730 \\ +18.0026 \\ \hline 23.7756 \end{array}$$

d.
$$\begin{array}{r} 6.204 \\ +23.248 \\ \hline 29.452 \end{array}$$

e.
$$\begin{array}{r} 9.24 \\ +14.28 \\ \hline 23.52 \end{array}$$

4. a.
$$\begin{array}{r} 35.60 \\ + 3.76 \\ \hline 39.36 \end{array}$$

b.
$$\begin{array}{r} 43.234 \\ + 4.800 \\ \hline 48.034 \end{array}$$

c.
$$\begin{array}{r} 22.232 \\ + 5.430 \\ \hline 27.662 \end{array}$$

d.
$$\begin{array}{r} 33.23 \\ + 7.89 \\ \hline 41.12 \end{array}$$

e.
$$\begin{array}{r} 39.4217 \\ + 8.3400 \\ \hline 47.7617 \end{array}$$

5. a.
$$\begin{array}{r} {}^{2\;11\;14\;4\;10} \\ 3\cancel{3}\cancel{2}.4\cancel{5}0 \\ - \quad 17.649 \\ \hline 314.801 \end{array}$$

b.
$$\begin{array}{r} {}^{3\;11\;13} \\ 3\cancel{4}\cancel{2}.34 \\ - \quad 18.90 \\ \hline 323.44 \end{array}$$

c.
$$\begin{array}{r} {}^{2\;11\;13\;2\;11\;10} \\ \cancel{3}\cancel{2}\cancel{3}.3\cancel{2}0 \\ - \quad 45.045 \\ \hline 278.275 \end{array}$$

d.
$$\begin{array}{r} {}^{5\;14\;13} \\ 3\cancel{6}\cancel{5}.35 \\ - \quad 17.80 \\ \hline 347.55 \end{array}$$

e.
$$\begin{array}{r} {}^{3\;12\;15} \\ \cancel{4}\cancel{3}.56 \\ - \quad 19.90 \\ \hline 23.66 \end{array}$$

6. a.
$$\begin{array}{r} 3.14 \\ \times 0.012 \\ \hline 628 \\ 314 \\ \hline 0.03768 \end{array}$$

b.
$$\begin{array}{r} 2.34 \\ \times 0.14 \\ \hline 936 \\ 234 \\ \hline 0.3276 \end{array}$$

c.
$$\begin{array}{r} 0.9615 \\ \times \quad 3.45 \\ \hline 48075 \\ 38460 \\ 28845 \\ \hline 3.317175 \end{array}$$

d.
$$\begin{array}{r} 2.345 \\ \times 0.016 \\ \hline 14070 \\ 2345 \\ \hline 0.037520 = 0.03752 \end{array}$$

e.
$$\begin{array}{r} 3.42 \\ \times \quad 0.3 \\ \hline 1.026 \end{array}$$

7. a. $0.37 \cdot 1000 = 370$
b. $0.049 \cdot 100 = 4.9$
c. $0.25 \cdot 10 = 2.5$
d. $4.285 \cdot 1000 = 4285$
e. $0.945 \cdot 1000 = 945$

8. a. $\dfrac{21.35}{0.35} = \dfrac{2135}{35} = 35\overline{)2135}$ $\boxed{61}$

$$\begin{array}{r} \underline{210} \\ 35 \\ \underline{35} \\ 0 \end{array}$$

b. $\dfrac{57.33}{0.91} = \dfrac{5733}{91} = 91\overline{)5733}$ $\boxed{63}$

$$\begin{array}{r} \underline{546} \\ 273 \\ \underline{273} \\ 0 \end{array}$$

c. $\dfrac{3.864}{0.042} = \dfrac{3864}{42} = 42\overline{)3864}$ $\boxed{92}$

$$\begin{array}{r} \underline{378} \\ 84 \\ \underline{84} \\ 0 \end{array}$$

d. $\dfrac{2.9052}{0.36} = \dfrac{290.52}{36} = 36\overline{)290.52}$ $\boxed{8.07}$

$$\begin{array}{r} \underline{288} \\ 2\,52 \\ \underline{2\,52} \\ 0 \end{array}$$

e. $\dfrac{3.7228}{0.041} = \dfrac{3722.8}{41} = 41\overline{)3722.8}$ $\boxed{90.8}$

$$\begin{array}{r} \underline{369} \\ 32\,8 \\ \underline{32\,8} \\ 0 \end{array}$$

9. a. $329.\underline{6}7 \to 329.7$

   b. $238.\underline{3}4 \to 238.3$

   c. $887.\underline{3}62 \to 887.4$

   d. $459.\underline{4}3 \to 459.4$

   e. $348.\underline{3}44 \to 348.3$

10. a. $80 \div 15 = 15\overline{)80.000}$ $\boxed{5.33}$

$$\begin{array}{r} \underline{75} \\ 5\,0 \\ \underline{4\,5} \\ 5 \end{array}$$

b. $90 \div 16 = 15\overline{)90.000} \approx \boxed{5.63}$ 5.625

$$\begin{array}{r} \underline{80} \\ 10\,0 \\ \underline{9\,6} \\ 40 \\ \underline{32} \\ 80 \\ \underline{80} \\ 0 \end{array}$$

c. $48 \div 7 = 7\overline{)48.000} \approx \boxed{6.86}$ 6.857

$$\begin{array}{r} \underline{42} \\ 6\,0 \\ \underline{5\,6} \\ 40 \\ \underline{35} \\ 50 \\ \underline{49} \\ 1 \end{array}$$

d. $84 \div 13 = 13\overline{)84.000} \approx \boxed{6.46}$ 6.461

$$\begin{array}{r} \underline{78} \\ 6\,0 \\ \underline{5\,2} \\ 80 \\ \underline{78} \\ 20 \\ \underline{13} \\ 7 \end{array}$$

e. $97 \div 14 = 14\overline{)97.000} \approx \boxed{6.93}$ 6.928

$$\begin{array}{r} \underline{84} \\ 13\,0 \\ \underline{12\,6} \\ 40 \\ \underline{28} \\ 120 \\ \underline{112} \\ 8 \end{array}$$

11. a. $3.12 \div 100 = 0.0312$

   b. $4.18 \div 1000 = 0.00418$

   c. $32.1 \div 100 = 0.321$

   d. $82.15 \div 10 = 8.215$

   e. $472.3 \div 100 = 4.723$

12. a. $\dfrac{\$1.92}{6} = 6\overline{)1.92}$

$$\begin{array}{r} 0.32 \\ \hline 1.92 \\ 18 \\ \hline 12 \\ 12 \\ \hline 0 \end{array}$$

The price per can is \$0.32 or 32¢.

b. $\dfrac{\$2.04}{6} = 6\overline{)2.04}$

$$\begin{array}{r} 0.34 \\ \hline 2.04 \\ 18 \\ \hline 24 \\ 24 \\ \hline 0 \end{array}$$

The price per can is \$0.34 or 34¢.

c. $\dfrac{\$1.80}{6} = 6\overline{)1.80}$

$$\begin{array}{r} 0.30 \\ \hline 1.80 \\ 18 \\ \hline 00 \end{array}$$

The price per can is \$0.30 or 30¢.

d. $\dfrac{\$1.68}{6} = 6\overline{)1.68}$

$$\begin{array}{r} 0.28 \\ \hline 1.68 \\ 12 \\ \hline 48 \\ 48 \\ \hline 0 \end{array}$$

The price per can is \$0.28 or 28¢.

e. $\dfrac{\$1.62}{6} = 6\overline{)1.62}$

$$\begin{array}{r} 0.27 \\ \hline 1.62 \\ 12 \\ \hline 42 \\ 42 \\ \hline 0 \end{array}$$

The price per can is \$0.27 or 27¢.

13. a. $\dfrac{3}{5} = 5\overline{)3.0}$ so $\dfrac{3}{5} = 0.6$

$$\begin{array}{r} 0.6 \\ \hline 3.0 \\ 30 \\ \hline 0 \end{array}$$

b. $\dfrac{9}{10} = 9 \div 10 = 0.9$

c. $\dfrac{5}{2} = 2\dfrac{1}{2} = 2 + \dfrac{1}{2} = 2 + 0.5 = 2.5$

d. $\dfrac{3}{16} = 16\overline{)3.0000}$ so $\dfrac{3}{16} = 0.1875$

$$\begin{array}{r} 0.1875 \\ \hline 3.0000 \\ 16 \\ \hline 140 \\ 128 \\ \hline 120 \\ 112 \\ \hline 80 \end{array}$$

e. $\dfrac{7}{8} = 8\overline{)7.000}$ so $\dfrac{7}{8} = 0.875$

$$\begin{array}{r} 0.875 \\ \hline 7.000 \\ 64 \\ \hline 60 \\ 56 \\ \hline 40 \\ 40 \\ \hline 0 \end{array}$$

14. a. $\dfrac{1}{3} = 3\overline{)1.00}$ so $\dfrac{1}{3} = 0.\overline{3}$

$$\begin{array}{r} 0.3\overline{3} \\ \hline 1.00 \\ 9 \\ \hline 10 \end{array}$$

b. $\dfrac{5}{6} = 6\overline{)5.000}$ so $\dfrac{5}{6} = 0.8\overline{3}$

$$\begin{array}{r} 0.83\overline{3} \\ \hline 5.000 \\ 48 \\ \hline 20 \\ 18 \\ \hline 20 \end{array}$$

c. $\dfrac{2}{3} = 3\overline{)2.00}$ so $\dfrac{2}{3} = 0.\overline{6}$

$$\begin{array}{r} 0.6\overline{6} \\ \hline 2.00 \\ 18 \\ \hline 20 \end{array}$$

d. $\dfrac{2}{7} = 2 \div 7 = 0.\overline{285714}$

$$\begin{array}{r} 0.285714 \\ \hline 2.000000 \\ 14 \\ \hline 60 \\ 56 \\ \hline 40 \\ 35 \\ \hline 50 \\ 49 \\ \hline 10 \\ 7 \\ \hline 30 \\ 28 \\ \hline 2 \end{array}$$

e. $\dfrac{1}{9} = 9\overline{)1.00}$ so $\dfrac{1}{9} = 0.\overline{1}$

$$\begin{array}{r} 0.1\overline{1} \\ \hline 1.00 \\ 9 \\ \hline 10 \end{array}$$

15. a. $0.38 = \dfrac{38}{100} = \dfrac{\cancel{2} \cdot 19}{\cancel{2} \cdot 50} = \dfrac{19}{50}$

   b. $0.41 = \dfrac{41}{100}$

   c. $0.6 = \dfrac{6}{10} = \dfrac{\cancel{2} \cdot 3}{\cancel{2} \cdot 5} = \dfrac{3}{5}$

   d. $0.03 = \dfrac{3}{100}$

   e. $0.333 = \dfrac{333}{1000}$

16. a. $2.33 = 2 + \dfrac{33}{100} = 2\dfrac{33}{100} = \dfrac{233}{100}$

   b. $3.47 = 3 + \dfrac{47}{100} = 3\dfrac{47}{100} = \dfrac{347}{100}$

   c. $6.55 = 6 + \dfrac{55}{100} = 6 + \dfrac{11}{20} = 6\dfrac{11}{20} = \dfrac{131}{20}$

   d. $1.37 = 1 + \dfrac{37}{100} = 1\dfrac{37}{100} = \dfrac{137}{100}$

   e. $2.134 = 2 + \dfrac{134}{1000} = 2\dfrac{67}{500} = \dfrac{1067}{500}$

17. a. $0.\overline{45} = \dfrac{45}{99} = \dfrac{\cancel{9} \cdot 5}{\cancel{9} \cdot 11} = \dfrac{5}{11}$

   b. $0.\overline{08} = \dfrac{8}{99}$

   c. $0.\overline{080} = \dfrac{080}{999} = \dfrac{80}{999}$

   d. $0.\overline{004} = \dfrac{004}{999} = \dfrac{4}{999}$

   e. $0.\overline{011} = \dfrac{011}{999} = \dfrac{11}{999}$

18. Let d = the decimal part.

   a. $d \times 16 = 4$
   $$\dfrac{d \times 16}{16} = \dfrac{4}{16}$$
   $$d = \dfrac{4}{16} = \dfrac{1}{4} = 0.25$$

   b. $d \times 5 = 3$
   $$\dfrac{d \times 5}{5} = \dfrac{3}{5}$$
   $$d = \dfrac{3}{5} = 0.6$$

   c. $d \times 12 = 6$
   $$\dfrac{d \times 12}{12} = \dfrac{6}{12}$$
   $$d = \dfrac{6}{12} = \dfrac{1}{2} = 0.5$$

   d. $d \times 8 = 6$
   $$\dfrac{d \times 8}{8} = \dfrac{6}{8}$$
   $$d = \dfrac{6}{8} = \dfrac{3}{4} = 0.75$$

   e. $d \times 16 = 3$
   $$\dfrac{d \times 16}{16} = \dfrac{3}{16}$$
   $$d = \dfrac{3}{16}$$
   $$= 0.1875$$

$$\dfrac{3}{16} = 16\overline{)3.0000}$$

$$
\begin{array}{r}
0.1875 \\
16\,)\overline{3.0000} \\
\underline{16} \\
140 \\
\underline{128} \\
120 \\
\underline{112} \\
80
\end{array}
$$

19. The fraction for rent = $\dfrac{\text{amt. for rent}}{\text{total expenses}}$

   a. $\dfrac{3\cancel{00}}{15\cancel{00}} = \dfrac{3}{15} = \dfrac{1}{5}$

   b. $\dfrac{3\cancel{00}}{12\cancel{00}} = \dfrac{3}{12} = \dfrac{1}{4}$

   c. $\dfrac{3\cancel{00}}{21\cancel{00}} = \dfrac{3}{21} = \dfrac{1}{7}$

   d. $\dfrac{3\cancel{00}}{10\cancel{00}} = \dfrac{3}{10}$

   e. $\dfrac{3\cancel{00}}{17\cancel{00}} = \dfrac{3}{17}$

20. a. 1.032
   1.030
   1.003
   so we have: $1.032 > 1.03 > 1.003$

   b. 2.032
   2.030
   2.003
   so we have: $2.032 > 2.03 > 2.003$

   c. 3.033
   3.030
   3.032
   so we have: $3.033 > 3.032 > 3.03$

d. 4.050
4.052
4.055
so we have: $4.055 > 4.052 > 4.05$

e. 5.003
5.030
5.033
so we have: $5.033 > 5.03 > 5.003$

21. a. $1.2\overline{16} = 1.2166\cdots$
$1.216 = 1.2160$
$1.2\overline{16} = 1.21616\cdots$
so we have: $1.2\overline{16} > 1.2\overline{16} > 1.216$

b. $2.336 = 2.3360$
$2.33\overline{6} = 2.3366\cdots$
$2.3\overline{36} = 2.33636\cdots$
so we have: $2.33\overline{6} > 2.3\overline{36} > 2.336$

c. $3.2\overline{16} = 3.2166\cdots$
$3.2\overline{16} = 3.21616\cdots$
$3.216 = 3.2160$
so we have: $3.2\overline{16} > 3.2\overline{16} > 3.216$

d. $4.5\overline{42} = 4.54242\cdots$
$4.54\overline{2} = 4.5422\cdots$
$4.542 = 4.5420$
so we have: $4.5\overline{42} > 4.54\overline{2} > 4.542$

e. $5.1\overline{23} = 5.12323\cdots$
$5.123 = 5.1230$
$5.12\overline{3} = 5.1233\cdots$
so we have: $5.12\overline{3} > 5.1\overline{23} > 5.123$

22. a. $\dfrac{1}{11} = 0.0909\cdots \approx 0.091$ so $\dfrac{1}{11} > 0.09$

b. $\dfrac{2}{11} = 0.1818\cdots \approx 0.181$ so $\dfrac{2}{11} > 0.18$

c. $\dfrac{3}{11} = 0.2727\cdots \approx 0.27$ so $\dfrac{3}{11} < 0.28$

d. $\dfrac{4}{11} = 0.3636\cdots \approx 0.36$ so $\dfrac{4}{11} < 0.37$

e. $\dfrac{5}{11} = 0.4545\cdots \approx 0.455$ so $\dfrac{5}{11} > 0.45$

23. a. $\text{BMI} = \dfrac{150}{60^2} \times 703$

$= \dfrac{\cancel{150}\,^{5}}{_{2}\cancel{60} \cdot \cancel{60}\,_{12}} \times 703 = \dfrac{703}{24}$

$\begin{array}{r} 29.2 \\ 24\overline{)703.0} \\ \underline{48} \\ 223 \\ \underline{216} \\ 7\,0 \\ \underline{48} \\ 2\,2 \end{array}$ Hence, the BMI is 29.

b. $\text{BMI} = \dfrac{150}{65^2} \times 703$

$= \dfrac{\cancel{150}\,^{30\ 6}}{_{13}\cancel{65} \cdot \cancel{65}\,_{13}} \times 703 = \dfrac{4218}{169}$

$\begin{array}{r} 24.9 \\ 169\overline{)4218.0} \\ \underline{338} \\ 838 \\ \underline{676} \\ 162\,0 \\ \underline{152\,0} \\ 10 \end{array}$ Hence, the BMI is 25.

c. $\text{BMI} = \dfrac{200}{65^2} \times 703$

$= \dfrac{\cancel{200}\,^{40\ 8}}{_{13}\cancel{65} \cdot \cancel{65}\,_{13}} \times 703 = \dfrac{5624}{169}$

$\begin{array}{r} 33.2 \\ 169\overline{)5624.0} \\ \underline{507} \\ 554 \\ \underline{507} \\ 47\,0 \\ \underline{33\,8} \\ 13\,2 \end{array}$ Hence, the BMI is 33.

d. $\text{BMI} = \dfrac{200}{60^2} \times 703$

$= \dfrac{\cancel{200}\,^{10\ 1}}{_{3}\cancel{60} \cdot \cancel{60}\,_{6}} \times 703 = \dfrac{703}{18}$

$\begin{array}{r} 39.0 \\ 18\overline{)703.0} \\ \underline{54} \\ 163 \\ \underline{162} \\ 1\,0 \end{array}$ Hence, the BMI is 39.

e. $\text{BMI} = \dfrac{160}{70^2} \times 703$

$$= \dfrac{\overset{16\,8}{\cancel{160}}}{{}_7\cancel{70} \cdot \cancel{70}{}_{35}} \times 703 = \dfrac{5624}{245}$$

$$245\overline{\smash)5624.0} \quad \text{Hence, the BMI is 23.}$$

$$\begin{array}{r} 22.9 \\ \underline{490} \\ 724 \\ \underline{490} \\ 234\,0 \\ \underline{220\,5} \\ 13\,5 \end{array}$$

24. a.
$$x + 3.6 = 7.9$$
$$x + 3.6 - 3.6 = 7.9 - 3.6$$
$$x = 4.3$$

b.
$$x + 4.6 = 6.9$$
$$x + 4.6 - 4.6 = 6.9 - 4.6$$
$$x = 2.3$$

c.
$$x + 5.4 = 5.9$$
$$x + 5.4 - 5.4 = 5.9 - 5.4$$
$$x = 0.5$$

d.
$$x + 6.3 = 9.9$$
$$x + 6.3 - 6.3 = 9.9 - 6.3$$
$$x = 3.6$$

e.
$$x + 7.2 = 9.9$$
$$x + 7.2 - 7.2 = 9.9 - 7.2$$
$$x = 2.7$$

25. a.
$$y - 1.4 = 5.9$$
$$y - 1.4 + 1.4 = 5.9 + 1.4$$
$$y = 7.3$$

b.
$$y - 1.5 = 6.2$$
$$y - 1.5 + 1.5 = 6.2 + 1.5$$
$$y = 7.7$$

c.
$$y - 7.42 = 5.9$$
$$y - 7.42 + 7.42 = 5.9 + 7.42$$
$$y = 13.32$$

d.
$$y - 4.2 = 5.8$$
$$y - 4.2 + 4.2 = 5.8 + 4.2$$
$$y = 10$$

e.
$$y - 7.8 = 3.7$$
$$y - 7.8 + 7.8 = 3.7 + 7.8$$
$$y = 11.5$$

26. a.
$$4.5 = 0.9y$$
$$\dfrac{4.5}{0.9} = \dfrac{\cancel{0.9}y}{\cancel{0.9}}$$
$$\dfrac{45}{9} = y$$
$$5 = y$$

b.
$$5.6 = 0.8y$$
$$\dfrac{5.6}{0.8} = \dfrac{\cancel{0.8}y}{\cancel{0.8}}$$
$$\dfrac{56}{8} = y$$
$$7 = y$$

c.
$$3.6 = 0.9y$$
$$\dfrac{3.6}{0.9} = \dfrac{\cancel{0.9}y}{\cancel{0.9}}$$
$$\dfrac{36}{9} = y$$
$$4 = y$$

d.
$$7.2 = 0.9y$$
$$\dfrac{7.2}{0.9} = \dfrac{\cancel{0.9}y}{\cancel{0.9}}$$
$$\dfrac{72}{9} = y$$
$$8 = y$$

e.
$$4.8 = 0.6y$$
$$\dfrac{4.8}{0.6} = \dfrac{\cancel{0.6}y}{\cancel{0.6}}$$
$$\dfrac{48}{6} = y$$
$$8 = y$$

27. a.
$$6 = \dfrac{z}{4.1}$$
$$4.1 \cdot 6 = \dfrac{z}{\cancel{4.1}} \cdot \cancel{4.1}$$
$$24.6 = z$$

b.
$$7 = \dfrac{z}{5.1}$$
$$5.1 \cdot 7 = \dfrac{z}{\cancel{5.1}} \cdot \cancel{5.1}$$
$$35.7 = z$$

c.
$$2 = \dfrac{z}{5.4}$$
$$5.4 \cdot 2 = \dfrac{z}{\cancel{5.4}} \cdot \cancel{5.4}$$
$$10.8 = z$$

d.
$$6.2 = \dfrac{z}{7.1}$$
$$7.1 \cdot 6.2 = \dfrac{z}{\cancel{7.1}} \cdot \cancel{7.1}$$
$$44.02 = z$$

e.
$$7 = \dfrac{z}{8.1}$$
$$8.1 \cdot 7 = \dfrac{z}{\cancel{8.1}} \cdot \cancel{8.1}$$
$$56.7 = z$$

28. Let p = the price before the decrease.

  a. $\qquad p - 4.28 = 39.95$

    $p - 4.28 + 4.28 = 39.95 + 4.28$

                 $p = 44.23$

    The price of the item was \$44.23.

    Verify: \$44.23 − \$4.28 = \$39.95✓

  b. $\qquad p - 3.01 = 39.95$

    $p - 3.01 + 3.01 = 39.95 + 3.01$

                 $p = 42.96$

    The price of the item was \$42.96.

    Verify: \$42.96 − \$3.01 = \$39.95✓

  c. $\qquad p - 7.73 = 39.95$

    $p - 7.73 + 7.73 = 39.95 + 7.73$

                 $p = 47.68$

    The price of the item was \$47.68.

    Verify: \$47.68 − \$7.73 = \$39.95✓

  d. $\qquad p - 6.83 = 39.95$

    $p - 6.83 + 6.83 = 39.95 + 6.83$

                 $p = 46.78$

    The price of the item was \$46.78.

    Verify: \$46.78 − \$6.83 = \$39.95✓

  e. $\qquad p - 9.55 = 39.95$

    $p - 9.55 + 9.55 = 39.95 + 9.55$

                 $p = 49.50$

    The price of the item was \$49.50.

    Verify: \$49.50 − \$9.55 = \$39.95✓

Cumulative Review Chapters 1–3

1. $300 + 90 + 4 = 394$

2. 3210

3. $2\underline{|20}$ The prime factors of 20 are 2 and 5.
   $2\underline{|10}$
     5

4. $2\underline{|60}$ Therefore, $60 = 2^2 \times 3 \times 5$.
   $2\underline{|30}$
   $3\underline{|15}$
     5

5. $2^2 \times 5 \times 5^0 = 4 \times 5 \times 1 = 20$

6. $49 \div 7 \cdot 7 + 8 - 5 = 7 \cdot 7 + 8 - 5$
                     $= 49 + 8 - 5$
                     $= 57 - 5$
                     $= 52$

7. Since the numerator is larger than the denominator, $\dfrac{5}{4}$ is improper.

8. $\dfrac{11}{2} \rightarrow 2\overline{)11}$ so $\dfrac{11}{2} = 5$ r 1; hence $\dfrac{11}{2} = 5\dfrac{1}{2}$.
          $\underline{10}$
           1

9. $5\dfrac{3}{8} = \dfrac{8 \times 5 + 3}{8} = \dfrac{40 + 3}{8} = \dfrac{43}{8}$

10. $\dfrac{2}{3} = \dfrac{?}{27}$; $\dfrac{2}{3} = \dfrac{2 \cdot 9}{3 \cdot 9} = \dfrac{18}{27}$. The missing number is 18.

11. $\dfrac{5}{7} = \dfrac{25}{?}$; $\dfrac{5}{7} = \dfrac{5 \cdot 5}{7 \cdot 5} = \dfrac{25}{35}$. The missing number is 35.

12. $\dfrac{1}{2} \cdot 5\dfrac{1}{6} = \dfrac{1}{2} \cdot \dfrac{31}{6} = \dfrac{31}{12}$ or $2\dfrac{7}{12}$

13. $\left(\dfrac{7}{2}\right)^2 \cdot \dfrac{1}{49} = \dfrac{\cancel{49}}{4} \cdot \dfrac{1}{\cancel{49}} = \dfrac{1}{4}$

14. $\dfrac{45}{4} \div 2\dfrac{7}{9} = \dfrac{45}{4} \div \dfrac{25}{9}$

       $= \dfrac{\cancel{45}^{\,9}}{4} \cdot \dfrac{9}{\cancel{25}_{\,5}} = \dfrac{81}{20}$ or $4\dfrac{1}{20}$

15. $6\dfrac{2}{3} + 8\dfrac{3}{8} = 6\dfrac{2 \cdot 8}{3 \cdot 8} + 8\dfrac{3 \cdot 3}{8 \cdot 3}$

            $= 6\dfrac{16}{24} + 8\dfrac{9}{24}$

            $= 14\dfrac{25}{24}$

            $= 14 + 1\dfrac{1}{24}$

            $= 15\dfrac{1}{24}$

16. $13\dfrac{1}{3} - 1\dfrac{3}{4} = \dfrac{40}{3} - \dfrac{7}{4}$

           $= \dfrac{40 \cdot 4}{3 \cdot 4} - \dfrac{7 \cdot 3}{4 \cdot 3}$

           $= \dfrac{160}{12} - \dfrac{21}{12} = \dfrac{139}{12} = 11\dfrac{7}{12}$

17.
$$z - \frac{6}{7} = \frac{3}{5}$$
$$z - \frac{6}{7} + \frac{6}{7} = \frac{3}{5} + \frac{6}{7}$$
$$z = \frac{3}{5} + \frac{6}{7}$$
$$z = \frac{21}{35} + \frac{30}{35} = \frac{51}{35}$$

18. Let n = the number.
$$\frac{7}{8}n = 3\frac{1}{2}$$
$$\frac{7}{8}n = \frac{7}{2}$$
$$\frac{\frac{7}{8}n}{\frac{7}{8}} = \frac{\frac{7}{2}}{\frac{7}{8}}$$
$$n = \frac{\cancel{7}^1}{\cancel{2}_1} \cdot \frac{\cancel{8}^4}{\cancel{7}_1} = 4$$

The number is 4.

19. $3\frac{1}{2}$ lb cost 49 cents so 1 lb costs

$$49 \div 3\frac{1}{2} = 49 \div \frac{7}{2} = \frac{\cancel{49}^7}{1} \cdot \frac{2}{\cancel{7}_1} = 14 \text{ cents.}$$

Thus 7 lb costs $7 \cdot 14 = 98$ cents.

20. 135.64: One hundred thirty-five and sixty-four hundredths

21. $94.478 = 90 + 4 + \frac{4}{10} + \frac{7}{100} + \frac{8}{1000}$

22.
$$\begin{array}{r} 46.654 \\ +\ \ 9.690 \\ \hline 56.344 \end{array}$$

23.
$$\begin{array}{r} {}^{3\ 10\ 14} \\ 24\cancel{1}.42 \\ -\ \ \ 12.50 \\ \hline 228.92 \end{array}$$

24.
$$\begin{array}{r} 5.98 \\ \times\ \ 1.9 \\ \hline 5382 \\ 598\ \ \\ \hline 11.362 \end{array}$$

25. $\dfrac{663}{0.39} = \dfrac{66{,}300}{39} = 39\overline{)66300}$

$$39\overline{)66300} \quad \boxed{1700}$$
$$\begin{array}{r} \underline{39} \\ 273 \\ \underline{273} \\ 00 \end{array}$$

26. $2\underline{4}9.851 \rightarrow 250$

27. $10 \div 0.13 = 0.13\overline{)10} \rightarrow 13\overline{)1000.000}$

$$\begin{array}{r} 76.922 \\ 13\overline{)1000.000} \\ \underline{91} \\ 90 \\ \underline{78} \\ 120 \\ \underline{117} \\ 30 \\ \underline{26} \\ 40 \end{array}$$

Thus $10 \div 0.13 \approx 76.92$.

28. $\$3.04 \div 8 = 8\overline{)3.04}$

$$\begin{array}{r} 0.38 \\ 8\overline{)3.04} \\ \underline{24} \\ 64 \\ \underline{64} \\ 0 \end{array}$$

The cost per bottle was $0.38, or 38¢.

29. $\dfrac{5}{6} = 5 \div 6 = 6\overline{)5.000}$

$$\begin{array}{r} \boxed{0.83\overline{3}} \\ 6\overline{)5.000} \\ \underline{48} \\ 20 \\ \underline{18} \\ 20 \end{array}$$

30. $0.35 = \dfrac{35}{100} = \dfrac{\cancel{5} \cdot 7}{\cancel{5} \cdot 20} = \dfrac{7}{20}$

31. $0.\overline{78} = \dfrac{78}{99} = \dfrac{\cancel{3} \cdot 26}{\cancel{3} \cdot 33} = \dfrac{26}{33}$

32. Let d = the decimal part.
$$d \times 15 = 3$$
$$\frac{d \times \cancel{15}}{\cancel{15}} = \frac{3}{15}$$
$$d = \frac{1}{5} = 0.2$$

33. The fraction for rent is

$$\frac{\text{rent amt.}}{\text{monthly expenses}} = \frac{2\cancel{00}}{20\cancel{00}} = \frac{2}{20} = \frac{1}{10}.$$

34. $5.314 = 5.3140$

$5.31\overline{4} = 5.3144\cdots$

$5.3\overline{14} = 5.31414\cdots$

so we have: $5.31\overline{4} > 5.3\overline{14} > 5.314$

35. $\dfrac{13}{20} = \dfrac{13 \cdot 5}{20 \cdot 5} = \dfrac{65}{100} = 0.65$ which is greater

than 0.26. Thus $0.26 < \dfrac{13}{20}$.

36. $\quad x + 2.1 = 9.4$

$x + 2.1 - 2.1 = 9.4 - 2.1$

$\quad\quad\quad x = 7.3$

37. $\quad 3.2 = 0.4y$

$\dfrac{3.2}{0.4} = \dfrac{\cancel{0.4}y}{\cancel{0.4}}$

$\dfrac{32}{4} = y$

$\quad 8 = y$

38. $\quad 7 = \dfrac{z}{4.8}$

$4.8 \cdot 7 = \dfrac{z}{\cancel{4.8}} \cdot \cancel{4.8}$

$\quad 33.6 = z$

Chapter 4

Ratio, Rate, and Proportion

Section 4.1 – Ratio and Proportion

Problems

1. a. 1 to 7 is written as $\frac{1}{7}$.

   b. 3 to 15 is written as $\frac{3}{15} = \frac{1}{5}$.

   c. 4 to 18 is written as $\frac{4}{18} = \frac{2}{9}$.

2. $\frac{375}{1000} = \frac{25 \cdot 15}{25 \cdot 40} = \frac{15}{40} = \frac{3}{8}$

3. $\frac{10.78}{3.43} = \frac{10.78 \cdot 100}{3.43 \cdot 100} = \frac{1078}{343} = \frac{22}{7}$

4. a. $\frac{58}{100} = \frac{2 \cdot 29}{2 \cdot 50} = \frac{29}{50}$

   b. $\frac{52}{100} = \frac{4 \cdot 13}{4 \cdot 25} = \frac{13}{25}$

5. a. $\frac{4}{12} = \frac{1}{3}$

   b. $\frac{5}{8} = \frac{10}{y}$

6. $\frac{10}{19} \stackrel{?}{=} \frac{255}{505} \rightarrow 10 \cdot 505 \stackrel{?}{=} 19 \cdot 255$

   $5050 \neq 4845$

   No, because $10 \cdot 505 \neq 19 \cdot 255$.

7. a. $\frac{8}{x} = \frac{4}{5}$

   $8 \cdot 5 = 4 \cdot x$

   $40 = 4x$

   $\frac{40}{4} = x$

   $10 = x$

   The solution is $x = 10$.

b. $\frac{x}{6} = \frac{7}{8}$

   $8 \cdot x = 6 \cdot 7$

   $8x = 42$

   $x = \frac{42}{8} = \frac{21}{4}$

   The solution is $x = \frac{21}{4}$.

c. $\frac{6}{7} = \frac{x}{14}$

   $6 \cdot 14 = 7 \cdot x$

   $84 = 7x$

   $\frac{84}{7} = x$

   $12 = x$

   The solution is $x = 12$.

8. $\frac{x}{20} = \frac{32}{5}$

   $5x = 20 \cdot 32$

   $5x = 640$

   $x = \frac{640}{5} = 128$

   He can produce 128 parts in 20 hours.

9. Since Mini Me is said to be $\frac{3}{7}$ of Dr. Evil's size, the scale is 3 inches = 7 inches. Thus

   $\frac{3 \text{ inches}}{32 \text{ inches}} = \frac{7 \text{ inches}}{E}$

   $3 \cdot E = 32 \cdot 7$

   $E = \frac{224}{3} = 74\frac{2}{3}$.

   Dr. Evil is $74\frac{2}{3}$ inches tall.

Exercises 4.1

1. 3 to 8 is written as $\frac{3}{8}$.

3. 5 to 35 is written as $\frac{5}{35} = \frac{1}{7}$.

5. 32 to 4 is written as $\frac{32}{4} = \frac{8}{1}$.

7. 11 to 3 is written as $\frac{11}{3}$.

9. 0.5 to 0.15 is written as
$$\frac{0.5}{0.15} = \frac{0.5 \cdot 100}{0.15 \cdot 100} = \frac{50}{15} = \frac{10}{3}.$$

11. $\frac{1}{4} = \frac{5}{20}$

13. $\frac{a}{3} = \frac{b}{7}$

15. $\frac{a}{6} = \frac{b}{18}$

17. $\frac{3}{a} = \frac{12}{b}$

19. $\frac{3}{4} \stackrel{?}{=} \frac{5}{6}$; $3 \cdot 6 \stackrel{?}{=} 4 \cdot 5$
$$18 \neq 20$$
No, since $3 \cdot 6 \neq 4 \cdot 5$.

21. $\frac{6}{9} \stackrel{?}{=} \frac{8}{12}$; $6 \cdot 12 \stackrel{?}{=} 9 \cdot 8$
$$72 = 72$$
Yes, since $6 \cdot 12 = 9 \cdot 8$.

23. $\frac{0.3}{5} \stackrel{?}{=} \frac{3}{50}$; $0.3 \cdot 50 \stackrel{?}{=} 5 \cdot 3$
$$15 = 15$$
Yes, since $0.3 \cdot 50 = 5 \cdot 3$.

25. $\frac{6}{1.5} \stackrel{?}{=} \frac{8}{2}$; $6 \cdot 2 \stackrel{?}{=} 1.5 \cdot 8$
$$12 = 12$$
Yes, since $6 \cdot 2 = 1.5 \cdot 8$.

27. $\frac{3}{1.2} \stackrel{?}{=} \frac{5}{0.2}$; $3 \cdot 0.2 \stackrel{?}{=} 1.2 \cdot 5$
$$0.6 \neq 6$$
No, since $3 \cdot 0.2 \neq 1.2 \cdot 5$.

29. $\frac{3}{4} = \frac{6}{x}$
$$3x = 4 \cdot 6$$
$$3x = 24$$
$$x = \frac{24}{3} = 8$$

31. $\frac{9}{10} = \frac{18}{x}$
$$9x = 10 \cdot 18$$
$$9x = 180$$
$$x = \frac{180}{9} = 20$$

33. $\frac{8}{5} = \frac{x}{30}$
$$5x = 8 \cdot 30$$
$$5x = 240$$
$$x = \frac{240}{5} = 48$$

35. $\frac{12}{x} = \frac{4}{5}$
$$4x = 12 \cdot 5$$
$$4x = 60$$
$$x = \frac{60}{4} = 15$$

37. $\frac{20}{x} = \frac{4}{5}$
$$4x = 20 \cdot 5$$
$$4x = 100$$
$$x = \frac{100}{5} = 25$$

39. $\frac{x}{21} = \frac{2}{3}$
$$3x = 21 \cdot 2$$
$$3x = 42$$
$$x = \frac{42}{3} = 14$$

41. $\frac{x}{16} = \frac{3}{4}$
$$4x = 16 \cdot 3$$
$$4x = 48$$
$$x = \frac{48}{4} = 12$$

43. $\dfrac{15}{x} = \dfrac{5}{3}$

$5x = 15 \cdot 3$

$5x = 45$

$x = \dfrac{45}{5} = 9$

45. $\dfrac{8}{9} = \dfrac{x}{16}$

$9x = 8 \cdot 16$

$9x = 128$

$x = \dfrac{128}{9}$

47. $\dfrac{14}{22} = \dfrac{7}{x}$

$14x = 22 \cdot 7$

$14x = 154$

$x = \dfrac{154}{14} = 11$

49. $\dfrac{3.5}{4} = \dfrac{x}{7}$

$4x = 3.5 \cdot 7$

$4x = 24.5$

$x = \dfrac{24.5}{4}$

$= \dfrac{245}{40} = 6\dfrac{5}{40} = 6\dfrac{1}{8} = 6.125$

51. $\dfrac{2}{26} = \dfrac{8}{d}$

$2d = 26 \cdot 8$

$2d = 208$

$d = 104$

53. $\dfrac{5}{4} = \dfrac{8}{x}$

$5x = 32$

$x = \dfrac{32}{5} = 6\dfrac{2}{5} = 6.40$

55. $\dfrac{4.38}{4} = \dfrac{y}{6}$

$4y = 4.38 \cdot 6$

$4y = 26.28$

$y = \dfrac{26.28}{4} = 6.57$

57. $\dfrac{9}{3} = \dfrac{z}{5}$

$3z = 9 \cdot 5$

$3z = 45$

$z = 15$

59. $\dfrac{6.95}{5} = \dfrac{5.56}{x}$

$6.95x = 5 \cdot 5.56$

$6.95x = 27.8$

$x = \dfrac{27.8}{6.95} = 4$

61. $\dfrac{34\cancel{0}}{100\cancel{0}} = \dfrac{34}{100} = \dfrac{\cancel{2} \cdot 17}{\cancel{2} \cdot 50} = \dfrac{17}{50}$

63. $\dfrac{114}{10,000} = \dfrac{\cancel{2} \cdot 57}{\cancel{2} \cdot 5000} = \dfrac{57}{5000}$

65. $\dfrac{6.15}{2.61} = \dfrac{615}{261} = \dfrac{\cancel{3} \cdot 205}{\cancel{3} \cdot 87} = \dfrac{205}{87}$

67. Let x = number of teachers.

$\dfrac{18}{1} = \dfrac{900}{x}$

$18x = 900$

$x = \dfrac{900}{18} = 50$

It has 50 teachers.

69. Let t = tax.

$\dfrac{9}{1000} = \dfrac{t}{40,000}$

$1000t = 9 \cdot 40,000$

$1000t = 360,000$

$t = 360$

The tax is $360.

71. Let x = actual wingspan of a 747.

$\dfrac{1}{144} = \dfrac{40}{x}$

$x = 144 \cdot 40 = 5760$

The actual wingspan is 5760 cm or 57.60 m.

73. Let $x = $ # of stocks advanced in price.

$$\frac{5}{2} = \frac{300,000}{x}$$
$$5x = 2 \cdot 300,000$$
$$5x = 600,000$$
$$x = 120,000$$

There were 120,000 stocks that advanced in price.

75. **a.** 85 out of 100 is $\dfrac{85}{100} = \dfrac{\cancel{5} \cdot 17}{\cancel{5} \cdot 20} = \dfrac{17}{20}$

b. Let $x = $ number having no trouble.

$$\frac{17}{20} = \frac{x}{500}$$
$$20x = 17 \cdot 500$$
$$20x = 8500$$
$$x = \frac{8500}{20} = 425$$

You would expet 425 calls to have no trouble.

c. $15\% = \dfrac{15}{100} = \dfrac{3}{20}$ is the ratio

experiencing trouble. If $x = $ number of

calls, then $\dfrac{3}{20} = \dfrac{102}{x}$
$$3x = 20 \cdot 102$$
$$3x = 2040$$
$$x = \frac{2040}{3} = 680.$$

Thus, 680 calls were made to 911.

77. Let $x = $ number of people.

$$\frac{80}{1} = \frac{x}{50}$$
$$x = 80 \cdot 50 = 4000$$

You would expect 4000 people.

79. Let $x = $ # of people per sq. km (density).

$$\frac{x}{1} = \frac{446,250}{21}$$
$$21x = 445,250$$
$$x = \frac{445,250}{21} = 21,250$$

The density in Macau is 21,250 people/km^2.

81. $\dfrac{350}{1000} = \dfrac{35}{100} = \dfrac{\cancel{5} \cdot 7}{\cancel{5} \cdot 20} = \dfrac{7}{20}$

83. $\dfrac{4.8}{3.2} = \dfrac{48}{32} = \dfrac{\cancel{16} \cdot 3}{\cancel{16} \cdot 2} = \dfrac{3}{2}$

85. current ratio $= \dfrac{15,\cancel{000}}{8\cancel{000}} = \dfrac{15}{8}$

87. net profit margin $= \dfrac{8\cancel{000}}{50,\cancel{000}} = \dfrac{8}{50} = \dfrac{4}{25}$

89. Answers may vary.

Mastery Test 4.1

91. $\dfrac{17\cancel{0,000}}{21\cancel{0,000}} = \dfrac{17}{21}$

93. **a.** $\dfrac{2}{5} = \dfrac{4}{10}$ **b.** $\dfrac{5}{8} = \dfrac{15}{x}$

95. $\dfrac{10}{19} \overset{?}{=} \dfrac{50}{95}$; $\;10 \cdot 95 \overset{?}{=} 19 \cdot 50$
$$950 = 950$$

Yes, since $10 \cdot 95 = 19 \cdot 50$.

97. Let $h = $ number of hours.

$$\frac{10}{3} = \frac{30}{h}$$
$$10h = 90$$
$$h = 9$$

It would take 9 hours.

Section 4.2 – Rates

Problems

1. $\dfrac{564}{41} = 41\overline{)564.0}$

$$\begin{array}{r} 13.7 \\ 41\overline{)564.0} \\ \underline{41} \\ 154 \\ \underline{123} \\ 310 \\ \underline{287} \\ 23 \end{array}$$

This is about 14 books per year.

2. $\dfrac{1900}{62} = 62\overline{)1900.0}$

$$\begin{array}{r} 30.6 \\ 62\overline{)1900.0} \\ \underline{186} \\ 40\,0 \\ \underline{37\,2} \\ 2\,8 \end{array}$$

Their rate was about 31 miles per gallon.

3. $\dfrac{6}{18} = \dfrac{1}{3} = 0.33\overline{3}$

The cost is \$0.33 per pound.

4. **a.** $\dfrac{\$612}{2 \text{ weeks}} = \$306$ per week

   b. $\dfrac{\$612}{72 \text{ hours}} = \dfrac{\$17}{2 \text{ hours}} = \$8.50$ per hour

5. $\dfrac{\$1.89}{10 \text{ oz}} = \dfrac{189 \text{ cents}}{10 \text{ oz}} = 18.9$ cents per oz

6. $\dfrac{\$1.09}{29 \text{ oz}} = \dfrac{109 \text{ cents}}{29 \text{ oz}} = 29\overline{)109.00}$ so the 29-

$$\begin{array}{r} 3.75 \\ 29\overline{)109.00} \\ \underline{87} \\ 220 \\ \underline{203} \\ 170 \\ \underline{145} \\ 25 \end{array}$$

oz bottle cost about 3.75¢ per oz.

$\dfrac{\$0.89}{24 \text{ oz}} = \dfrac{89 \text{ cents}}{24 \text{ oz}} = 24\overline{)89.00}$

$$\begin{array}{r} 3.70 \\ 24\overline{)89.00} \\ \underline{72} \\ 170 \\ \underline{168} \\ 20 \end{array}$$

The 24 oz bottle cost about 3.70¢ per oz. Thus the 24 oz bottle is the better buy.

Exercises 4.2

1. $\dfrac{\$38}{8 \text{ hr}} = \dfrac{19}{4} \dfrac{\text{dollars}}{\text{hr}} = \$4.75$ per hr

3. $\dfrac{258 \text{ miles}}{12 \text{ gal}} = \dfrac{\cancel{6} \cdot 43 \text{ miles}}{\cancel{6} \cdot 2 \text{ gal}}$
$= \dfrac{43 \text{ miles}}{2 \text{ gal}} = 21.5$ mi/gal

5. $\dfrac{840 \text{ miles}}{1.5 \text{ hr}} = \dfrac{8400 \text{ miles}}{15 \text{ hr}} = 560$ mi/hr

7. $\dfrac{2459 \text{ miles}}{2 \text{ hr}} = 1229.5$ mi/hr

9. $\dfrac{74}{30} = 30\overline{)74.00}$ This is 2.5 tortillas/min.

$$\begin{array}{r} 2.46 \\ 30\overline{)74.00} \\ \underline{60} \\ 14\,0 \\ \underline{12\,0} \\ 2\,00 \\ \underline{1\,80} \\ 20 \end{array}$$

11. $\dfrac{263 \text{ calories}}{100 \text{ g}} = 2.63$ calories/g

13. $\dfrac{90 \text{ cars}}{1 \text{ day}} = \dfrac{90 \text{ cars}}{7.5 \text{ hr}} = \dfrac{900 \text{ cars}}{75 \text{ hr}} = 12$ cars/hr

15. $\dfrac{36 \text{ in.}}{24 \text{ hr}} = \dfrac{\cancel{12} \cdot 3 \text{ in.}}{\cancel{12} \cdot 2 \text{ hr}} = \dfrac{3 \text{ in.}}{2 \text{ hr}} = 1.5$ in./hr

17. $\dfrac{1110 \text{ calories}}{1.5 \text{ lb}} = \dfrac{11{,}100 \text{ calories}}{15 \text{ lb}}$
$= 740 \text{ calories/lb}$

19. $\dfrac{31{,}419}{1045} = 1045\overline{)31419.00}$

$\begin{array}{r} 30.06 \\ \underline{3135} \\ 69\,00 \\ \underline{62\,00} \\ 7\,00 \end{array}$

His scoring rate was 30.1 points/game.

21. $\dfrac{872}{84} = 84\overline{)872.000} \approx \$10.38 \text{ per hr}$

$\begin{array}{r} 10.380 \\ \underline{84} \\ 32\,0 \\ 25\,2 \\ \overline{6\,80} \\ \underline{6\,72} \\ 80 \end{array}$

23. $\dfrac{87}{4} = 4\overline{)87.00}$; $\$21.75 \text{ per hr}$

$\begin{array}{r} 21.75 \\ \underline{8} \\ 7 \\ \underline{4} \\ 3\,0 \\ \underline{2\,8} \\ 20 \\ \underline{20} \\ 0 \end{array}$

25. $\dfrac{895}{35} = 35\overline{)895.000} \approx \$25.57 \text{ per hr}$

$\begin{array}{r} 25.571 \\ \underline{70} \\ 195 \\ \underline{175} \\ 20\,0 \\ 175 \\ \overline{2\,50} \\ 2\,45 \\ \overline{50} \\ \underline{35} \\ 15 \end{array}$

27. $\dfrac{1400 \text{ lb}}{1 \text{ week}} = \dfrac{1400 \text{ lb}}{7 \text{ days}} = 200 \text{ lb/day}$

29. $\dfrac{300\cancel{0} \text{ teeth}}{4\cancel{0} \text{ yr}} = \dfrac{300 \text{ teeth}}{4 \text{ yr}} = 75 \text{ teeth/yr}$

31. $\dfrac{\$2.29}{6 \text{ oz}} = \dfrac{229 \text{ cents}}{6 \text{ oz}} = 6\overline{)229.0}$

$\begin{array}{r} 38.1 \\ \underline{18} \\ 49 \\ \underline{48} \\ 1\,0 \\ \underline{6} \\ 4 \end{array}$

The 6-oz tube cost 38¢ per oz.

$\dfrac{\$1.59}{4 \text{ oz}} = \dfrac{159 \text{ cents}}{4 \text{ oz}} = 4\overline{)159.0}$ so the 4-oz

$\begin{array}{r} 39.7 \\ \underline{12} \\ 39 \\ \underline{36} \\ 30 \\ \underline{2\,8} \\ 2 \end{array}$

The 4-oz tube cost 40¢ per oz. The first tube has the lowest per unit price.

33. $\dfrac{\$1.99}{26 \text{ oz}} = \dfrac{199 \text{ cents}}{26 \text{ oz}} = 26\overline{)199.0}$

$\begin{array}{r} 7.6 \\ \underline{182} \\ 170 \\ \underline{156} \\ 14 \end{array}$

The first item cost 8¢ per oz.
$\dfrac{\$0.99}{11 \text{ oz}} = \dfrac{99 \text{ cents}}{11 \text{ oz}} = 9¢ \text{ per oz.}$
The First item has the lowest per unit price.

35. $\dfrac{\$4.49}{2.5 \text{ oz}} = \dfrac{\$44.9}{25 \text{ oz}} = 25\overline{)44.900}$

$\begin{array}{r} 1.797 \\ \underline{25} \\ 19\,9 \\ 175 \\ \overline{240} \\ \underline{225} \\ 150 \\ \underline{150} \\ 0 \end{array}$

The first item cost $1.80 per oz.

$$\frac{\$3.79}{1.5 \text{ oz}} = \frac{\$37.9}{15 \text{ oz}} = 15\overline{)37.900}$$

$$\begin{array}{r} 2.526 \\ \underline{30} \\ 79 \\ \underline{75} \\ 40 \\ \underline{30} \\ 100 \\ \underline{90} \\ 10 \end{array}$$

The second item cost $2.53 per oz. Hence the first item has the lowest per unit cost.

37. a. $\dfrac{\$1.31}{22 \text{ oz}} = \dfrac{131 \text{ cents}}{22 \text{ oz}} = 22\overline{)131.0}$

$$\begin{array}{r} 5.9 \\ \underline{110} \\ 21\,0 \\ \underline{19\,8} \\ 12 \end{array}$$

The cost is about 6¢ per oz.

b. $\dfrac{\$1.75}{32 \text{ oz}} = \dfrac{175 \text{ cents}}{32 \text{ oz}} = 32\overline{)175.0}$

$$\begin{array}{r} 5.4 \\ \underline{160} \\ 15\,0 \\ \underline{12\,8} \\ 2\,2 \end{array}$$

The cost is about 5¢ per oz.

c. White magic is the better buy.

39. $18 = 3x$

$\dfrac{18}{3} = x$

$6 = x$

41. $6 \cdot 5 = 14x$

$\dfrac{\overset{3}{\cancel{6}} \cdot 5}{\underset{7}{\cancel{14}}} = x$

$\dfrac{15}{7} = x$

43. $6x = 9 \cdot 10$

$x = \dfrac{\overset{3}{\cancel{9}} \cdot \overset{5}{\cancel{10}}}{\underset{\cancel{2}}{\cancel{6}}}$

$x = 15$

45. a. $\dfrac{\$348}{10 \text{ days}} = \$34.80/\text{day}$

b. $\dfrac{\$34.80}{10} = \$3.48$ per line per day

47. No; answers may vary. Sample answer: a map might read 3 inches = 50 miles giving the ratio $\dfrac{3 \text{ inches}}{50 \text{ miles}}$ which is not a unit rate.

49. Answers may vary.

Mastery Test 4.2

51. $\dfrac{\$2.30}{6 \text{ oz}} = \dfrac{230 \text{ cents}}{6 \text{ oz}} = 6\overline{)230.0}$

$$\begin{array}{r} 38.\overline{3} \\ \underline{18} \\ 50 \\ \underline{48} \\ 20 \\ \underline{18} \\ 2 \end{array}$$

The unit price is 38.3¢ per oz.

53. a. $\dfrac{\$508.40}{2 \text{ weeks}} = 2\overline{)508.40}$

$$\begin{array}{r} 254.20 \\ \underline{4} \\ 10 \\ \underline{10} \\ 08 \\ \underline{8} \\ 0\,4 \\ \underline{4} \\ 00 \end{array}$$

This is $254.20 per week.

b. $\dfrac{\$508.40}{82 \text{ hr}} = 82\overline{)508.40}$

$$\begin{array}{r} 6.20 \\ \underline{492} \\ 16\,4 \\ \underline{16\,4} \\ 00 \end{array}$$

The pay rate was $6.20 per hour.

55. $\dfrac{280 \text{ mi}}{13 \text{ gal}} = 13\overline{)280.0} \approx 22 \text{ mi/gal}$

$$\begin{array}{r} 21.5 \\ \underline{26} \\ 20 \\ \underline{13} \\ 70 \\ \underline{65} \\ 5 \end{array}$$

Section 4.3 – Word Problems Involving Proportions

Problems

1. Let p = number of pounds needed.

$$\frac{1}{100} = \frac{p}{4500}$$
$$100p = 1 \cdot 4500$$
$$p = \frac{4500}{100} = 45$$

You need 45 pounds.

2. Let f = # of families that ran into trouble.

$$\frac{2}{5} = \frac{300}{f}$$
$$2f = 5 \cdot 300$$
$$f = \frac{1500}{2} = 750$$

There were 750 people surveyed.

3. Let t = number of tablespoons.

$$\frac{2}{8} = \frac{t}{56}$$
$$8t = 2 \cdot 56$$
$$t = \frac{2 \cdot \cancel{56}^{7}}{\cancel{8}} = 14$$

He needs 14 tablespoons.

4. Let n = number of shares.

$$\frac{50}{137.50} = \frac{n}{1100}$$
$$137.50n = 50 \cdot 1100$$
$$n = \frac{55,000}{137.50} = 400$$

You can buy 400 shares with $1000.

5. Let c = dose for a child.

$$\frac{W_c}{W_a} = \frac{c}{a} \text{ gives } \frac{75}{150} = \frac{c}{3}$$
$$150c = 75 \cdot 3$$
$$c = \frac{\cancel{75} \cdot 3}{\cancel{150}_2} = \frac{3}{2} = 1\frac{1}{2}$$

The dose for a 75-lb child is 1.5 pills.

6. Let W = weight that can be carried.

$$\frac{30}{25,500} = \frac{90}{W}$$
$$30W = 90 \cdot 25,500$$
$$n = \frac{\overset{3}{\cancel{90}} \cdot 25,500}{\cancel{30}} = 76,500$$

He can carry 76,500 kg.

Exercises 4.3

1. Let p = number of pounds needed.

$$\frac{1}{200} = \frac{p}{3000}$$
$$200p = 1 \cdot 3000$$
$$p = \frac{3000}{200} = 15$$

You need 15 pounds.

3. Let n = number advanced in price.

$$\frac{2}{5} = \frac{n}{1900}$$
$$5n = 2 \cdot 1900$$
$$n = \frac{2 \cdot \cancel{1900}^{380}}{\cancel{5}} = 760$$

There were 760 stocks that advanced in price.

5. Let c = amount of cheddar cheese needed.

$$\frac{2}{14} = \frac{c}{56}$$
$$14c = 2 \cdot 56$$
$$c = \frac{2 \cdot \cancel{56}^{4}}{\cancel{14}} = 8$$

You need 8 ounces of cheddar cheese.

7. Let n = number of shares.

$$\frac{50}{112.50} = \frac{n}{562.50}$$
$$112.50n = 50 \cdot 562.50$$
$$n = \frac{50 \cdot \cancel{562.50}^{5}}{\cancel{112.50}} = 250$$

You can buy 250 shares with $562.50.

9. Let m = length of shower.

$$\frac{10}{1.5} = \frac{m}{6}$$

$$1.5m = 10 \cdot 6$$

$$m = \frac{10 \cdot \cancel{6}^{\,4}}{\cancel{1.5}} = 40$$

You have to shower 40 minutes.

11. Let v = number of votes.

$$\frac{10}{100} = \frac{770}{v}$$

$$10v = 100 \cdot 770$$

$$v = \frac{100 \cdot 77\cancel{0}}{\cancel{10}} = 7700$$

They can expect 7700 votes.

13. Let g = number of gulps.

2 hr = 2 × 60 min = 120 min

$$\frac{4}{20} = \frac{g}{120}$$

$$20g = 4 \cdot 120$$

$$g = \frac{4 \cdot \cancel{120}^{\,6}}{\cancel{20}} = 24$$

You need 24 gulps.

15. Let p = number of pages.

$$\frac{1}{330} = \frac{p}{6600}$$

$$330p = 1 \cdot 6600$$

$$p = \frac{6600}{330} = 20$$

There would be 20 pages in a double-spaced 6600-word paper.

17. Let a = amount added.

$$\frac{1}{2.3} = \frac{2.5}{a}$$

$$a = 2.3 \cdot 2.5$$

$$a = 5.75$$

There would be $5.75 billion added to federal spending.

19. Let n = number of pounds of water.

$$\frac{1725}{1} = \frac{n}{50}$$

$$n = 50 \cdot 1725$$

$$n = 86,250$$

A 50-lb bale needs 86,250 pounds of water.

21. Let f = number of grams of fat.

$$\frac{\frac{1}{4}}{13} = \frac{\frac{3}{4}}{f}$$

$$\frac{1}{4}f = 13 \cdot \frac{3}{4}$$

$$f = 13 \cdot \frac{3}{4} \cdot 4 = 39$$

There are 39 grams of fat in $\frac{3}{4}$ of the pizza.

23. Let f = number of grams of fat.

$$\frac{\frac{1}{4}}{7} = \frac{1}{f}$$

$$\frac{1}{4}f = 7$$

$$f = 7 \cdot 4 = 28$$

There are 28 grams of fat in the whole pizza.

25. Let g = # in global Internet population.

$$\frac{\frac{1}{4}}{175} = \frac{1}{g}$$

$$\frac{1}{4}g = 175$$

$$g = 175 \cdot 4 = 700$$

About 700 million (700,000,000) people are in the global Internet population.

27. Let c = number of carbon atoms.

$$\frac{3}{6} = \frac{c}{720}$$

$$6c = 3 \cdot 720$$

$$c = \frac{3 \cdot \cancel{720}^{\,120}}{\cancel{6}} = 360$$

You need 360 carbon atoms.

29. Let d = distance between the cities.

$$\frac{1}{20} = \frac{3.5}{d}$$

$$d = 20 \cdot 3.5$$

$$d = 70$$

The cities are 70 miles apart.

31. $245.\underline{9}2 \to 245.9$

33. $2\underline{4}5.92 \to 250$

35. Let n = total number of birds.

$$\frac{12}{980} = \frac{300}{n}$$

$$12n = 980 \cdot 300$$

$$n = \frac{980 \cdot \overset{25}{\cancel{300}}}{\cancel{12}} = 24{,}500$$

There were about 24,500 birds in the Vienna Woods.

37. No; answers may vary.

Mastery Test 4.3

39. $\dfrac{W_c}{W_a} = \dfrac{c}{a}$

$$\frac{50}{150} = \frac{c}{6}$$

$$150c = 50 \cdot 6$$

$$c = \frac{300}{150} = 2 \text{ teaspoons/day}$$

41. Let n = number of shares.

$$\frac{50}{212.50} = \frac{n}{850}$$

$$212.50n = 50 \cdot 850$$

$$n = \frac{50 \cdot \overset{4}{\cancel{850}}}{\cancel{212.50}} = 200 \text{ shares}$$

43. Let n = number passed.

$$\frac{2}{3} = \frac{n}{600}$$

$$3n = 2 \cdot 600$$

$$n = \frac{2 \cdot \overset{200}{\cancel{600}}}{\cancel{3}} = 400 \text{ students}$$

Review Exercises – Chapter 4

1. a. 1 to 10 is written as $\dfrac{1}{10}$.

   b. 2 to 10 is written as $\dfrac{2}{10} = \dfrac{1}{5}$.

   c. 3 to 10 is written as $\dfrac{3}{10}$.

   d. 4 to 10 is written as $\dfrac{4}{10} = \dfrac{2}{5}$.

   e. 5 to 10 is written as $\dfrac{5}{10} = \dfrac{1}{2}$.

2. a. $\dfrac{425}{1000} = \dfrac{\cancel{25} \cdot 17}{\cancel{25} \cdot 40} = \dfrac{17}{40}$

   b. $\dfrac{325}{1000} = \dfrac{\cancel{25} \cdot 13}{\cancel{25} \cdot 40} = \dfrac{13}{40}$

   c. $\dfrac{3\cancel{00}}{10\cancel{00}} = \dfrac{3}{10}$

   d. $\dfrac{15\cancel{0}}{100\cancel{0}} = \dfrac{15}{100} = \dfrac{\cancel{5} \cdot 3}{\cancel{5} \cdot 20} = \dfrac{3}{20}$

   e. $\dfrac{4\cancel{0}}{100\cancel{0}} = \dfrac{4}{100} = \dfrac{1}{25}$

3. a. $\dfrac{14.52}{4.62} = \dfrac{1452}{462} = \dfrac{22}{7}$

   b. $\dfrac{14.58}{4.64} = \dfrac{1458}{464} = \dfrac{729}{232}$

   c. $\dfrac{14.60}{4.65} = \dfrac{1460}{465} = \dfrac{292}{93}$

   d. $\dfrac{14.64}{4.66} = \dfrac{1464}{466} = \dfrac{732}{233}$

   e. $\dfrac{14.68}{4.67} = \dfrac{1468}{467}$

4. a. $\dfrac{58}{42} = \dfrac{29}{21}$ **b.** $\dfrac{55}{45} = \dfrac{11}{9}$

   c. $\dfrac{52}{48} = \dfrac{13}{12}$ **d.** $\dfrac{48}{52} = \dfrac{12}{13}$

   e. $\dfrac{46}{54} = \dfrac{23}{27}$

5. a. $\dfrac{3}{7} = \dfrac{6}{x}$ **b.** $\dfrac{4}{7} = \dfrac{12}{x}$

   c. $\dfrac{5}{7} = \dfrac{x}{21}$ **d.** $\dfrac{6}{7} = \dfrac{33}{x}$

   e. $\dfrac{7}{35} = \dfrac{5}{x}$

6. a. $\dfrac{2}{3} \overset{?}{=} \dfrac{4}{5}$; No, since $2 \cdot 5 \neq 3 \cdot 4$

   b. $\dfrac{8}{10} \overset{?}{=} \dfrac{4}{5}$; Yes, since $8 \cdot 5 = 10 \cdot 4$

   c. $\dfrac{5}{6} \overset{?}{=} \dfrac{12}{15}$; No, since $5 \cdot 15 \neq 6 \cdot 12$

d. $\dfrac{12}{18} \overset{?}{=} \dfrac{2}{3}$; Yes, since $12 \cdot 3 = 18 \cdot 2$

e. $\dfrac{9}{12} \overset{?}{=} \dfrac{3}{4}$; Yes, since $9 \cdot 4 = 12 \cdot 3$

7. a. $\dfrac{x}{4} = \dfrac{1}{2}$
$2x = 4$
$x = 2$

b. $\dfrac{x}{6} = \dfrac{1}{2}$
$2x = 6$
$x = 3$

c. $\dfrac{x}{8} = \dfrac{1}{2}$
$2x = 8$
$x = 4$

d. $\dfrac{x}{10} = \dfrac{1}{2}$
$2x = 10$
$x = 5$

e. $\dfrac{x}{12} = \dfrac{1}{2}$
$2x = 12$
$x = 6$

8. a. $\dfrac{2}{x} = \dfrac{2}{5}$
$2x = 2 \cdot 5$
$x = 5$

b. $\dfrac{4}{x} = \dfrac{2}{5}$
$2x = 4 \cdot 5$
$x = 10$

c. $\dfrac{6}{x} = \dfrac{2}{5}$
$2x = 6 \cdot 5$
$x = 15$

d. $\dfrac{10}{x} = \dfrac{2}{5}$
$2x = 10 \cdot 5$
$x = 25$

e. $\dfrac{12}{x} = \dfrac{2}{5}$
$2x = 12 \cdot 5$
$x = 30$

9. a. $\dfrac{x}{4} = \dfrac{9}{2}$
$2x = 4 \cdot 9$
$x = 18$

b. $\dfrac{x}{4} = \dfrac{9}{12}$
$12x = 4 \cdot 9$
$x = 3$

c. $\dfrac{x}{4} = \dfrac{9}{18}$
$18x = 4 \cdot 9$
$x = 2$

d. $\dfrac{x}{4} = \dfrac{9}{36}$
$36x = 4 \cdot 9$
$x = 1$

e. $\dfrac{x}{4} = \dfrac{9}{6}$
$6x = 4 \cdot 9$
$x = 6$

10. a. $\dfrac{5}{7} = \dfrac{x}{7}$
$7x = 5 \cdot 7$
$x = 5$

b. $\dfrac{5}{7} = \dfrac{x}{14}$
$7x = 5 \cdot 14$
$x = 10$

c. $\dfrac{5}{7} = \dfrac{x}{28}$
$7x = 5 \cdot 28$
$x = 20$

d. $\dfrac{5}{7} = \dfrac{x}{35}$
$7x = 5 \cdot 35$
$x = 25$

e. $\dfrac{5}{7} = \dfrac{x}{42}$
$7x = 5 \cdot 42$
$x = 30$

11. Let n = number of parts.

a. $\dfrac{3}{27} = \dfrac{1}{n}$
$3n = 27$
$n = 9$ parts

b. $\dfrac{3}{27} = \dfrac{2}{n}$
$3n = 27 \cdot 2$
$n = 18$ parts

c. $\dfrac{3}{27} = \dfrac{4}{n}$
$3n = 27 \cdot 4$
$n = 36$ parts

d. $\dfrac{3}{27} = \dfrac{5}{n}$
$3n = 27 \cdot 5$
$n = 45$ parts

e. $\dfrac{3}{27} = \dfrac{6}{n}$
$3n = 27 \cdot 6$
$n = 54$ parts

12. a. $\dfrac{\$12,000}{2,000 \text{ words}} = \$6 / \text{word}$

b. $\dfrac{\$12,000}{3,000 \text{ words}} = \$4 / \text{word}$

c. $\dfrac{\$12,000}{4,000 \text{ words}} = \$3 / \text{word}$

d. $\dfrac{\$12,000}{6,000 \text{ words}} = \$2 / \text{word}$

e. $\dfrac{\$12,000}{8,000 \text{ words}} = \dfrac{\$3}{2 \text{ words}} = \$1.50 / \text{word}$

13. a. $\dfrac{400 \text{ miles}}{18 \text{ gal}} = 18 \overline{)400.0} \approx 22 \text{ mi/gal}$

$$\begin{array}{r} 22.2 \\ 18\overline{)400.0} \\ \underline{36} \\ 40 \\ \underline{36} \\ 4 \end{array}$$

b. $\dfrac{400 \text{ miles}}{19 \text{ gal}} = 19\overline{)400.0} \approx 21 \text{ mi/gal}$

$\begin{array}{r} 21.0 \\ \hline 38 \\ \hline 20 \\ 19 \\ \hline 10 \end{array}$

c. $\dfrac{40\!\!\!/0 \text{ miles}}{2\!\!\!/0 \text{ gal}} = 20 \text{ mi/gal}$

d. $\dfrac{400 \text{ miles}}{21 \text{ gal}} = 21\overline{)400.0} \approx 19 \text{ mi/gal}$

$\begin{array}{r} 19.0 \\ \hline 21 \\ \hline 190 \\ 189 \\ \hline 10 \end{array}$

e. $\dfrac{400 \text{ miles}}{22 \text{ gal}} = 22\overline{)400.0} \approx 18 \text{ mi/gal}$

$\begin{array}{r} 18.1 \\ \hline 22 \\ \hline 180 \\ 176 \\ \hline 4\,0 \\ 2\,2 \\ \hline 8 \end{array}$

14. a. $\dfrac{500\!\!\!/0 \text{ ft}^2}{2\!\!\!/0 \text{ lb}} = 250 \text{ ft}^2/\text{lb}$

b. $\dfrac{5000 \text{ ft}^2}{22 \text{ lb}} = 22\overline{)5000.0} \approx 227 \text{ ft}^2/\text{lb}$

$\begin{array}{r} 227.2 \\ \hline 44 \\ \hline 60 \\ 44 \\ \hline 160 \\ 154 \\ \hline 6\,0 \\ 4\,4 \\ \hline 16 \end{array}$

c. $\dfrac{5000 \text{ ft}^2}{24 \text{ lb}} = 24\overline{)5000.0} \approx 208 \text{ ft}^2/\text{lb}$

$\begin{array}{r} 208.3 \\ \hline 48 \\ \hline 200 \\ 192 \\ \hline 8\,0 \\ 7\,2 \\ \hline 8 \end{array}$

d. $\dfrac{5000 \text{ ft}^2}{25 \text{ lb}} = 200 \text{ ft}^2/\text{lb}$

e. $\dfrac{500\!\!\!/0 \text{ ft}^2}{5\!\!\!/0 \text{ lb}} = 100 \text{ ft}^2/\text{lb}$

15. a. $\dfrac{\$2.39}{24 \text{ oz}} = \dfrac{239 \text{ cents}}{24 \text{ oz}} = 24\overline{)239.0} \approx 10\text{¢/oz}$

$\begin{array}{r} 9.9 \\ \hline 216 \\ \hline 23\,0 \\ 216 \\ \hline 14 \end{array}$

b. $\dfrac{\$2.39}{16 \text{ oz}} = \dfrac{239 \text{ cents}}{16 \text{ oz}} = 16\overline{)239.0} \approx 15\text{¢/oz}$

$\begin{array}{r} 14.9 \\ \hline 16 \\ \hline 79 \\ 64 \\ \hline 15\,0 \\ 144 \\ \hline 6 \end{array}$

c. $\dfrac{\$2.39}{32 \text{ oz}} = \dfrac{239 \text{ cents}}{32 \text{ oz}} = 32\overline{)239.0} \approx 7\text{¢/oz}$

$\begin{array}{r} 7.4 \\ \hline 224 \\ \hline 15\,0 \\ 12\,8 \\ \hline 2\,2 \end{array}$

d. $\dfrac{\$2.39}{40 \text{ oz}} = \dfrac{239 \text{ cents}}{40 \text{ oz}} = 40\overline{)239.0} \approx 6\text{¢/oz}$

$\begin{array}{r} 5.9 \\ \hline 200 \\ \hline 39\,0 \\ 36\,0 \\ \hline 30 \end{array}$

e. $\dfrac{\$2.39}{48 \text{ oz}} = \dfrac{239 \text{ cents}}{48 \text{ oz}} = 48\overline{)239.0} \approx 5\text{¢/oz}$

$\begin{array}{r} 4.9 \\ \hline 192 \\ \hline 470 \\ 432 \\ \hline 3\,0 \end{array}$

16. The unit cost for the generic brand is

$\dfrac{\$1.39}{12 \text{ oz}} = \dfrac{139 \text{ cents}}{12 \text{ oz}} = 12\overline{)139.00} \approx 11.6\text{¢/oz.}$

$\begin{array}{r} 11.58 \\ \hline 12 \\ \hline 19 \\ 12 \\ \hline 7\,0 \\ 6\,0 \\ \hline 100 \\ 96 \\ \hline 4 \end{array}$

a. $\dfrac{275 \text{ cents}}{24 \text{ oz}} = 24\overline{)275.00} \approx 11.5¢/\text{oz}$

$$
\begin{array}{r}
11.45 \\
24\overline{)275.00} \\
24 \\
\hline
35 \\
24 \\
\hline
110 \\
96 \\
\hline
140 \\
120 \\
\hline
20
\end{array}
$$

Brand X is the better buy.

b. $\dfrac{276 \text{ cents}}{24 \text{ oz}} = 24\overline{)276.0} = 11.5¢/\text{oz}$

$$
\begin{array}{r}
11.5 \\
24\overline{)276.0} \\
24 \\
\hline
36 \\
24 \\
\hline
120 \\
120 \\
\hline
0
\end{array}
$$

Brand X is the better buy.

c. $\dfrac{277 \text{ cents}}{24 \text{ oz}} = 24\overline{)277.00} \approx 11.5¢/\text{oz}$

$$
\begin{array}{r}
11.54 \\
24\overline{)277.00} \\
24 \\
\hline
37 \\
24 \\
\hline
130 \\
120 \\
\hline
100 \\
96 \\
\hline
4
\end{array}
$$

Brand X is the better buy.

d. $\dfrac{274 \text{ cents}}{24 \text{ oz}} = 24\overline{)274.00} \approx 11.4¢/\text{oz}$

$$
\begin{array}{r}
11.41 \\
24\overline{)274.00} \\
24 \\
\hline
34 \\
24 \\
\hline
100 \\
96 \\
\hline
40 \\
26 \\
\hline
14
\end{array}
$$

Brand X is the better buy.

e. $\dfrac{279 \text{ cents}}{24 \text{ oz}} = 24\overline{)279.00} \approx 11.6¢/\text{oz}$

$$
\begin{array}{r}
11.62 \\
24\overline{)279.00} \\
24 \\
\hline
39 \\
24 \\
\hline
150 \\
144 \\
\hline
60 \\
48 \\
\hline
12
\end{array}
$$

The generic brand is the better buy.

17. Let p = number of pounds needed.

a. $\dfrac{1}{120} = \dfrac{p}{1800}$
$120p = 1800$
$p = 15$ pounds

b. $\dfrac{1}{120} = \dfrac{p}{2100}$
$120p = 2100$
$p = 17.5$ pounds

c. $\dfrac{1}{120} = \dfrac{p}{2160}$
$120p = 2160$
$p = 18$ pounds

d. $\dfrac{1}{120} = \dfrac{p}{2700}$
$120p = 2700$
$p = 22.5$ pounds

e. $\dfrac{1}{120} = \dfrac{p}{3000}$
$120p = 3000$
$p = 25$ pounds

18. Let n = number having used this brand.

a. $\dfrac{3}{5} = \dfrac{n}{3000}$
$5n = 3 \cdot 3000$
$n = 1800$

b. $\dfrac{3}{5} = \dfrac{n}{4000}$
$5n = 3 \cdot 4000$
$n = 2400$

c. $\dfrac{3}{5} = \dfrac{n}{5000}$
$5n = 3 \cdot 5000$
$n = 3000$

d. $\dfrac{3}{5} = \dfrac{n}{6000}$
$5n = 3 \cdot 6000$
$n = 3600$

e. $\dfrac{3}{5} = \dfrac{n}{8000}$
$5n = 3 \cdot 8000$
$n = 4800$

19. Let n = number of ounces needed.

a. $\dfrac{3}{8} = \dfrac{n}{56}$
$8n = 3 \cdot 56$
$n = 21$

b. $\dfrac{4}{8} = \dfrac{n}{56}$
$8n = 4 \cdot 56$
$n = 28$

c. $\dfrac{5}{8} = \dfrac{n}{56}$
$8n = 5 \cdot 56$
$n = 35$

d. $\dfrac{6}{8} = \dfrac{n}{56}$
$8n = 6 \cdot 56$
$n = 42$

e. $\dfrac{7}{8} = \dfrac{n}{56}$
$8n = 7 \cdot 56$
$n = 49$

20. Let n = number of shares you can buy.

a. $\dfrac{50}{87.50} = \dfrac{n}{350}$

$87.50n = 50 \cdot 350$

$n = 200$

b. $\dfrac{50}{87.50} = \dfrac{n}{700}$

$87.50n = 50 \cdot 700$

$n = 400$

c. $\dfrac{50}{87.50} = \dfrac{n}{612.50}$

$87.50n = 50 \cdot 612.50$

$n = 350$

d. $\dfrac{50}{87.50} = \dfrac{n}{787.50}$

$87.50n = 50 \cdot 787.50$

$n = 450$

e. $\dfrac{50}{87.50} = \dfrac{n}{1050}$

$87.50n = 50 \cdot 1050$

$n = 600$

Cumulative Review Chapters 1–4

1. 9810

2. $56 = 8 \times 7 = 2^3 \times 7$ so the prime factors are 2 and 7.

3. $2^3 \times 7 \times 4^0 = 8 \times 7 \times 1 = 56$

4. $25 \div 5 \cdot 5 + 7 - 3 = 5 \cdot 5 + 7 - 3$
$= 25 + 7 - 3$
$= 32 - 3$
$= 29$

5. $\dfrac{9}{7}$ is improper, since the numerator is larger than the denominator.

6. $\dfrac{39}{4} = 9\dfrac{3}{4}$

7. $7\dfrac{2}{3} = \dfrac{3 \times 7 + 2}{3} = \dfrac{21 + 2}{3} = \dfrac{23}{3}$

8. $\dfrac{1}{2} \cdot 3\dfrac{1}{7} = \dfrac{1}{\cancel{2}} \cdot \dfrac{\cancel{22}^{11}}{7} = \dfrac{11}{7}$ or $1\dfrac{4}{7}$

9. $\left(\dfrac{3}{7}\right)^2 \cdot \dfrac{1}{9} = \dfrac{\cancel{9}}{49} \cdot \dfrac{1}{\cancel{9}} = \dfrac{1}{49}$

10. $\dfrac{8}{5} \div 2\dfrac{2}{7} = \dfrac{8}{5} \div \dfrac{16}{7} = \dfrac{\cancel{8}}{5} \cdot \dfrac{7}{\cancel{16}_2} = \dfrac{7}{10}$

11. $7\dfrac{2}{3} + 1\dfrac{3}{5} = 7\dfrac{2 \cdot 5}{3 \cdot 5} + 1\dfrac{3 \cdot 3}{5 \cdot 3}$
$= 7\dfrac{10}{15} + 1\dfrac{9}{15}$
$= 8\dfrac{19}{15}$
$= 8 + 1\dfrac{4}{15}$
$= 9\dfrac{4}{15}$

12. $5\dfrac{1}{4} - 1\dfrac{7}{8} = \dfrac{21}{4} - \dfrac{15}{8}$
$= \dfrac{42}{8} - \dfrac{15}{8} = \dfrac{27}{8} = 3\dfrac{3}{8}$

13. $c - \dfrac{7}{9} = \dfrac{1}{2}$

$c = \dfrac{1}{2} + \dfrac{7}{9}$

$c = \dfrac{23}{18}$

14. Let n = the number.

$\dfrac{11}{12}n = 7\dfrac{1}{10}$

$\dfrac{11}{12}n = \dfrac{71}{10}$

$\dfrac{\cancel{11}n}{\cancel{12}} = \dfrac{\frac{71}{10}}{\frac{11}{12}}$

$n = \dfrac{71}{_5\cancel{10}} \cdot \dfrac{\cancel{12}^6}{11} = \dfrac{426}{55} = 7\dfrac{41}{55}$

15. 241.35: Two hundred forty-one and thrity-five hundred

16. $44.874 = 40 + 4 + \dfrac{8}{10} + \dfrac{7}{100} + \dfrac{4}{1000}$

17.
$\begin{array}{r} 36.454 \\ +\ 9.690 \\ \hline 46.144 \end{array}$

18.

$$\begin{array}{r}
{\scriptstyle 3\ 11\ 14}\\
3\cancel{4}\,\cancel{2}.42\\
-\ \ 13.50\\
\hline
3\,2\,8.92
\end{array}$$

19.

$$\begin{array}{r}
0.554\\
\times\ \ 0.15\\
\hline
2770\\
554\ \ \ \\
\hline
0.08310
\end{array}$$

20. $\dfrac{135}{0.27} = \dfrac{13{,}500}{27} = 27)\overline{13500}$

$$\begin{array}{r}
\boxed{500}\\
27)\overline{13500}\\
135\ \ \ \\
\hline
00
\end{array}$$

21. $44\underline{9}.851 \rightarrow 450$

22. $10 \div 0.13 = \dfrac{10}{0.13} = \dfrac{1000}{13} = 13)\overline{1000.000}$

$$\begin{array}{r}
76.923\\
13)\overline{1000.000}\\
91\ \ \ \ \ \ \ \\
\hline
90\ \ \ \ \ \ \\
78\ \ \ \ \ \ \\
\hline
12\,0\ \ \ \ \\
11\,7\ \ \ \ \\
\hline
30\ \ \\
26\ \ \\
\hline
40\\
39\\
\hline
1
\end{array}$$

Thus $10 \div 0.13 \approx 7.92$.

23. $\dfrac{7}{12} = 12)\overline{7.000}$

$$\begin{array}{r}
\boxed{0.583}\\
12)\overline{7.000}\\
6\,0\ \ \ \\
\hline
1\,00\ \ \\
96\ \ \\
\hline
40\\
36\\
\hline
4
\end{array}$$

24. $0.15 = \dfrac{15}{100} = \dfrac{\cancel{5} \cdot 3}{\cancel{5} \cdot 20} = \dfrac{3}{20}$

25. $0.\overline{84} = \dfrac{84}{99} = \dfrac{\cancel{3} \cdot 28}{\cancel{3} \cdot 33} = \dfrac{28}{33}$

26. Let d = the decimal part.

$$d \times 12 = 9$$
$$d = \dfrac{9}{12} = \dfrac{3}{4} = 0.75$$

27. $6.435 = 6.4350$

$6.43\overline{5} = 6.4355\cdots$

$6.4\overline{35} = 6.43535\cdots$

Using the fourth decimal digit, we have
$6.43\overline{5} > 6.4\overline{35} > 6.435$.

28. $0.89 = \dfrac{89}{100}$ and $\dfrac{7}{20} = \dfrac{7 \cdot 5}{20 \cdot 5} = \dfrac{35}{100}$. Since

89 is greater 35, we have $0.89 > \dfrac{7}{20}$.

29. $x + 2.5 = 6.5$

$$x = 6.5 - 2.5$$
$$x = 4$$

30. $2.1 = 0.3y$

$$\dfrac{2.1}{0.3} = y$$
$$\dfrac{21}{3} = y$$
$$7 = y$$

31. $9 = \dfrac{z}{6.9}$

$$6.9 \cdot 9 = z$$
$$62.1 = z$$

32. $\dfrac{42\cancel{0}}{100\cancel{0}} = \dfrac{42}{100} = \dfrac{\cancel{2} \cdot 21}{\cancel{2} \cdot 50} = \dfrac{21}{50}$

33. $\dfrac{6}{2} = \dfrac{54}{x}$

34. $\dfrac{10}{19} \overset{?}{=} \dfrac{50}{97}$; $10 \cdot 97 = 970$ and $19 \cdot 50 = 950$.

Since $10 \cdot 97 \neq 19 \cdot 50$, the flag is not of the correct ratio.

35. $\dfrac{j}{5} = \dfrac{6}{150}$

$$150j = 5 \cdot 6$$
$$j = \dfrac{30}{150} = \dfrac{1}{5}$$

36. $\dfrac{20}{c} = \dfrac{4}{3}$

$4c = 20 \cdot 3$

$c = \dfrac{60}{4} = 15$

37. Let n = number of parts.

$\dfrac{9}{25} = \dfrac{36}{n}$

$9n = 25 \cdot 36$

$n = \dfrac{25 \cdot \cancel{36}^{4}}{\cancel{9}} = 100$

She can produce 100 parts in 36 hours.

38. $\dfrac{600}{17} = 17{\overline{\smash{\big)}\,600.0}}$

$\begin{array}{r} 35.2 \\ 17{\overline{\smash{\big)}\,600.0}} \\ \underline{51} \\ 90 \\ \underline{85} \\ 5\,0 \\ \underline{3\,4} \\ 16 \end{array}$

39. $\dfrac{\$2.89}{24\ \text{oz}} = 24{\overline{\smash{\big)}\,289.0}} \approx 12¢/\text{oz}$

$\begin{array}{r} 12.0 \\ 24{\overline{\smash{\big)}\,289.0}} \\ \underline{24} \\ 49 \\ \underline{48} \\ 1\,0 \end{array}$

40. Let n = number of pounds.

$\dfrac{1}{120} = \dfrac{n}{480}$

$120n = 1 \cdot 480$

$n = \dfrac{4800}{120} = 40$

You need 40 lb.

41. Let n = number of ounces.

$\dfrac{2}{4} = \dfrac{n}{48}$

$4n = 2 \cdot 48$

$n = \dfrac{2 \cdot \cancel{48}^{12}}{\cancel{4}} = 24$

You need 24 ounces.

42. Let n = number of shares.

$\dfrac{80}{87.50} = \dfrac{n}{875}$

$87.50n = 80 \cdot 875$

$n = \dfrac{80 \cdot \cancel{875}^{10}}{\cancel{87.50}} = 800$

You can buy 800 shares.

Chapter 5

Percent

Section 5.1 – Percent Notation

Problems

1. a. $47\% = .47 = 0.47$
  b. $493\% = 4.93$

2. a. $2\frac{1}{8}\% = 2.125\% = .02125 = 0.02125$
  b. $5\frac{1}{3}\% = 5.\overline{3}\% = .05\overline{3} = 0.05\overline{3}$

3. a. $0.07 = 007.\% = 7\%$
  b. $3.14 = 314.\% = 314\%$
  c. $71.8 = 7180.\% = 7180\%$

4. a. $41\% = \frac{41}{100}$
  b. $25\% = \frac{25}{100} = \frac{1}{4}$

5. a. $3\frac{1}{3}\% = \frac{3\frac{1}{3}}{100} = \frac{\frac{10}{3}}{100} = \frac{\cancel{10}}{3} \cdot \frac{1}{\cancel{100}} = \frac{1}{30}$
  b. $4.5\% = \frac{4.5}{100} = \frac{4.5 \times 10}{100 \times 10} = \frac{45}{1000} = \frac{9}{20}$

6. $\frac{2}{5} = \frac{2 \cdot 20}{5 \cdot 20} = \frac{40}{100} = 40\%$

7. $\frac{3}{16} = 16\overline{\smash{)}3.00}$

$\phantom{7.\quad \frac{3}{16} = 16)}0.18$
$\underline{1\ 6}$
$1\ 40$
$\underline{1\ 28}$
12

Thus $\frac{3}{16} = 0.18\frac{12}{16} = 0.18\frac{3}{4} = 18\frac{3}{4}\%$.

8. a. $30\% = \frac{30}{100} = \frac{3}{10}$
  b. $30\% = .30 = 0.30$

9. a. The fraction that can sit in the Guadalajara Room is $\frac{45}{200} = \frac{9}{40}$.

  b. $\frac{9}{40} = 40\overline{\smash{)}9.00}$

0.22
$\underline{8\ 0}$
$\overline{1\ 00}$
$\underline{80}$
$\overline{20}$

  Thus $\frac{9}{40} = 0.22\frac{20}{40}$
  $ = 0.22\frac{1}{2} = 22\frac{1}{2}\%$.

  c. $\frac{9}{40} = 22\frac{1}{2}\% = 22.5\% = 0.225$

Exercises 5.1

1. $3\% = .03 = 0.03$

3. $10\% = .10 = 0.10$

5. $300\% = 3.00 = 3$

7. $12\frac{1}{4}\% = 12.25\% = .1225 = 0.1225$

9. $11.5\% = .115 = 0.115$

11. $0.3\% = .003 = 0.003$

13. $0.04 = 4.\% = 4\%$

15. $0.813 = 81.3\%$

17. $3.14 = 314.\% = 314\%$

19. $1.00 = 100.\% = 100\%$

21. $0.002 = 000.2\% = 0.2\%$

23. $30\% = \dfrac{3\cancel{0}}{10\cancel{0}} = \dfrac{3}{10}$

25. $6\% = \dfrac{6}{100} = \dfrac{3}{50}$

27. $7\% = \dfrac{7}{100}$

29. $4\dfrac{1}{2}\% = \dfrac{4\frac{1}{2}}{100} = \dfrac{\frac{9}{2}}{100} = \dfrac{9}{2} \cdot \dfrac{1}{100} = \dfrac{9}{200}$

31. $1\dfrac{1}{3}\% = \dfrac{1\frac{1}{3}}{100} = \dfrac{\frac{4}{3}}{100} = \dfrac{\cancel{4}}{3} \cdot \dfrac{1}{\cancel{100}_{25}} = \dfrac{1}{75}$

33. $3.4\% = \dfrac{3.4}{100} = \dfrac{3.4 \times 10}{100 \times 10} = \dfrac{34}{1000} = \dfrac{17}{500}$

35. $10.5\% = \dfrac{10.5}{100} = \dfrac{105}{1000} = \dfrac{\cancel{5} \cdot 21}{\cancel{5} \cdot 200} = \dfrac{21}{200}$

37. $\dfrac{3}{5} = \dfrac{3 \cdot 20}{5 \cdot 20} = \dfrac{60}{100} = 60\%$

39. $\dfrac{1}{2} = \dfrac{1 \cdot 50}{2 \cdot 50} = \dfrac{50}{100} = 50\%$

41. $\dfrac{5}{6} = 6{\overline{\smash{\big)}\,5.00}}$

$$\begin{array}{r} 0.83 \\ 6{\overline{\smash{\big)}\,5.00}} \\ \underline{4\,8} \\ 20 \\ \underline{18} \\ 2 \end{array}$$

Thus $\dfrac{5}{6} = 0.83\dfrac{2}{6} = 0.83\dfrac{1}{3} = 83\dfrac{1}{3}\%$

43. $\dfrac{3}{8} = 8{\overline{\smash{\big)}\,3.00}}$

$$\begin{array}{r} 0.37 \\ 8{\overline{\smash{\big)}\,3.00}} \\ \underline{2\,4} \\ 60 \\ \underline{56} \\ 4 \end{array}$$

Thus $\dfrac{3}{8} = 0.37\dfrac{4}{8} = 0.37\dfrac{1}{2} = 37\dfrac{1}{2}\%$.

45. $\dfrac{4}{3} = 3{\overline{\smash{\big)}\,4.00}}$

$$\begin{array}{r} 1.33 \\ 3{\overline{\smash{\big)}\,4.00}} \\ \underline{3} \\ 1\,0 \\ \underline{9} \\ 10 \\ \underline{9} \\ 1 \end{array}$$

Thus $\dfrac{4}{3} = 1.33\dfrac{1}{3} = 133\dfrac{1}{3}\%$.

47. $\dfrac{81}{100} = 81\%$

49. $\dfrac{3}{20} = \dfrac{3 \cdot 5}{20 \cdot 5} = \dfrac{15}{100} = 15\%$

51. a. $41\% = \dfrac{41}{100}$ and $41\% = 0.41$

b. $33\% = \dfrac{33}{100}$ and $33\% = 0.33$

c. $32\% = \dfrac{32}{100} = \dfrac{8}{25}$ and $32\% = 0.32$

d. $31\% = \dfrac{31}{100}$ and $31\% = 0.31$

53. a. $44\% = \dfrac{44}{100} = \dfrac{11}{25}$ and $44\% = 0.44$

b. $20\% = \dfrac{20}{100} = \dfrac{1}{5}$ and $20\% = 0.20$

c. $15\% = \dfrac{15}{100} = \dfrac{3}{20}$ and $15\% = 0.15$

55. a. $\dfrac{165}{280} = \dfrac{\cancel{5} \cdot 33}{\cancel{5} \cdot 56} = \dfrac{33}{56}$

b. $\dfrac{33}{56} = 56{\overline{\smash{\big)}\,33.000}}$; $0.588 = 58.8\% \approx 59\%$

$$\begin{array}{r} 0.588 \\ 56{\overline{\smash{\big)}\,33.000}} \\ \underline{28\,0} \\ 5\,00 \\ \underline{4\,48} \\ 520 \\ \underline{448} \\ 72 \end{array}$$

c. $59\% = 0.59$

57. a. $\dfrac{66}{165} = \dfrac{2 \cdot \cancel{33}}{5 \cdot \cancel{33}} = \dfrac{2}{5}$

b. $\dfrac{2}{5} = \dfrac{2 \cdot 20}{5 \cdot 20} = \dfrac{40}{100} = 40\%$

c. $\dfrac{2}{5} = 0.4$

59. a. $\dfrac{1\cancel{00}}{48\cancel{00}} = \dfrac{1}{48}$

b. $\dfrac{1}{48} = 48\overline{)1.000}^{\;0.020} \approx 2\%$
$\underline{96}$
40

c. $2\% = 0.02$

61. a. $\dfrac{85}{4800} = \dfrac{\cancel{5} \cdot 17}{\cancel{5} \cdot 960} = \dfrac{17}{960}$

b. $\dfrac{17}{960} = 960\overline{)17.000}^{\;0.017} \approx 1.7\%$ or 2%
$\underline{9\,60}$
$7\,400$
$\underline{6\,720}$
680

c. $\dfrac{17}{960} = 0.017 \dfrac{680}{960} = 0.017 \dfrac{17}{24} \approx 0.018$

63. a. $\dfrac{42\cancel{00}}{14{,}0\cancel{00}} = \dfrac{42}{140} = \dfrac{3 \cdot \cancel{14}}{10 \cdot \cancel{14}} = \dfrac{3}{10}$

b. $\dfrac{3}{10} = \dfrac{3 \cdot 10}{10 \cdot 10} = \dfrac{30}{100} = 30\%$

c. $30\% = 0.30$

65. $0.30x = 12$
$x = \dfrac{12}{0.30}$
$x = \dfrac{120}{3.0} = \dfrac{120}{3} = 40$

67. $15 = 45x$
$\dfrac{15}{45} = x$
$\dfrac{1}{3} = x$

69. $\dfrac{7}{10} = \dfrac{70}{100} = 70\%$

71. a. $0.40 = \dfrac{40}{100} = 40\%$

b. $0.40 = \dfrac{4\cancel{0}}{10\cancel{0}} = \dfrac{4}{10} = \dfrac{2}{5}$

73. a. $49\% = \dfrac{49}{100}$

b. $49\% = .49 = 0.49$

Marital Status	Number	Percent
Single	25,480	49%
75. Remarried	2,600	5%
77. Widowed	520	1%

79. Answers may vary.

Mastery Test 5.1

81. a. $82\% = \dfrac{82}{100} = \dfrac{41}{100}$

b. $82\% = 0.82$

83. a. $6\dfrac{1}{2}\% = \dfrac{6\frac{1}{2}}{100} = \dfrac{\frac{13}{2}}{100} = \dfrac{13}{2} \cdot \dfrac{1}{100} = \dfrac{13}{200}$

b. $6.55\% = \dfrac{6.55}{100} = \dfrac{655}{10{,}000} = \dfrac{131}{2000}$

85. a. $0.06 = 6\%$
b. $6.19 = 619\%$
c. $42.2\% = 4220\%$

87. a. $38\% = 0.38$
b. $29.3\% = 0.293$

Section 5.2 – Percent Problems

Problems

1. Let P = the number.
$$70\% \times 40 = P$$
$$0.70 \times 40 = P$$
$$28 = P$$
Thus 70% of 40 is 28.

2. Let P = the number.
$$8\frac{1}{2}\% \times 60 = P$$
$$0.08\frac{1}{2} \times 60 = P$$
$$.085 \times 60 = P$$
$$5.1 = P$$
Thus $8\frac{1}{2}\%$ of 60 is 5.1.

3. Let P = the number.
$$P = 66\frac{2}{3}\% \times 90$$
$$P = \frac{2}{3} \times 90 = P$$
$$P = 60$$
Thus $66\frac{2}{3}\%$ of 90 is 60.

4. Let R = the percent.
$$R \times 30 = 3$$
$$R = \frac{3}{30} = \frac{1}{10} = \frac{1 \cdot 10}{10 \cdot 10} = \frac{10}{100} = 10\%$$
Thus 10% of 300 is 3.

5. Let B = the number.
$$50 = 20\% \times B$$
$$50 = 0.20 \times B$$
$$\frac{50}{0.20} = B$$
$$\frac{500}{2} = B$$
$$250 = B$$
Thus 50 is 20% of 250.

6. Let B = the price.
$$3.28 = 8\% \times B$$
$$3.28 = 0.08 \times B$$
$$\frac{3.28}{0.08} = B$$
$$\frac{328}{8} = B$$
$$41 = B$$
The price was $41.

7. a. $0.80 \times 7000 = 5600$
$7000 - 5600 = 1400$
The scholarship pays $5600 and Doug pays $1400.

   b. $\dfrac{10\cancel{00}}{14\cancel{00}} = \dfrac{5}{7} \approx 71\%$

   c. $25\% \times T = 1400$
$0.25 \times T = 1400$
$$T = \frac{1400}{0.25} = \frac{140,000}{25} = 5600$$
It costs the state $5600 per student.

Exercises 5.2

1. Let P = the number.
$$40\% \times 80 = P$$
$$0.40 \times 80 = P$$
$$32 = P$$
Thus 40% of 80 is 32.

3. Let P = the number.
$$150\% \times 8 = P$$
$$1.50 \times 8 = P$$
$$12 = P$$
Thus 150% of 8 is 12.

5. Let P = the amount.
$$20\% \times \$15 = P$$
$$0.20 \times 15 = P$$
$$3 = P$$
Thus 20% of $15 is $3.

7. Let P = the number.
$$60\% \times 48 = P$$
$$0.60 \times 48 = P$$
$$28.8 = P$$
Thus 60% of 48 is 28.8.

9. Let P = the number.
$$12\frac{1}{2}\% \times 40 = P$$
$$0.12\frac{1}{2} \times 40 = P$$
$$.125 \times 40 = P$$
$$5 = P$$
Thus $12\frac{1}{2}\%$ of 40 is 5.

11. Let P = the number.
$$3.5\% \times 60 = P$$
$$0.035 \times 60 = P$$
$$2.1 = P$$
Thus 3.5% of 60 is 2.1.

13. Let P = the number.
$$P = 16\frac{2}{3}\% \times 120$$
$$P = \frac{1}{6} \times 120 = P$$
$$P = 20$$
Thus $16\frac{2}{3}\%$ of 120 is 20.

15. Let R = the percent.
$$315 = R \times 3150$$
$$\frac{315}{3150} = R$$
$$\frac{1}{10} = R \text{ so } R = \frac{1 \cdot 10}{10 \cdot 10} = \frac{10}{100} = 10\%$$
Thus 315 is 10% of 3150.

17. Let R = the percent.
$$8 = R \times 4$$
$$\frac{8}{4} = R$$
$$2 = R \text{ so } R = 200\%$$
Thus 8 is 200% of 4.

19. Let R = the percent.
$$R \times 50 = 5$$
$$R = \frac{5}{50} = \frac{1}{10} = 10\%$$
Thus 10% of 50 is 5.

21. Let R = the percent.
$$R \times 40 = 5$$
$$R = \frac{5}{40} = \frac{1}{8} = 0.125 = 12\frac{1}{2}\%$$
Thus $12\frac{1}{2}\%$ of 40 is 5.

23. Let R = the percent.
$$5\frac{1}{2} = R \times 22$$
$$\frac{11}{2} = R \times 22$$
$$\frac{\frac{11}{2}}{22} = R \text{ so}$$
$$R = \frac{\cancel{11}}{2} \cdot \frac{1}{\cancel{22}_2} = \frac{1}{4} = 0.25 = 25\%$$
Thus $5\frac{1}{2}$ is 25% of 22.

25. Let R = the percent.
$$3 = R \times 5$$
$$\frac{3}{5} = R \text{ so } R = \frac{3 \cdot 20}{5 \cdot 20} = \frac{60}{100} = 60\%$$
Thus 3 is 60% of 5.

27. Let R = the percent.
$$50 = R \times 25$$
$$\frac{50}{25} = R \text{ so } R = 2 = 200\%$$
Thus 50 is 200% of 25.

29. Let R = the percent.
$$\$50 = R \times \$60$$
$$\frac{50}{60} = R$$
$$\frac{5}{6} = R \text{ so}$$
$$R = \frac{5}{6} = 0.8\overline{3} = 83\frac{1}{3}\%$$
Thus \$50 is $83\frac{1}{3}\%$ of \$60.

31. Let R = the percent.

$$\$0.75 = R \times \$4.50$$

$$\frac{0.75}{4.50} = R$$

$$\frac{75}{450} = R$$

$$\frac{1}{6} = R \text{ so}$$

$$R = \frac{1}{6} = 0.16\overline{6} = 16\frac{2}{3}\%$$

Thus $0.75 is $16\frac{2}{3}\%$ of $4.50.

33. Let B = the number.

$$30\% \times B = 60$$

$$0.30 \times B = 60$$

$$B = \frac{60}{0.30} = \frac{600}{3} = 200$$

The number is 200.

35. Let B = the number.

$$2\frac{1}{4}\% \times B = 9$$

$$0.0225 \times B = 9$$

$$B = \frac{9}{0.0225} = \frac{90,000}{225} = 400$$

The number is 400.

37. Let B = the number.

$$20 = 40\% \times B$$

$$20 = 0.40 \times B$$

$$\frac{20}{0.40} = B \text{ so } B = \frac{200}{4} = 50$$

The number is 50.

39. Let B = the number.

$$15 = 33\frac{1}{3}\% \times B$$

$$15 = \frac{1}{3} \times B$$

$$45 = B$$

The number is 45.

41. Let B = the number.

$$120\% \times B = 20$$

$$1.20 \times B = 20$$

$$B = \frac{20}{1.20} = \frac{200}{12} = \frac{50}{3} = 16\frac{2}{3}$$

The number is $16\frac{2}{3}$.

43. Let B = the number.

$$40\% \times B = 7\frac{1}{2}$$

$$0.40 \times B = 7.5$$

$$B = \frac{7.5}{0.40} = \frac{750}{4} = \frac{375}{2} = 18\frac{3}{4}$$

The number is $18\frac{3}{4}$.

45. Let B = the number.

$$4.75\% \times B = 38$$

$$0.0475 \times B = 38$$

$$B = \frac{38}{0.0475} = \frac{380,000}{475} = 800$$

The number is 800.

47. Let B = the number.

$$100\% \times B = 3$$

$$1 \times B = 3$$

$$B = 3$$

The number is 3.

49. Let B = the number.

$$12\frac{1}{2}\% \times B = 37.5$$

$$0.125 \times B = 37.5$$

$$B = \frac{37.5}{0.125} = \frac{37,500}{125} = 300$$

The number is 300.

51. a. $11\% \times 280 = 0.11 \times 280 = 30.8$

About 31 execs had cell phones in 2002.

b. $77\% \times 280 = 0.77 \times 280 = 215.6$

About 216 execs had cell phones in 1988.

53. $57\% \times 40 = 0.57 \times 40 = 22.8$

You would expect 23 of the 40 members to clip coupons.

55. Let B = the number of people surveyed.

$$38\% \times B = 95$$

$$0.38 \times B = 95$$

$$B = \frac{95}{0.38} = \frac{9500}{38} = 250$$

There were 250 people surveyed.

57. a. Let R = the percent.
$$R \times 280 = 112$$
$$R = \frac{112}{280} = \frac{2}{5} = 0.4 = 40\%$$
Thus 40% of the 280 is 112.
b. This is more than the graph predicted.

59. Let B = number of parents surveyed.
$$153 = 17\% \times B$$
$$153 = 0.17 \times B$$
$$\frac{153}{0.17} = B \text{ so } B = \frac{15,300}{17} = 900$$
There were 900 parents surveyed.

61. Let B = world's oil reserves.
$$5\% \times B = 50$$
$$0.05 \times B = 50$$
$$B = \frac{50}{0.05} = \frac{5000}{5} = 1000$$
The world's oil reserves are 1000 billion barrels.

63. Let B = world's coal reserves.
$$30\% \times B = 300,000$$
$$0.30 \times B = 300,000$$
$$B = \frac{300,000}{0.30}$$
$$= \frac{3,000,000}{3} = 1,000,000$$
The world's coal reserves are 1,000,000 short tons.

65. Let P = population increase.
$$1.8\% \times 1 = P$$
$$0.018 \times 1 = P$$
$$0.018 = P$$
The population would increase by 0.018 billion (18 million) people the first year.

67. Let R = the percent.
$$R \times 190 = 50$$
$$R = \frac{50}{190} = \frac{5}{19} \approx 0.263$$
That is about 26%.

69. $48\% \times 16 = 0.48 \times 16 = 7.68$ oz of sugar
$41\% \times 16 = 0.41 \times 16 = 6.56$ oz of water

71. $\dfrac{20}{100} = \dfrac{n}{80}$
$$100n = 20 \cdot 80$$
$$n = \frac{2\cancel{0} \cdot 8\cancel{0}}{1\cancel{0}\cancel{0}} = 16$$

73. $\dfrac{40}{100} = \dfrac{20}{W}$
$$40W = 100 \cdot 20$$
$$W = \frac{100 \cdot \cancel{20}}{\cancel{40}_2} = 50$$

75.

	Interest
Month 1	$0.03 \times \$200 = \$6.00$
Month 2	$0.03 \times\ \ 150 =\ \ 4.50$
Month 3	$0.03 \times\ \ 100 =\ \ 3.00$
Month 4	$0.03 \times\ \ \ \ 50 =\ \ 1.50$
	Total Interest $= \$15.00$

You pay $15 in total interest.

77. Let R = the percent.
$$R \times 200 = 15$$
$$R = \frac{15}{200} = \frac{3}{40} = 0.075 = 7.5\%$$
The total interest is $7\frac{1}{2}\%$ of the $200.

79. Three (percentage, base, and percent).
Answers may vary.

Mastery Test 5.2

81. $53\% \times 1500 = 0.53 \times 1500 = 795$
The amount paid by financial aid is $795.
$1500 - 795 = 705$
You have to pay $705.

83. Let B = the number.
$$20 = 40\% \times B$$
$$20 = 0.40 \times B$$
$$\frac{20}{0.40} = B \text{ so } B = \frac{200}{4} = 50$$
Thus 20 is 40% of 50.

85. $33\frac{1}{3}\% \times 180 = \frac{1}{3} \times 180 = 60$

87. Let P = the percentage.

$$12\frac{1}{2}\% \times 160 = P$$
$$0.125 \times 160 = P$$
$$20 = P$$

Thus $12\frac{1}{2}\%$ of 160 = 20.

Section 5.3 – Solving Percent Problems Using Proportions

Problems

1. $\dfrac{30}{100} = \dfrac{n}{90}$

$100n = 30 \cdot 90$

$n = \dfrac{2700}{100} = 27$

Thus, 30% of 90 is 27.

2. $\dfrac{\text{Percent}}{100} = \dfrac{15}{45}$

$\text{Percent} \times 45 = 100 \cdot 15$

$\text{Percent} = \dfrac{100 \cdot 15}{45_3} = \dfrac{100}{3} = 33\frac{1}{3}$

Thus, $33\frac{1}{3}\%$ of 45 is 15.

3. $\dfrac{40}{100} = \dfrac{50}{\text{Whole}}$

$40 \times \text{Whole} = 100 \cdot 50$

$\text{Whole} = \dfrac{\overset{5}{100} \cdot \overset{25}{50}}{\underset{2}{40}} = 125$

Thus, 50 is 40% of 125.

Exercises 5.3

1. $\dfrac{30}{100} = \dfrac{P}{80}$

$100P = 30 \cdot 80$

$P = \dfrac{2400}{100} = 24$

Thus 30% of 80 is 24.

3. $\dfrac{150}{100} = \dfrac{P}{20}$

$100P = 150 \cdot 20$

$P = \dfrac{3000}{100} = 30$

Thus 150% of 20 is 30.

5. $\dfrac{20}{100} = \dfrac{P}{30}$

$100P = 20 \cdot 30$

$P = \dfrac{600}{100} = 6$

Thus 20% of $30 is $6.

7. $\dfrac{60}{100} = \dfrac{P}{45}$

$100P = 60 \cdot 45$

$P = \dfrac{2700}{100} = 27$

Thus 60% of 45 is 27.

9. $\dfrac{12\frac{1}{2}}{100} = \dfrac{P}{60}$

$100P = 12\frac{1}{2} \cdot 60$

$100P = \dfrac{25}{2} \cdot \overset{30}{60}$

$100P = 750$

$P = \dfrac{750}{100} = 7.5$

Thus $12\frac{1}{2}\%$ of 60 is 7.5.

11. $\dfrac{3.5}{100} = \dfrac{P}{60}$

$100P = 3.5 \cdot 60$

$P = \dfrac{210}{100} = 2.1$

Thus 3.5% of 60 is 2.1.

13. $\dfrac{16\frac{2}{3}}{100} = \dfrac{P}{132}$

$100P = 16\frac{2}{3} \cdot 132$

$100P = \dfrac{50}{\cancel{3}} \cdot \cancel{132}^{44}$

$P = \dfrac{2200}{100} = 22$

Thus $16\frac{2}{3}$% of 132 is 22.

15. $\dfrac{R}{100} = \dfrac{325}{3250}$

$3250R = 100 \cdot 325$

$R = \dfrac{32,500}{3250} = 10$

Thus 325 is 10% of 3250.

17. $\dfrac{R}{100} = \dfrac{16}{4}$

$4R = 100 \cdot 16$

$R = \dfrac{1600}{4} = 400$

Thus 16 is 400% of 4.

19. $\dfrac{R}{100} = \dfrac{10}{50}$

$50R = 100 \cdot 10$

$R = \dfrac{^2\cancel{100} \cdot 10}{\cancel{50}} = 20$

Thus 10 is 20% of 50.

21. $\dfrac{R}{100} = \dfrac{8}{40}$

$40R = 100 \cdot 8$

$R = \dfrac{80\cancel{0}}{4\cancel{0}} = 20$

Thus 8 is 20% of 40.

23. $\dfrac{R}{100} = \dfrac{6\frac{1}{2}}{26}$

$26R = 100 \cdot 6\frac{1}{2}$

$26R = \cancel{100}^{50} \cdot \dfrac{13}{\cancel{2}}$

$R = \dfrac{50 \cdot \cancel{13}}{\cancel{26}_2} = 25$

Thus $6\frac{1}{2}$ is 25% of 26.

25. $\dfrac{R}{100} = \dfrac{4}{5}$

$5R = 400$

$R = \dfrac{400}{5} = 80$

Thus 4 is 80% of 5.

27. $\dfrac{R}{100} = \dfrac{50}{25}$

$25R = 100 \cdot 50$

$R = \dfrac{100 \cdot \cancel{50}^2}{\cancel{25}} = 200$

Thus 50 is 200% of 25.

29. $\dfrac{R}{100} = \dfrac{50}{60}$

$60R = 100 \cdot 50$

$R = \dfrac{^5\cancel{100} \cdot 50}{\cancel{60}_3} = \dfrac{250}{3} = 83\frac{1}{3}$

Thus $50 is $83\frac{1}{3}$% of $60.

31. $\dfrac{R}{100} = \dfrac{0.75}{4.50}$

$4.50R = 100 \cdot 0.75$

$R = \dfrac{75}{4.50} = \dfrac{750}{45} = \dfrac{50}{3} = 16\frac{2}{3}$

Thus $0.75 is $16\frac{2}{3}$% of $4.50.

33. $\dfrac{30}{100} = \dfrac{90}{B}$

$30B = 100 \cdot 90$

$B = \dfrac{9000}{30} = 300$

Thus 30% of 300 is 90.

35. $\dfrac{2\frac{1}{4}}{100} = \dfrac{18}{B}$

$2\frac{1}{4} \cdot B = 100 \cdot 18$

$\dfrac{9}{4} \cdot B = 1800$

$B = \cancel{1800}^{\,200} \cdot \dfrac{4}{\cancel{9}} = 800$

Thus $2\frac{1}{4}\%$ of 800 is 18.

37. $\dfrac{40}{100} = \dfrac{60}{B}$

$40B = 100 \cdot 60$

$B = \dfrac{6000}{40} = 150$

Thus 60 is 40% of 150.

39. $\dfrac{33\frac{1}{3}}{100} = \dfrac{30}{B}$

$33\frac{1}{3} \cdot B = 100 \cdot 30$

$\dfrac{100}{3} B = 3000$

$B = 3000 \cdot \dfrac{3}{100} = 90$

The number is 90.

41. $\dfrac{120}{100} = \dfrac{42}{B}$

$120B = 100 \cdot 42$

$B = \dfrac{4200}{120} = 35$

The number is 35.

43. $\dfrac{40}{100} = \dfrac{7\frac{1}{2}}{B}$

$40B = 100 \cdot 7\frac{1}{2}$

$40B = \cancel{100}^{\,50} \cdot \dfrac{15}{\cancel{2}}$

$B = \dfrac{750}{40} = \dfrac{75}{4} = 18.75$

Thus 40% of $18\frac{3}{4}$ is $7\frac{1}{2}$.

45. $\dfrac{4.75}{100} = \dfrac{76}{B}$

$4.75B = 100 \cdot 76$

$B = \dfrac{7600}{4.75} = \dfrac{760{,}000}{475} = 1600$

Thus 4.75% of 1600 is 76.

47. $\dfrac{100}{100} = \dfrac{49}{B}$

$100B = 100 \cdot 49$

$B = 49$

Thus 100% of 49 is 49.

49. $\dfrac{12\frac{1}{2}}{100} = \dfrac{37.5}{B}$

$12\frac{1}{2} \cdot B = 100 \cdot 37.5$

$\dfrac{25}{2} B = 3750$

$B = \cancel{3750}^{\,150} \cdot \dfrac{2}{\cancel{25}} = 300$

Thus $12\frac{1}{2}\%$ of 300 is 37.5.

51. $5\frac{1}{2}\% = 5.5\% = \dfrac{5.5}{100} = 0.055$

53. $16\frac{2}{3}\% = 16.\overline{6}\% = \dfrac{16.\overline{6}}{100} = 0.16\overline{6}$

55. $\dfrac{24}{h} > \dfrac{8}{10}$

$24 \cdot 10 > 8h$

$\dfrac{\overset{3}{\cancel{24}} \cdot 10}{\cancel{8}} > h$

$30 > h$

A hip size less than 30 inches will make her risk of heart disease increase.

57. $\dfrac{32}{h} > \dfrac{1}{1}$

$32 > h$

A hip size less than 32 inches will make his risk of heart disease increase.

59. Answers may vary.

Mastery Test 5.3

61.
$$\frac{40}{100} = \frac{P}{80}$$
$$100P = 40 \cdot 80$$
$$P = \frac{3200}{100} = 32$$
Thus 40% of 80 is 32.

63.
$$\frac{3\frac{1}{2}}{100} = \frac{21}{B}$$
$$3\frac{1}{2} \cdot B = 100 \cdot 21$$
$$\frac{7}{2}B = 2100$$
$$B = \cancel{2100}^{\,300} \cdot \frac{2}{\cancel{7}} = 600$$

Section 5.4 – Taxes, Interest, Commissions, and Discounts

Problems

1. $5\% \times 33 = 0.05 \times 33 = 1.65$ is the tax;
$33 + 1.65 = 34.65$ is total cost.
You would pay $34.65.

2. $I = P \times R \times T$
$$= 1200 \times 8.5\% \times \frac{1}{3}$$
$$= 1200 \times 0.085 \times \frac{1}{3}$$
$$= \$34$$

3. Here P = \$100, r = 6%, and the
semiannual rate is $\frac{6\%}{2} = 3\% = 0.03$.

$100 \times .03 = \$3.00$
$103 \times .03 = \$3.09$
$106.09 \times .03 = \$3.1827$
$109.27 \times .03 = \$3.2781$
The total compound interest is
$\$3 + \$3.09 + \$3.18 + \$3.28 = \$12.55$, and
the final compound amount is
$\$100 + \$12.55 = \$112.55$.

4. $P = \$10,000,\ i = \frac{8\%}{2} = 4\% = 0.03,$ and
$n = 2 \cdot 2 = 4$ periods.
$A = \$10,000(1 + 0.04)^4$
$= \$10,000(1.04)^4$
$= \$10,000(1.04 \times 1.04 \times 1.04 \times 1.04)$
$= \$10,000(1.1698586)$
$= \$11,698.59$
The interest is $I = A - P$
$= \$11,698.59 - \$10,000$
$= \$1,698.59$.

5. Comission $= 6\% \times 1599.95$
$= 0.06 \times 1599.95$
$= \$95.997$
$= \$96$

6. Discount: $20\% \times 8.50 = 0.20 \times 8.50$
$= \$1.70$
Sale price: $\$8.50 - \$1.70 = \$6.80$

7. $0.20 \times 60.70 = \$12.14$

Exercises 5.4

1. Tax $= 0.04 \times 3500 = \$140$
Total cost $= 3500 + 140 = \$3640$

3. Tax $= 0.04 \times 12.5 = \$0.50$
Total cost $= 12.50 + 0.50 = \$13.00$

5. $0.05 \times 13,200 = 660$
The tax was $660.

7. Let R = sales tax rate.
$R \times 1500 = 60$
$$R = \frac{60}{1500} = \frac{1}{25} = \frac{4}{100} = 4\%$$
The sales tax rate was 4%.

In Problems 9-16, use the formula
$I = P \times R \times T$ **to obtain the interest**
answers.

	Principal	Annual Interest Rate	Time	*Interest*
9.	$200	12%	1 yr	*$24*
11.	$400	15%	$\frac{1}{2}$ yr	*$30*
13.	$1500	15%	2 mo	*$37.50*
15.	$300	9%	60 days	*$4.50*

17. a. $I = P \times R \times T$
$= 500 \times 0.20 \times 1 = \$100$
You would earn $100 interest.
b. $I = P \times R \times T$
$= 2500 \times 0.20 \times 2 = \$1000$
You would pay $1000 interest.

19. $I = P \times R \times T$
$= 1000 \times 0.08 \times 2 = \$160$
She received $160 interest in two years.

21. $P = \$5000, \; i = \dfrac{10\%}{2} = 5\% = 0.05,$ and
there are $2 \cdot 2 = 4$ periods.
$A = P(1 + i)^n$
$= 5000(1 + 0.05)^4$
$= 5000(1.05)^4$
$= 5000(1.05 \times 1.05 \times 1.05 \times 1.05)$
$= 5000(1.2155062)$
$= 6077.53125$
The amount at the end of the period will
be $6077.53.

23. $P = \$50,000, \; i = \dfrac{10\%}{2} = 5\% = 0.05,$ and
there are $2 \cdot 2 = 4$ periods.
$A = P(1 + i)^n$
$= 50,000(1 + 0.05)^4$
$= 50,000(1.05)^4$
$= 50,000(1.05 \times 1.05 \times 1.05 \times 1.05)$
$= 50,000(1.2155062)$
$= 60775.3125$
He should receive $60,775.31.

25. $P = \$500, \; i = \dfrac{12\%}{12} = 1\% = 0.01, \; n = 2$
$A = 500(1 + 0.01)^2$
$= 500(1.01)^2$
$= 500(1.0201) = \$510.05$
His balance was $510.05.

27. Commission $= 8\% \times \$3500$
$= 0.08 \times \$3500 = \$280$

29. Commission $= 2\% \times \$6300$
$= 0.02 \times \$6300 = \$126$
Total Salary $= \$500 + \$126 = \$626$

31. Discount $= 10\% \times \$14.50$
$= 0.10 \times \$14.50 = \$1.45$
Sale price $= \$14.50 - \$1.45 = \$13.05$
You have to pay $13.05 for it now.

33. Discount $= 10\% \times \$460$
$= 0.10 \times \$460 = \$46$
New price $= \$460 - \$46 = \$414$

35. Discount $= 2\% \times \$250$
$= 0.02 \times \$250 = \$5$

37. $25\% \times \$80.30 = 0.25 \times \$80.30 = \$20.075$
The tip was $20.08.

39. $15\% \times \$82.50 = 0.15 \times \$82.50 = \$12.375$
The service charge is $12.38.

41. Let $R =$ the percent.
$R \times 48 = 80$
$R = \dfrac{80}{48} = \dfrac{5}{3} = 1\dfrac{2}{3} = 1.6\overline{6} = 166.\overline{6}\%$
Thus about 167% of 48 is 80.

43. Let $R =$ the percent.
$R \times 72 = 40$
$R = \dfrac{40}{72} = \dfrac{5}{9} = 0.55\overline{5} = 55.\overline{5}\%$
Thus about 56% of 72 is 40.

45. Let $R =$ the percent.
$25 = R \times 30$
$\dfrac{25}{30} = R$ so $R = \dfrac{5}{6} = 0.83\overline{3} = 83.\overline{3}\%$
Thus 25 is about 83% of 30.

47. Tax $= \$1400 + 0.15 \times (25,000 - 14,000)$
$= \$1400 + 0.15 \times \$11,000$
$= \$1400 + \$1650$
$= \$3050$

49. Tax $= \$7820 + 0.25 \times (60,000 - 56,800)$
$= \$7820 + 0.25 \times \$3200$
$= \$7820 + \$800$
$= \$8620$

51. Tax $= \$7820 + 0.25 \times (100,000 - 56,800)$
$= \$7820 + 0.25 \times \$43,200$
$= \$7820 + \$10,800$
$= \$18,620$

53. Taking 20% off would be better. Answers may vary.

55. a. Move the decimal point left one place.
b. Answers may vary.
c. Answers may vary.

Mastery Test 5.4

57. Discount $= 12\% \times \$80$
$= 0.12 \times \$80 = \$9.60$
Sale price $= \$80 - \$9.60 = \$70.40$

59. $P = \$10,000, i = \dfrac{4\%}{2} = 2\% = 0.02,$ and
there are $2 \cdot 2 = 4$ periods.
$A = P(1 + i)^n$
$= 10,000(1 + 0.02)^4$
$= 10,000(1.02)^4$
$= 10,000(1.02 \times 1.02 \times 1.02 \times 1.02)$
$= 10,000(1.0824322)$
$= 10,824.322$
and $I = A - P$
$= \$10,824.32 - \$10,000$
$= \$824.32$
The compound amount is \$10,824.32 and the compound interest is \$824.32.

61. Tax $= 7.25\% \times \$120$
$= 0.0725 \times \$120 = \$8.70$
Total cost $= \$120 + \$8.70 = \$128.70$

Section 5.5 – Applications: Percent of Increase or Decrease

Problems

1. Amount of increase: $4.3 - 3 = 1.3$

$$\dfrac{1.3}{3} = 3\overline{\smash{)}1.300}$$
$$\begin{array}{r} 0.433 \\ \underline{1\,2} \\ \overline{10} \\ \underline{9} \\ \overline{10} \\ \underline{9} \\ \overline{1} \end{array}$$

The percent of increase is about 43%.

2. Amount of decrease: $9606 - 8921 = 685$

$$\dfrac{685}{9606} = 9606\overline{\smash{)}685.000}$$
$$\begin{array}{r} 0.071 \\ \underline{672\,42} \\ \overline{12\,580} \\ \underline{9\,606} \\ \overline{2\,974} \\ \underline{9} \\ \overline{1} \end{array}$$

The market declined by 7%.

3. $95 + 21\% \times 95 = 95 + 0.21 \times 95$
$= 95 + 19.95$
$= 114.95$
About \$115 billion will be spent in two years.

4. Amount of increase:
$54,500 - 18,900 = 35,600$

$$\frac{35,600}{18,900} = \frac{356}{189} = 189\overline{)356.000}$$

$$\begin{array}{r} 1.883 \\ \hline 189 \\ \hline 167\,0 \\ 151\,2 \\ \hline 15\,80 \\ 15\,12 \\ \hline 680 \\ 567 \\ \hline 113 \end{array}$$

The percent increase in pay for a master's degree over not having a high school diploma is about 188%.

5. Increase $= 27,700 - 25,200 = 2500$

$$\frac{2500}{25,200} = \frac{25}{252} = 252\overline{)25.000}$$

$$\begin{array}{r} 0.099 \\ \hline 22\,68 \\ \hline 2\,320 \\ 2\,268 \\ \hline 52 \\ 9 \\ \hline 1 \end{array}$$

The percent increase over the base price is about 10%.

6. Decrease $= 20,000 - 15,000 = 5000$

$$\frac{5000}{20,000} = \frac{1}{4} = 0.25 = 25\%$$

The depreciation after 500 miles of use is 25%.

Exercises 5.5

1. Increase $= 4000$

$$\frac{4000}{20,000} = \frac{1}{5} = 0.20 = 20\%$$

This was a 20% increase in salary.

3. Increase $= 70 - 60 = 10$

$$\frac{10}{60} = \frac{1}{6} = 0.16\overline{6} = 16.\overline{6}\%$$

The percent of increase in households reached was 17%.

5. Increase $= 73 - 72 = 1$

$$\frac{1}{72} = 72\overline{)1.000}$$

$$\begin{array}{r} 0.013 \\ \hline 72 \\ \hline 280 \\ 216 \\ \hline 64 \end{array}$$

The predicted percent of increase in passengers is 1%.

7. Increase $= 19.02 - 18.27 = 0.75$

$$\frac{0.75}{18.27} = \frac{75}{1827} = 1827\overline{)75.000}$$

$$\begin{array}{r} 0.041 \\ \hline 73\,08 \\ \hline 1\,920 \\ 1\,827 \\ \hline 93 \end{array}$$

That was a 4% increase.

9. Increase $= 10,400 - 10,000 = 400$

$$\frac{400}{10,000} = \frac{1}{25} = \frac{4}{100} = 4\%$$

That was a 4% increase.

11. Decrease $= 4.5 - 3.85 = 0.65$

$$\frac{0.65}{4.5} = \frac{65}{450} = \frac{13}{90} = 90\overline{)13.000}$$

$$\begin{array}{r} 0.144 \\ \hline 9\,0 \\ \hline 4\,00 \\ 3\,60 \\ \hline 400 \\ 360 \\ \hline 40 \end{array}$$

This is a 14% decrease in complaints.

13. Decrease $= 5.07 - 3.66 = 1.41$

$$\frac{1.41}{5.07} = \frac{141}{507} = \frac{47}{169} = 169\overline{)47.000}$$

$$\begin{array}{r} 0.278 \\ \hline 33\,8 \\ \hline 13\,20 \\ 11\,83 \\ \hline 1\,370 \\ 1\,352 \\ \hline 18 \end{array}$$

This is a 28% decrease in complaints.

15. Increase = 9100 − 3650 = 5450

$$\frac{5450}{3650} = \frac{109}{73} = 73\overline{)109.000}$$

$$\begin{array}{r} 1.493 \\ \underline{73} \\ 36\,0 \\ \underline{29\,2} \\ 6\,80 \\ \underline{6\,57} \\ 230 \\ \underline{219} \\ 11 \end{array}$$

There was a 149% increase of dollar stores during this time.

17. Increase = 98 − 38 = 60

$$\frac{60}{38} = \frac{30}{19} = 19\overline{)30.000}$$

$$\begin{array}{r} 1.578 \\ \underline{19} \\ 11\,0 \\ \underline{95} \\ 15\,0 \\ \underline{13\,3} \\ 1\,70 \\ \underline{1\,52} \\ 18 \end{array}$$

That is a 158% percent of increase.

19. a. Increase = 13 − 10 = 3

$$\frac{3}{10} = \frac{30}{100} = 30\%$$

This is a 30% increase.

b. $300 + 300 \times 30\% = 300 + 300 \times 0.30$
$$= 300 + 90$$
$$= 390$$

The number of customers next year is expected to be 390.

21. a. Increase = 10.3 − 3 = 7.3

$$\frac{7.3}{3} = \frac{73}{30} = 30\overline{)73.000}$$

$$\begin{array}{r} 2.433 \\ \underline{60} \\ 13\,0 \\ \underline{12\,0} \\ 100 \\ \underline{90} \\ 100 \\ \underline{90} \\ 10 \end{array}$$

This is a 243% increase.

b. $100,000 + 100,000 \times 243\%$
$$= 100,000 + 100,000 \times 2.43$$
$$= 100,000 + 243,000$$
$$= 343,000$$

The number of visitors next year is expected to be 343,000.

23. a. Increase = 4.8 − 3.7 = 1.1

$$\frac{1.1}{3.7} = \frac{11}{37} = 37\overline{)11.000}$$

$$\begin{array}{r} 0.297 \\ \underline{74} \\ 3\,60 \\ \underline{3\,33} \\ 270 \\ \underline{259} \\ 11 \end{array}$$

This is a 30% increase in visitors.

b. $6000 + 6000 \times 30\%$
$$= 6000 + 6000 \times 0.30$$
$$= 6000 + 1800$$
$$= 7800$$

You should expect 7800 visitors during the week ending November 10.

25. a. $10,000 + 10,000 \times 1.5\%$
$$= 10,000 + 10,000 \times 0.015$$
$$= 10,000 + 150$$
$$= 10,150$$

The investment will be worth $10,150 at the end of one year.

b. $10,150 - 10,150 \times 28\%$
$$= 10,150 - 10,150 \times 0.28$$
$$= 10,150 - 2842$$
$$= 7308$$

The investment will be worth $7308.

27. a. Increase = 1750 − 400 = 1350

$$\frac{1350}{400} = \frac{27}{8} = 3.375$$

This is a 338% increase.

b. Decrease = 1750 − 1700 = 50

$$\frac{50}{1750} = \frac{1}{35} \approx 0.0285$$

There is a 3% decrease in price between Friday and Saturday.

c. Decrease = 1700 − 1500 = 200

$$\frac{200}{1700} = \frac{2}{17} \approx 0.1176$$

There is a 12% decrease in price between Saturday and game day.

29. a. Decrease = 6040 − 5584 = 456

$$\frac{456}{6040} = \frac{57}{755} \approx 0.075$$

There is an 8% decrease in the interest paid.

b. Increase = 17.3 − 10 = 7.3

$$\frac{7.3}{10} = \frac{73}{100} = 73\%$$

There is a 73% increase in the number of years between the loans.

31. $I = P \times R \times T$

$$= 3550 \times 18\% \times \frac{1}{12}$$

$$= 3550 \times 0.18 \times \frac{1}{12} = \$53.25$$

33. $1.5\% \times \$260 = 0.015 \times \$260 = \$3.90$

35. $P = \$1000,\ i = \frac{8\%}{12} = \frac{0.08}{12} = 0.00667,$

and there are $1 \cdot 2 = 2$ periods.

$$A = P(1+i)^n$$

$$= 1000(1 + 0.00667)^2$$

$$= 1000(1.00667)^2$$

$$= 1000(1.00667 \times 1.00667)$$

$$= 1000(1.013384)$$

$$= 1013.384$$

$$= \$1013.38$$

37. Difference = 16,900 − 7400 = 9500

$$\frac{9500}{7400} = \frac{95}{74} = 74\overline{\smash{)}95.000}$$

$$\begin{array}{r} 1.283 \\ 74 \\ \hline 21\,0 \\ 14\,8 \\ \hline 6\,20 \\ 5\,92 \\ \hline 280 \\ 222 \\ \hline 58 \end{array}$$

There was a 128% difference.

39. Difference = 14,200 − 7400 = \$6800

$6800 \times 2 = \$13,600 < \$14,200$ so no, the cost did not double during this time period.

41. Total payment for the 5% loan is $\$106 \times 12 \times 10 = \$12,720$ so the interest paid is $12,720 - 10,000 = \$2720$. Total payment for the 3% loan is $\$100 \times 12 \times 10 = \$12,000$ so the interest paid is $12,000 - 10,000 = \$2000$. Thus, the 3% loan is better since you are paying $2720 - 2000 = \$720$ less interest.

43. Answers may vary.

Mastery Test 5.5

45. Depreciation = 25,000 − 22,000 = 3000

$$\frac{3000}{25,000} = \frac{3}{25} = \frac{12}{100} = 12\%$$

The car depreciates 12% after just using it for 100 miles.

47. Increase = 33,000 − 26,000 = 7000

$$\frac{7000}{26000} = \frac{7}{26} = 26\overline{\smash{)}7.000}$$

$$\begin{array}{r} 0.269 \\ 5\,2 \\ \hline 1\,80 \\ 1\,56 \\ \hline 240 \\ 234 \\ \hline 6 \end{array}$$

There is a 27% increase in annual salary between a high school graduate and a person with an associate's degree.

49. Decrease = 9000 − 8000 = 1000

$$\frac{1000}{9000} = \frac{1}{9} = 0.11\overline{1}$$

The DJIA declined 11%.

Section 5.6 – Consumer Credit

Problems

1. The increased rate means that you will be $(14\% - 10\%) \times \$1000 = 4\% \times \$1000 = \$40$ in additional interest. However, you will save the $50 annual feeyou're your total saving would be $\$50 - \$40 = \$10$.

2. First card: $12\% \times 1000 + 50 = \$170$
Second card: $18\% \times 1000 = \$180$
It is better to stay with the first card.

3. a. Minimum payment: $2\% \times \$3000 = \$60$
   b. Amount of interest:

$$0.15 \times 3000 \times \frac{1}{12} = \$37.50$$

   Amount to principal:
$$\$60 - \$37.50 = \$22.50$$

   c. Interest: $0.10 \times 3000 \times \frac{1}{12} = \$25$

   Amt. to principal: $\$60 - \$25 = \$35$

   d. The principal is reduce by $22.50 with the 15% card and by $35 with the 10% card.

4. a. Unpaid balance: $\$350 - \$50 = \$300$
   b. Finance charge: $1.5\% \times 300 = .015 \times 300$
$$= \$4.50$$

   c. New balance: $\$300 + \$4.50 + \$200$
$$= \$504.50$$
   d. Minimum payment: $5\% \times \$504.50$
$$= 0.05 \times 504.50$$
$$= \$25.23$$

5. a. The interest is
$$I = P \times R \times T$$
$$= 15,000 \times 0.04 \times \frac{1}{12} = \$50$$

   and the princpal payment is
$$\$151.87 - \$50 = \$101.87.$$

   b. $15,000 \times 0.0825 \times \frac{1}{12} = \$103.13$

   so $\$103.13 - \$50 = \$53.13$ more would have gone to interest.

c. The amount paid at 4% over the 10 years is $\$151.87 \times 120 = \$18,224.40$ so $\$18,224.40 - \$15,000 = \$3224.40$ is the interest.

The amount paid at 8.25% over the 10 years is $\$183.98 \times 120 = \$22,077.60$ so $\$22,077.60 - \$15,000 = \$7077.60$ is the interest. Thus $\$7077.60 - \$3224.40 = \$3853.20$ more interest would be paid with the 8.25% loan.

6. a. The interest is
$$I = P \times R \times T$$
$$= 100,000 \times 0.06 \times \frac{1}{12} = \$500$$

   and the princpal payment is
$$\$151.87 - \$50 = \$101.87.$$

   b. $\$600 - \$500 = \$100$ would go toward paying the principal.

Exercises 5.6

	Unpaid Balance	Finance Charge (1.5%)	New Balance
1.	$90	$1.35	$141.35
3.	$109.39	$1.64	$185.01
5.	$303.93	$4.56	$557.48
7.	$200	$3	$453

9. a. Finance charge: $0.015 \times \$85 = \$1.28$
   b. New balance: $\$85 + \$1.28 + \$150$
$$= \$236.28$$
   c. Minimum monthly payment:
$$0.05 \times \$236.28 = \$11.81$$

11. a. Finance charge: $0.015 \times \$344 = \$5.16$
   b. New balance: $\$344 + \$5.16 + \$60$
$$= \$409.16$$
   c. Minimum monthly payment:
$$0.05 \times \$409.16 = \$20.46$$

13. a. Finance charge: $0.015 \times \$80.45 = \$1.21$
   b. New balance: $\$80.45 + \$1.21 + \$98.73$
                $= \$180.39$
   c. Minimum monthly payment: $10

15. a. Finance charge: $0.015 \times \$55.90 = \$0.84$
   b. New balance: $\$55.90 + \$0.84 + \$35.99$
                $= \$92.73$
   c. Minimum monthly payment: $10

17. a. The interest is
$$I = P \times R \times T$$
$$= 20,000 \times 0.03 \times \frac{1}{12} = \$50$$
and the princpal payment is
$\$193.12 - \$50 = \$143.12$.
   b. $20,000 \times 0.0825 \times \dfrac{1}{12} = \$137.50$
so $\$137.50 - \$50 = \$87.50$ more would have gone to interest.
   c. The amount paid at 3% over the 10 years is $\$193.12 \times 120 = \$23,174.40$ so $\$23,174.40 - \$20,000 = \$3174.40$ is the interest.

The amount paid at 8.25% over the 10 years is $\$245.31 \times 120 = \$29,437.20$ so $\$29,437.20 - \$20,000 = \$9437.20$ is the interest. Thus $\$9437.20 - \$3174.40 = \$6262.80$ more interest would be paid with the 8.25% loan.

19. a. The first month interest is
$$I = P \times R \times T$$
$$= 10,000 \times 0.06 \times \frac{1}{12} = \$50.$$
   b. $\$111 - \$50 = \$61$ would go toward paying the principal.
   c. The total amount paid is $\$111 \times 120 = \$13,320$ so the total amount of interest that will be paid is $\$13,320 - \$10,000 = \$3320$.

21. a. The first month interest is
$$I = P \times R \times T$$
$$= 20,000 \times 0.06 \times \frac{1}{12} = \$100.$$
   b. $\$222 - \$100 = \$122$ would go toward paying the principal.
   c. The total amount paid is $\$222 \times 120 = \$26,640$ so the total amount of interest that will be paid is $\$26,640 - \$20,000 = \$6640$.

23. a. The amount of the loan would be $0.95 \times 50,000 = \$47,500$.
   b. The down payment would be $\$50,000 - \$47,500 = \$2500$.
   c. The amount of the first payment that is interest is
$$I = P \times R \times T$$
$$= 47,500 \times 0.07 \times \frac{1}{12}$$
$$= \$277.08$$
so the amount that will be principal is $\$316 - \$277.08 = \$38.92$.

25.

No.	Interest Paid: 6%	Principal Applied	New Balance
1	$500	$99.55	$99,900.45
2	$499.50	$100.05	$99,800.40

27.

No.	Interest Paid: 7%	Principal Applied	New Balance
1	$875	$122.95	$149,877.05
2	$874.28	$123.67	$199,753.38

29. $58.1 > 41.9$

31. $39.4 = 39.40$
$39.\overline{4} = 39.4\overline{4}$
Since $4 > 0$ in the last decimal digit shown, we have $39.4 < 39.\overline{4}$.

33. $I = P \times R \times T$
$$= 1000 \times 0.06 \times \frac{1}{12} = \$5$$

35. $T = \dfrac{1}{12}$; answers may vary.

37. $T = \dfrac{90}{360} = \dfrac{1}{4}$; answers may vary.

Mastery Test 5.6

39. a. The first month interest would be
$$I = P \times R \times T$$
$$= 200{,}000 \times 0.06 \times \dfrac{1}{12} = \$1000.$$

b. The amount going toward the principal is $\$1200 - \$1000 = \$200$.

41. a. Unpaid balance: $\$200 - \$50 = \$150$
b. Finance charge: $1.5\% \times 150 = .015 \times 150$
$$= \$2.25$$
c. New balance: $\$150 + \$2.25 + \$60$
$$= \$212.25$$
d. Minimum payment: $5\% \times \$212.25$
$$= 0.05 \times 212.25$$
$$= \$10.61$$

Review Exercises – Chapter 5

1. a. $39\% = .39 = 0.39$
b. $1\% = .01 = 0.01$
c. $13\% = .13 = 0.13$
d. $101\% = 1.01$
e. $207\% = 2.07$

2. a. $3.2\% = .032 = 0.032$
b. $11.2\% = .112 = 0.112$
c. $71.4\% = .714 = 0.714$
d. $17.51\% = .1751 = 0.1751$
e. $142.5\% = 1.425$

3. a. $6\dfrac{1}{4}\% = 6.25\% = .0625 = 0.0625$

b. $71\dfrac{1}{2}\% = 71.5\% = .715 = 0.715$

c. $5\dfrac{3}{8}\% = 5.375\% = .05375 = 0.05375$

d. $14\dfrac{1}{4}\% = 17.25\% = .1725 = 0.1725$

e. $52\dfrac{1}{8}\% = 52.125\% = .52125 = 0.52125$

4. a. $6\dfrac{1}{3}\% = 6.33\overline{3}\% = 0.063\overline{3}$

b. $8\dfrac{2}{3}\% = 8.66\overline{6}\% = .0.086\overline{6}$

c. $1\dfrac{1}{6}\% = 1.16\overline{6}\% = 0.011\overline{6}$

d. $18\dfrac{5}{6}\% = 18.83\overline{3}\% = 0.1883\overline{3}$

e. $20\dfrac{1}{9}\% = 20.11\overline{1}\% = 0.201\overline{1}$

5. a. $0.01 = 1.\% = 1\%$
b. $0.07 = 7.\% = 7\%$
c. $0.17 = 17.\% = 17\%$
d. $0.91 = 91.\% = 91\%$
e. $0.83 = 83.\% = 83\%$

6. a. $3.2 = 320.\% = 320\%$
b. $1.1 = 110.\% = 110\%$
c. $7.9 = 790.\% = 790\%$
d. $9.1 = 910.\% = 910\%$
e. $4.32 = 432.\% = 432\%$

7. a. $17\% = \dfrac{17}{100}$ **b.** $23\% = \dfrac{23}{100}$

c. $51\% = \dfrac{51}{100}$ **d.** $111\% = \dfrac{111}{100}$

e. $201\% = \dfrac{201}{100}$

8. a. $10\% = \dfrac{10}{100} = \dfrac{1}{10}$

b. $40\% = \dfrac{40}{100} = \dfrac{4}{10} = \dfrac{2}{5}$

c. $15\% = \dfrac{15}{100} = \dfrac{\cancel{5} \cdot 3}{\cancel{5} \cdot 20} = \dfrac{3}{20}$

d. $35\% = \dfrac{35}{100} = \dfrac{\cancel{5} \cdot 7}{\cancel{5} \cdot 20} = \dfrac{7}{20}$

e. $42\% = \dfrac{42}{100} = \dfrac{\cancel{2} \cdot 21}{\cancel{2} \cdot 50} = \dfrac{21}{50}$

9. a. $16\dfrac{2}{3}\% = \dfrac{16\dfrac{2}{3}}{100}$

$= \dfrac{\dfrac{50}{3}}{100} = \dfrac{\cancel{50}}{3} \cdot \dfrac{1}{\cancel{100}_2} = \dfrac{1}{6}$

b. $33\dfrac{1}{3}\% = \dfrac{33\dfrac{1}{3}}{100}$

$= \dfrac{\dfrac{100}{3}}{100} = \dfrac{\cancel{100}}{3} \cdot \dfrac{1}{\cancel{100}} = \dfrac{1}{3}$

c. $62\dfrac{1}{2}\% = \dfrac{62\dfrac{1}{2}}{100}$

$= \dfrac{\dfrac{125}{2}}{100} = \dfrac{^5\cancel{125}}{2} \cdot \dfrac{1}{\cancel{100}_4} = \dfrac{5}{8}$

d. $83\dfrac{1}{3}\% = \dfrac{83\dfrac{1}{3}}{100}$

$= \dfrac{\dfrac{250}{3}}{100} = \dfrac{^5\cancel{250}}{3} \cdot \dfrac{1}{\cancel{100}_2} = \dfrac{5}{6}$

e. $87\dfrac{1}{2}\% = \dfrac{87\dfrac{1}{2}}{100}$

$= \dfrac{\dfrac{175}{2}}{100} = \dfrac{^7\cancel{175}}{2} \cdot \dfrac{1}{\cancel{100}_4} = \dfrac{7}{8}$

10. a. $\dfrac{3}{8} = 8\overline{)3.00}$
$\dfrac{0.37}{}$
$2\,4$
$\overline{60}$
$\dfrac{56}{4}$

Thus $\dfrac{3}{8} = 0.37\dfrac{4}{8} = 0.37\dfrac{1}{2} = 37\dfrac{1}{2}\%$.

b. $\dfrac{5}{8} = 8\overline{)5.00}$
$\dfrac{0.62}{}$
$4\,8$
$\overline{20}$
$\dfrac{16}{4}$

Thus $\dfrac{5}{8} = 0.62\dfrac{4}{8} = 0.62\dfrac{1}{2} = 62\dfrac{1}{2}\%$.

c. $\dfrac{1}{16} = 16\overline{)1.00}$
$\dfrac{0.06}{}$
$\dfrac{96}{4}$

Thus $\dfrac{1}{16} = 0.06\dfrac{4}{16} = 0.06\dfrac{1}{4} = 6\dfrac{1}{4}\%$.

d. $\dfrac{3}{16} = 16\overline{)3.00}$
$\dfrac{0.18}{}$
16
$\overline{140}$
$\dfrac{128}{12}$

Thus $\dfrac{3}{16} = 0.18\dfrac{12}{16} = 0.18\dfrac{3}{4} = 18\dfrac{3}{4}\%$.

e. $\dfrac{5}{16} = 16\overline{)5.00}$
$\dfrac{0.31}{}$
$4\,8$
$\overline{20}$
$\dfrac{16}{4}$

Thus $\dfrac{5}{16} = 0.31\dfrac{4}{16} = 0.31\dfrac{1}{4} = 31\dfrac{1}{4}\%$.

11. a. $60\% \times 30 = 0.60 \times 30 = 18$
b. $70\% \times 140 = 0.70 \times 140 = 98$
c. $40\% \times 80 = 0.40 \times 80 = 32$
d. $30\% \times 90 = 0.30 \times 90 = 27$
e. $35\% \times 105 = 0.35 \times 105 = 36.75$

12. a. $12\dfrac{1}{2}\% \times 80 = 0.125 \times 80 = 10$

b. $40\dfrac{1}{2}\% \times 60 = 0.405 \times 60 = 24.3$

c. $15\dfrac{1}{2}\% \times 250 = 0.155 \times 250 = 38.75$

d. $10\dfrac{2}{3}\% \times 300 = \dfrac{\dfrac{32}{2}}{100} \times 300$

$= \dfrac{32}{\cancel{3}} \cdot \dfrac{1}{\cancel{100}} \times \cancel{300} = 32$

e. $24\dfrac{3}{4}\% \times 7000 = 0.2475 \times 7000$

$= 1732.5$

13. Let R = the percent.

   a. $R \times 20 = 5$

$$R = \frac{5}{20} = \frac{1}{5} = 0.2 = 20\%$$

   b. $R \times 50 = 10$

$$R = \frac{10}{50} = \frac{1}{5} = 0.2 = 20\%$$

   c. $R \times 60 = 20$

$$R = \frac{20}{60} = \frac{1}{3} = 0.33\overline{3} = 33\frac{1}{3}\%$$

   d. $R \times 80 = 160$

$$R = \frac{160}{80} = 2 = 200\%$$

   e. $R \times 5 = 1$

$$R = \frac{1}{5} = 0.2 = 20\%$$

14. Let R = the percent.

   a. $R \times 40 = 30$

$$R = \frac{30}{40} = \frac{3}{4} = 0.75 = 75\%$$

   b. $R \times 50 = 20$

$$R = \frac{20}{50} = \frac{2}{5} = 0.4 = 40\%$$

   c. $R \times 20 = 40$

$$R = \frac{40}{20} = 2 = 200\%$$

   d. $R \times 30 = 20$

$$R = \frac{20}{30} = \frac{2}{3} = 0.66\overline{6} = 66\frac{2}{3}\%$$

   e. $R \times 60 = 10$

$$R = \frac{10}{60} = \frac{1}{6} = 0.16\overline{6} = 16\frac{2}{3}\%$$

15. Let R = the percent.

   a. $20 = R \times 40$

$$\frac{20}{40} = R \text{ so } R = \frac{1}{2} = 0.5 = 50\%$$

   b. $30 = R \times 90$

$$\frac{30}{90} = R \text{ so } R = \frac{1}{3} = 0.33\overline{3} = 33\frac{1}{3}\%$$

   c. $60 = R \times 80$

$$\frac{60}{80} = R \text{ so } R = \frac{6}{8} = \frac{3}{4} = 0.75 = 75\%$$

   d. $90 = R \times 60$

$$\frac{90}{60} = R \text{ so } R = \frac{9}{6} = \frac{3}{2} = 1.5 = 150\%$$

   e. $30 = R \times 80$

$$\frac{30}{80} = R \text{ so } R = \frac{3}{8} = 0.375 = 37\frac{1}{2}\%$$

16. Let B = the number.

   a. $30 = 50\% \times B$

$$30 = 0.50 \times B$$
$$\frac{30}{0.50} = B \text{ so } B = \frac{300}{5} = 60$$

   b. $20 = 40\% \times B$

$$20 = 0.40 \times B$$
$$\frac{20}{0.40} = B \text{ so } B = \frac{200}{4} = 50$$

   c. $15 = 75\% \times B$

$$15 = 0.75 \times B$$
$$\frac{15}{0.75} = B \text{ so } B = \frac{1500}{75} = 20$$

   d. $20 = 60\% \times B$

$$20 = 0.60 \times B$$
$$\frac{20}{0.60} = B \text{ so } B = \frac{200}{6} = 33\frac{1}{3}$$

   e. $60 = 90\% \times B$

$$60 = 0.90 \times B$$
$$\frac{60}{0.90} = B \text{ so } B = \frac{600}{9} = 66\frac{2}{3}$$

17. Let B = the number.

   a. $60 = 40\% \times B$

$$60 = 0.40 \times B$$
$$\frac{60}{0.40} = B \text{ so } B = \frac{600}{4} = 150$$

   b. $90 = 30\% \times B$

$$90 = 0.30 \times B$$
$$\frac{90}{0.30} = B \text{ so } B = \frac{900}{3} = 300$$

   c. $40 = 80\% \times B$

$$40 = 0.80 \times B$$
$$\frac{40}{0.80} = B \text{ so } B = \frac{400}{8} = 50$$

   d. $20 = 60\% \times B$

$$20 = 0.60 \times B$$
$$\frac{20}{0.60} = B \text{ so } B = \frac{200}{6} = 33\frac{1}{3}$$

   e. $42 = 50\% \times B$

$$42 = 0.50 \times B$$
$$\frac{42}{0.50} = B \text{ so } B = \frac{420}{5} = 84$$

18. Let B = the number.

a. $40\% \times B = 10$

$0.40 \times B = 10$

$B = \dfrac{10}{0.40} = \dfrac{100}{4} = 25$

b. $50\% \times B = 5$

$0.50 \times B = 5$

$B = \dfrac{5}{0.50} = \dfrac{50}{5} = 10$

c. $70\% \times B = 140$

$0.70 \times B = 140$

$B = \dfrac{140}{0.70} = \dfrac{1400}{7} = 200$

d. $65\% \times B = 195$

$0.65 \times B = 195$

$B = \dfrac{195}{0.65} = \dfrac{19,500}{65} = 300$

e. $16\% \times B = 40$

$0.16 \times B = 40$

$B = \dfrac{40}{0.16} = \dfrac{4000}{16} = 250$

19. Let P = the part.

a. $\dfrac{30}{100} = \dfrac{P}{40}$

$100P = 30 \cdot 40$

$P = \dfrac{1200}{100} = 12$

Thus 12 is 30% of 40.

b. $\dfrac{40}{100} = \dfrac{P}{72}$

$100P = 40 \cdot 72$

$P = \dfrac{2880}{100} = 28.8$

Thus 28.8 is 40% of 72.

c. $\dfrac{50}{100} = \dfrac{P}{94}$

$100P = 50 \cdot 94$

$P = \dfrac{4700}{100} = 47$

Thus 47 is 50% of 94.

d. $\dfrac{60}{100} = \dfrac{P}{50}$

$100P = 60 \cdot 50$

$P = \dfrac{3000}{100} = 30$

Thus 30 is 60% of 50.

e. $\dfrac{70}{100} = \dfrac{P}{70}$

$100P = 70 \cdot 70$

$P = \dfrac{4900}{100} = 49$

Thus 49 is 70% of 70.

20. Let R = the percent.

a. $\dfrac{R}{100} = \dfrac{80}{800}$

$800R = 100 \cdot 80$

$R = \dfrac{8000}{800} = 10$

Thus 80 is 10% of 800.

b. $\dfrac{R}{100} = \dfrac{22}{110}$

$110R = 100 \cdot 22$

$R = \dfrac{2200}{110} = 20$

Thus 22 is 20% of 110.

c. $\dfrac{R}{100} = \dfrac{28}{70}$

$70R = 100 \cdot 28$

$R = \dfrac{2800}{70} = 40$

Thus 28 is 40% of 70.

d. $\dfrac{R}{100} = \dfrac{90}{180}$

$180R = 100 \cdot 90$

$R = \dfrac{9000}{180} = 50$

Thus 90 is 50% of 180.

e. $\dfrac{R}{100} = \dfrac{80}{40}$

$40R = 100 \cdot 80$

$R = \dfrac{8000}{40} = 200$

Thus 80 is 200% of 40.

21. Let B = the whole.

a. $\dfrac{10}{100} = \dfrac{20}{B}$

$10B = 100 \cdot 20$

$B = \dfrac{2000}{10} = 200$

Thus 20 is 10% of 200.

b. $\dfrac{12}{100} = \dfrac{30}{B}$

$12B = 100 \cdot 30$

$B = \dfrac{3000}{12} = 250$

Thus 30 is 12% of 250.

c. $\dfrac{90}{100} = \dfrac{45}{B}$

$90B = 100 \cdot 45$

$B = \dfrac{4500}{90} = 50$

Thus 45 is 90% of 50.

d. $\dfrac{25}{100} = \dfrac{50}{B}$

$25B = 100 \cdot 50$

$B = \dfrac{5000}{25} = 200$

Thus 50 is 25% of 200.

e. $\dfrac{18}{100} = \dfrac{63}{B}$

$18B = 100 \cdot 63$

$B = \dfrac{6300}{18} = 350$

Thus 63 is 18% of 350.

22. a. Total price $= 20 + 20 \times 0.06 = 20 + 1.2$
$$= \$21.20$$

b. Total price $= 50 + 50 \times 0.04 = 50 + 2$
$$= \$52$$

c. Total price $= 18 + 18 \times 0.05 = 18 + 0.90$
$$= \$18.90$$

d. $80 + 80 \times 0.065 = 80 + 5.20$
$$= \$85.20$$

e. Total price $= 300 + 300 \times 0.045$
$$= 300 + 13.50 = \$313.50$$

23. a. $I = P \times R \times T = 100 \times 0.10 \times 2 = \$20$

b. $I = P \times R \times T = 250 \times 0.12 \times 3 = \$90$

c. $I = P \times R \times T = 300 \times 0.15 \times 2 = \$90$

d. $I = P \times R \times T = 600 \times 0.09 \times 2 = \$108$

e. $I = P \times R \times T = 3000 \times 0.085 \times 3 = \$765$

24. a. $I = P \times R \times T = 600 \times 0.10 \times \dfrac{2}{12} = \$10$

b. $I = P \times R \times T = 600 \times 0.12 \times \dfrac{3}{12} = \$18$

c. $I = P \times R \times T = 250 \times 0.08 \times \dfrac{6}{12} = \$10$

d. $I = P \times R \times T = 450 \times 0.09 \times \dfrac{8}{12} = \$27$

e. $I = P \times R \times T = 300 \times 0.13 \times \dfrac{10}{12} = \$32.50$

25. $P = \$10,000$ and there are $2 \cdot 2 = 4$ periods in each case.

a. $i = \dfrac{4\%}{2} = 2\% = 0.02$

$A = P(1+i)^n$
$= 10,000(1 + 0.02)^4$
$= 10,000(1.02)^4$
$= 10,000(1.02 \times 1.02 \times 1.02 \times 1.02)$
$= 10,000(1.0824322)$
$= 10,824.322$

and $I = A - P$
$= \$10,824.32 - \$10,000$
$= \$824.32$

The compound amount is $10,824.32 and the compound interest is $824.32.

b. $i = \dfrac{6\%}{2} = 3\% = 0.03$

$A = P(1+i)^n$
$= 10,000(1 + 0.03)^4$
$= 10,000(1.03)^4$
$= 10,000(1.03 \times 1.03 \times 1.03 \times 1.03)$
$= 10,000(1.1255088)$
$= 11,255.088$

and $I = A - P$
$= \$11,255.09 - \$10,000$
$= \$1225.09$

The compound amount is $11,255.09 and the compound interest is $1255.09

c. $i = \dfrac{8\%}{2} = 4\% = 0.04$

$A = P(1+i)^n$
$= 10,000(1 + 0.04)^4$
$= 10,000(1.04)^4$
$= 10,000(1.04 \times 1.04 \times 1.04 \times 1.04)$
$= 10,000(1.1698586)$
$= 11698.586$

and $I = A - P$
$= \$11,698.59 - \$10,000$
$= \$1698.59$

The compound amount is $11,698.59 and the compound interest is $1698.59.

d. $i = \dfrac{10\%}{2} = 5\% = 0.05$

$A = P(1+i)^n$
$= 10,000(1+0.05)^4$
$= 10,000(1.05)^4$
$= 10,000(1.05 \times 1.05 \times 1.05 \times 1.05)$
$= 10,000(1.2155062)$
$= 12,155.063$

and $I = A - P$
$= \$12,155.06 - \$10,000$
$= \$2155.06$

The compound amount is \$12,155.06 and the compound interest is \$2155.06.

e. $i = \dfrac{12\%}{2} = 6\% = 0.06$

$A = P(1+i)^n$
$= 10,000(1+0.06)^4$
$= 10,000(1.06)^4$
$= 10,000(1.06 \times 1.06 \times 1.06 \times 1.06)$
$= 10,000(1.262477)$
$= 12,624.770$

and $I = A - P$
$= \$12,624.77 - \$10,000$
$= \$2624.77$

The compound amount is \$12,624.77 and the compound interest is \$2624.77.

26. a. Commission $= 0.08 \times 100 = \$8$
b. Commission $= 0.06 \times 250 = \$15$
c. Commission $= 0.07 \times 17,500 = \$1225$
d. Commision $= 0.065 \times 300 = \$19.50$
e. Commission $= 0.055 \times 700 = \$38.50$

27. a. Sale price $= 35 - 0.40 \times 35 = 35 - 14$
$\qquad\qquad = \$21$

b. Sale price $= 40 - 0.50 \times 40 = 40 - 20$
$\qquad\qquad = \$20$

c. Sale price $= 60 - \dfrac{1}{3} \times 60 = 60 - 20$
$\qquad\qquad = \$40$

d. Sale price $= 90 - 0.10 \times 90 = 90 - 9$
$\qquad\qquad = \$81$

e. Sale price $= 540 - 0.15 \times 540 = 540 - 20$
$\qquad\qquad = \$81$

28. a. Increase $= 3 - 2 = 1$ billion dollars
$\dfrac{1}{2} = 0.5 = 50\%$
Sales increased by 50%.

b. Increase $= 3.1 - 2 = 1.1$ billion dollars
$\dfrac{1.1}{2} = \dfrac{11}{20} = \dfrac{55}{100} = 55\%$
Sales increased by 55%.

c. Increase $= 3.2 - 2 = 1.2$ billion dollars
$\dfrac{1.2}{2} = \dfrac{12}{20} = \dfrac{60}{100} = 60\%$
Sales increased by 60%.

d. Increase $= 3.3 - 2 = 1.3$ billion dollars
$\dfrac{1.3}{2} = \dfrac{13}{20} = \dfrac{65}{100} = 65\%$
Sales increased by 65%.

e. Increase $= 3.4 - 2 = 1.4$ billion dollars
$\dfrac{1.4}{2} = \dfrac{14}{20} = \dfrac{70}{100} = 70\%$
Sales increased by 70%.

29. a. Decrease $= 400 - 395 = 5$
$\dfrac{5}{400} = \dfrac{5 \div 4}{400 \div 4} = \dfrac{1.25}{100} = 1.25\%$
The price decreased by 1%.

b. Decrease $= 400 - 390 = 10$
$\dfrac{10}{400} = \dfrac{10 \div 4}{400 \div 4} = \dfrac{2.5}{100} = 2.5\%$
The price decreased by 3%.

c. Decrease $= 400 - 385 = 15$
$\dfrac{15}{400} = \dfrac{15 \div 4}{400 \div 4} = \dfrac{3.75}{100} = 3.75\%$
The price decreased by 4%.

d. Decrease $= 400 - 380 = 20$
$\dfrac{20}{400} = \dfrac{20 \div 4}{400 \div 4} = \dfrac{5}{100} = 5\%$
The price decreased by 5%.

e. Increase $= 400 - 375 = 25$
$\dfrac{25}{400} = \dfrac{25 \div 4}{400 \div 4} = \dfrac{6.25}{100} = 6.25\%$
The price decreased by 6%.

30. a. $48 + 16\dfrac{2}{3}\% \times 48 = 48 + \dfrac{1}{6} \times 48$
$\qquad\qquad\qquad = 48 + 8$
$\qquad\qquad\qquad = \$56$ million

b. $48 + 33\dfrac{1}{3}\% \times 48 = 48 + \dfrac{1}{3} \times 48$
$\qquad\qquad\qquad = 48 + 16$
$\qquad\qquad\qquad = \$64$ million

c. $48 + 37\frac{1}{2}\% \times 48 = 48 + 0.375 \times 48$

$\phantom{48 + 37\frac{1}{2}\% \times 48} = 48 + 18$

$\phantom{48 + 37\frac{1}{2}\% \times 48} = \$66 \text{ million}$

d. $48 + 66\frac{2}{3}\% \times 48 = 48 + \frac{2}{3} \times 48$

$\phantom{48 + 66\frac{2}{3}\% \times 48} = 48 + 32$

$\phantom{48 + 66\frac{2}{3}\% \times 48} = \$80 \text{ million}$

e. $48 + 83\frac{1}{3}\% \times 48 = 48 + \frac{5}{6} \times 48$

$\phantom{48 + 83\frac{1}{3}\% \times 48} = 48 + 40$

$\phantom{48 + 83\frac{1}{3}\% \times 48} = \$88 \text{ million}$

31. a. Unpaid balance $= 100 - 20 = \$80$
Finance charge $= 0.015 \times 80 = \$1.20$
New balance $= 80 + 1.20 + 50 = \$131.20$
b. Unpaid balance $= 150 - 30 = \$120$
Finance charge $= 0.015 \times 120 = \$1.80$
New balance $= 150 + 1.80 + 60 = \$181.80$
c. Unpaid balance $= 200 - 40 = \$160$
Finance charge $= 0.015 \times 160 = \$2.40$
New balance $= 160 + 2.40 + 70 = \$232.40$
d. Unpaid balance $= 300 - 100 = \$200$
Finance charge $= 0.015 \times 200 = \$3$
New balance $= 200 + 3 + 100 = \$303$
e. Unpaid balance $= 500 - 200 = \$300$
Finance charge $= 0.015 \times 300 = \$4.50$
New balance $= 300 + 4.50 + 70 = \$604.50$

32. a. The interest is
$I = P \times R \times T$

$ = 5000 \times 0.03 \times \frac{1}{12} = \$12.50$

and principal payment is
$\$48.28 - \$12.50 = \$35.78$.

At 8.25%, the interest is
$5000 \times 0.0825 \times \frac{1}{12} = \$34.38$ so

$\$34.38 - \$12.50 = \$21.88$ more would
have gone to interest.

The amount paid at 3% over the 10
years is $\$48.28 \times 120 = \$5793.60$ so
$\$5793.60 - \$5000 = \$793.60$ would be
the interest. The amount paid at 8.25%
over the 10 years is
$\$61.33 \times 120 = \$7359.60$ so
$\$7359.60 - \$5000 = \$2359.60$ would be

the interest. Thus
$2359.60 - 793.60 = \boxed{\$1566}$ more
interest would be paid with the 8.25%
loan.
b. The interest is
$I = P \times R \times T$

$ = 10,000 \times 0.03 \times \frac{1}{12} = \$25$

and principal payment is
$\$96.56 - \$25 = \$71.56$.

At 8.25%, the interest is
$10,000 \times 0.0825 \times \frac{1}{12} = \$68.75$ so

$\$68.75 - \$25 = \$43.75$ more would
have gone to interest.

The amount paid at 3% over the 10
years is $\$96.56 \times 120 = \$11,587.20$ so
$\$11,587.20 - \$10,000 = \$1587.20$
would be the interest. The amount paid
at 8.25% over the 10 years is
$\$122.66 \times 120 = \$14,719.20$ so
$\$14,719.20 - \$10,000 = \$4719.20$
would be the interest. Thus
$4719.20 - 1587.20 = \boxed{\$3132}$ more
interest would be paid with the 8.25%
loan.
c. The interest is
$I = P \times R \times T$

$ = 15,000 \times 0.03 \times \frac{1}{12} = \$37.50$

and principal payment is
$\$144.84 - \$37.50 = \$107.34$.

At 8.25%, the interest is
$15,000 \times 0.0825 \times \frac{1}{12} = \$103.13$ so

$\$103.13 - \$37.50 = \$65.63$ more would
have gone to interest.

The amount paid at 3% over the 10
years is $\$144.84 \times 120 = \$17,380.80$ so
$\$17,380.80 - \$15,000 = \$2380.80$
would be the interest. The amount paid
at 8.25% over the 10 years is
$\$183.99 \times 120 = \$22,078.80$ so
$\$22,078.80 - \$15,000 = \$7078.80$

would be the interest. Thus
$7078.80 - 2380.80 = \boxed{\$4698}$ more
interest would be paid with the 8.25%
loan.

d. The interest is
$$I = P \times R \times T$$
$$= 20,000 \times 0.03 \times \frac{1}{12} = \$50$$
and principal payment is
$\$193.12 - \$50 = \$143.12$.

At 8.25%, the interest is
$20,000 \times 0.0825 \times \frac{1}{12} = \$137.50$ so

$\$137.50 - \$50 = \$87.50$ more would
have gone to interest.

The amount paid at 3% over the 10
years is $\$193.12 \times 120 = \$23,174.40$ so
$\$23,174.40 - \$20,000 = \$3174.40$
would be the interest. The amount paid
at 8.25% over the 10 years is
$\$245.32 \times 120 = \$29,438.40$ so
$\$29,438.40 - \$20,000 = \$9438.40$
would be the interest. Thus
$9438.40 - 3174.40 = \boxed{\$6264}$ more
interest would be paid with the 8.25%
loan.

e. The interest is
$$I = P \times R \times T$$
$$= 25,000 \times 0.03 \times \frac{1}{12} = \$62.50$$
and principal payment is
$\$241.40 - \$62.50 = \$178.90$.

At 8.25%, the interest is
$25,000 \times 0.0825 \times \frac{1}{12} = \$171.88$ so

$\$171.88 - \$62.50 = \$109.38$ more
would have gone to interest.

The amount paid at 3% over the 10
years is $\$241.40 \times 120 = \$28,968$ so
$\$28,968 - \$25,000 = \$3968$ would be
the interest. The amount paid at 8.25%
over the 10 years is
$\$306.65 \times 120 = \$36,798$ so
$\$36,798 - \$25,000 = \$11,798$ would be

the interest. Thus
$11,798 - 3968 = \$7830$ more interest
would be paid with the 8.25% loan.

Cumulative Review Chapters 1–5

1. 6510

2. $36 \div 6 \cdot 6 + 7 - 3 = 6 \cdot 6 + 7 - 3$
$$= 36 + 7 - 3$$
$$= 43 - 3$$
$$= 40$$

3. $\dfrac{7}{6}$ is improper, since the numerator is
larger than the denominator.

4. $\dfrac{31}{4} = 7\dfrac{3}{4}$

5. $6\dfrac{4}{9} = \dfrac{9 \times 6 + 4}{9} = \dfrac{54 + 4}{9} = \dfrac{58}{9}$

6. $\left(\dfrac{5}{6}\right)^2 \cdot \dfrac{1}{25} = \left(\dfrac{5}{6} \cdot \dfrac{5}{6}\right) \cdot \dfrac{1}{25} = \dfrac{\cancel{25}}{36} \cdot \dfrac{1}{\cancel{25}} = \dfrac{1}{36}$

7. $\dfrac{10}{7} \div 8\dfrac{1}{3} = \dfrac{10}{7} \div \dfrac{25}{3} = \dfrac{\overset{2}{\cancel{10}}}{7} \cdot \dfrac{3}{\underset{5}{\cancel{25}}} = \dfrac{6}{35}$

8. $\quad x - \dfrac{3}{4} = \dfrac{1}{3}$
$$x - \dfrac{3}{4} + \dfrac{3}{4} = \dfrac{1}{3} + \dfrac{3}{4}$$
$$x = \dfrac{4}{12} + \dfrac{9}{12} = \dfrac{13}{12}$$

9. Let n = the number.
$$\dfrac{10}{11}n = 7\dfrac{1}{9}$$
$$\dfrac{10}{11}n = \dfrac{64}{9}$$
$$\dfrac{\cancel{10}n}{\cancel{11}} = \dfrac{\frac{64}{9}}{\frac{10}{11}}$$
$$n = \dfrac{\overset{32}{\cancel{64}}}{9} \cdot \dfrac{11}{\underset{5}{\cancel{10}}} = \dfrac{352}{45} = 7\dfrac{37}{45}$$

10. 342.41: Three hundred forty-two and forty-one hundredths

11. $37.773 = 30 + 7 + \dfrac{7}{10} + \dfrac{7}{100} + \dfrac{3}{1000}$

12.
$$\begin{array}{r} {\scriptstyle 3\ 10\ 14} \\ 6\cancel{4}\cancel{1}.42 \\ -\ \ 14.50 \\ \hline 626.92 \end{array}$$

13.
$$\begin{array}{r} {\scriptstyle 4\ 2} \\ 5.94 \\ \times\ \ \ 1.5 \\ \hline 2970 \\ 594\ \ \\ \hline 8.910 \end{array}$$

14. $\dfrac{189}{0.27} = \dfrac{18,900}{27} = 27\overline{)18900}$
$$\begin{array}{r} \boxed{700} \\ 27\overline{)18900} \\ \underline{189}\ \ \ \\ 000 \end{array}$$

15. $749.\underline{8}51 \to 749.9$

16. $\dfrac{10}{0.13} = \dfrac{1000}{13} = 13\overline{)1000.000} = 76.92$
$$\begin{array}{r} 76.923 \\ 13\overline{)1000.000} \\ \underline{91}\ \ \ \ \ \ \ \\ 90\ \ \ \ \ \\ \underline{78}\ \ \ \ \ \\ 120\ \ \ \\ \underline{117}\ \ \ \\ 30\ \ \\ \underline{26}\ \ \\ 40\ \\ \underline{39}\ \\ 1 \end{array}$$

17. $0.\overline{12} = \dfrac{12}{99} = \dfrac{4 \times \cancel{3}}{33 \times \cancel{3}} = \dfrac{4}{33}$

18. Let d = the decimal part.
$$d \times 12 = 3$$
$$d = \dfrac{3}{12} = \dfrac{1}{4} = 0.25$$

19. $4.293 = 4.2930$
$4.29\overline{3} = 4.2933\cdots$
$4.2\overline{93} = 4.29393\cdots$

Looking at the fourth decimal digit, we have $4.29\overline{3} > 4.2\overline{93} > 4.293$.

20. $0.25 = \dfrac{1}{4} = \dfrac{5}{20}$. Thus $0.25 < \dfrac{13}{20}$.

21. $x + 2.3 = 6.2$
$x = 6.2 - 2.3 = 3.9$

22. $1.4 = 0.2y$
$$\dfrac{1.4}{0.2} = y$$
$$\dfrac{14}{2} = y$$
$$7 = y$$

23. $5 = \dfrac{z}{4.2}$
$$4.2 \times 5 = z$$
$$21 = z$$

24. $\dfrac{25\cancel{0}}{100\cancel{0}} = \dfrac{25}{100} = \dfrac{1}{4}$

25. $\dfrac{5}{9} = \dfrac{40}{x}$

26. $\dfrac{10}{19} \overset{?}{=} \dfrac{50}{97}$; $10 \cdot 97 \overset{?}{=} 19 \cdot 50$
$$970 \neq 950$$

Thus the flag is <u>not</u> in the correct ratio.

27. $\dfrac{f}{4} = \dfrac{5}{80}$
$$80f = 4 \cdot 5$$
$$f = \dfrac{20}{80} = \dfrac{1}{4}$$

28. $\dfrac{12}{f} = \dfrac{2}{3}$
$$2f = 12 \cdot 3$$
$$f = \dfrac{36}{2} = 18$$

29. $\dfrac{300 \text{ miles}}{16 \text{ gal}} = \dfrac{75}{4} = 18\dfrac{3}{4} = 18.75 \text{ mi/gal}$

30. $\dfrac{\$2.79}{24 \text{ oz}} = 24\overline{)2.790} = 12¢/\text{oz}$

$$\begin{array}{r} 0.116 \\ \hline 2.790 \\ 2\,4 \\ \hline 39 \\ 24 \\ \hline 150 \\ 144 \\ \hline 6 \end{array}$$

31. Let n = number of pounds.

$$\frac{1}{3500} = \frac{n}{14,000}$$
$$3500n = 14,000$$
$$n = \frac{14,000}{3500} = \frac{140}{35} = 4$$

32. Let n = number of ounces.

$$\frac{4}{2} = \frac{n}{28}$$
$$2n = 4 \cdot 28$$
$$n = \frac{\overset{2}{\cancel{4}} \cdot 28}{\cancel{2}} = 56$$

33. $12\% = .12 = 0.12$

34. $7\dfrac{1}{4}\% = 7.25\% = .0725 = 0.0725$

35. $0.03 = \dfrac{3}{100} = 3\%$

36. Let P = the number.
$$10\% \times 80 = P$$
$$0.10 \times 80 = P$$
$$8 = P$$

37. $66\dfrac{2}{3}\% \times 54 = \dfrac{2}{3} \times 54 = 36$

Thus $66\dfrac{2}{3}\%$ of 54 is 36.

38. Let R = the percent.
$$R \times 32 = 16$$
$$R = \frac{16}{32} = 0.5 = 50\%$$

39. Let B = the number.
$$12 = 60\% \times B$$
$$12 = 0.60 \times B$$
$$\frac{12}{0.60} = B \text{ so } B = \frac{120}{6} = 20$$

40. Total price $= 16 + 0.05 \times 16$
$$= 16 + 0.80$$
$$= \$16.80$$

41. $I = P \times R \times T$
$$= 400 \times 0.085 \times 2 = \$68$$
The interest earned is $68.

Chapter 6

Statistics and Graphs

Section 6.1 – Tables and Pictographs

Problems

1. **a.** 37.9%
   b. 41.6%
   c. 71.5% of males play video games and 28.5% of females play video games. Thus, males play more video games.

2. **a.** Mount Sinai has 5 (red) infected people.
   b. York Central Hospital has 2 (gray) known dead.

3. 2 whole cans = $2 \times 100 = 200$ students and $\frac{1}{2}$ can = 50 students. Thus 2 and $\frac{1}{2}$ cans = 200 + 50 = 250 students chose ACE cola as their number-one choice at the game.

4. **a.** 3 pictures = $3 \times 10 = 30$ middle managers
   b. Executives and business owners

Exercises 6.1

1. Wendy's (with 75% favorable)

3. 69%

5. 22 students pay every time

7. 94 professors pay every time

9. They are the same in the "Most times" category.

Table for Problems 11-16.

Type of Spam	July	August	Difference
Internet	14	22	22 – 14 = 8
Other	28	32	32 – 28 = 4
Scams	18	20	20 – 18 = 2
Products	40	40	40 – 40 = 0
Spiritual	2	2	2 – 2 = 0
Financial	30	28	28 – 30 = –2
Leisure	16	14	14 – 16 = –2
Adult	28	24	24 – 28 = –4
Health	24	18	18 – 24 = –6

11. Internet (8 more in August than in July)

13. Products and Spiritual spams stayed the same

15. Health (6 fewer in August than in July)

17. 5 clocks = $5 \times 10 = 50$ hours per week

19. 2 and about $\frac{1}{4}$ clocks = $2.25 \times 10 \approx 22$ hours per week

21. Administrative worked the fewest.

23. Microsoft had the most users.

25. 7 symbols = $7 \times 10 = 70$ million users

27. 3 and $\frac{1}{2}$ symbols = $3.5 \times 10 = 35$ million users

29. AOL: 75 million users; Yahoo: 70 million users. Thus the difference in the number of users is 75 – 70 = 5 million.

31. $2\% \times 725 = 0.02 \times 725 = 14.5$

33. $70\% \times 725 = 0.70 \times 725 = 507.5$

35. $34\% - 9\% = 25\%$

37. Wendy's has the fewest calories with 310.

39. Jack in the Box (since it has the least amount of sodium)

41. Answers may vary.

43. White

45. $526.0 - 344.2 = \$181.8$ billion

47. 7 students come by car.

49. Number by cycle – Number who walk
$= 7 - 4 = 3.$

Section 6.2 – Bar and Line Graphs

Problems

1. a. Looking at the red bar, we see that about 34 women reported blood clots while taking the medicine.
b. Looking at the blue bar, we see that about 17 women reported blood clots while taking the placebo.
c. 34 – 17 = 17 more women reported blood clots while taking the medicine.

2. a. The second most frequent destination is beach or lake, and 67% of the people select it.
b. The second least destination is the Big city, and 60% of the people select it.

3.

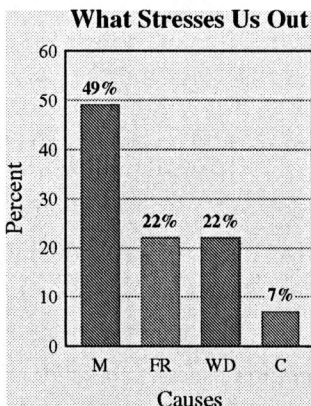

What Stresses Us Out

4.

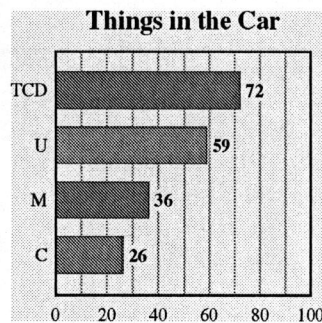

Things in the Car

5. a. The placebo group had fewer strokes from years 1 to 7.
b. The placebo group had fewer breast cancers from years 4 to 7.
c. Breast cancer patients were better off when taking the medicine in years 1 to 4.

6.

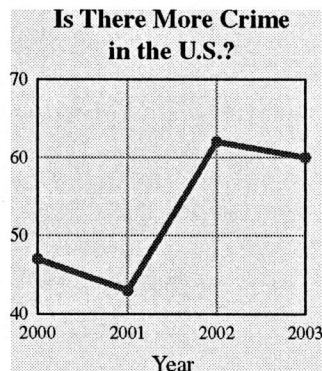

Is There More Crime in the U.S.?

Exercises 6.2

1. a. 10
   b. 16
   c. Taking the medicine would be better, since fewer women reported colorectal cancer while taking it.

3. a. About 4
   b. About 5
   c. Taking the medicine would be better, since fewer women reported endometrial cancer while taking it.

5. College spends the most; $413

7. $413 - 119 = 294$; The difference is $294.

9. a. 59%
   b. 20%
   c. $20\% \times 725 = 0.20 \times 725 = 145$
   This is 145 of the 725 employees.

11. a. The 20 to 29 age group has the most fatalities (40).
   b. The 13 to 15 age group has the least fatalities (other than the <1 age group); answers may vary.
   c. Less than 50:
   $0 + 5 + 1 + 31 + 40 + 22 + 27 = 126$
   More than 50: $28 + 9 + 19 + 17 + 2 = 75$
   Thus, there are more fatalities involving people that are less than 50 years old.
   d. The 90+ age group; this population is smaller than the other age groups.

13. a. 39
   b. $13 + 14 + 2 = 29$ were legally drunk
   c. 0.20–0.29; 14 had this BAL

15. a.

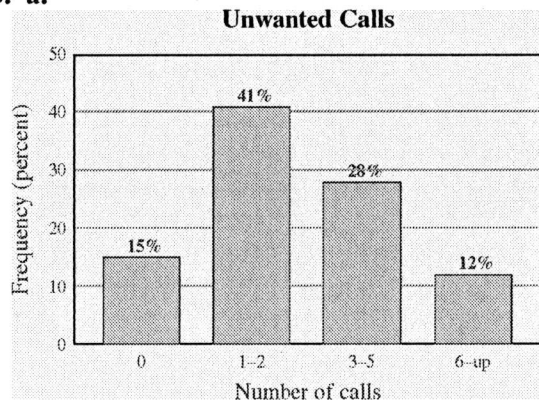

b. 1–2 calls is most common.
c. 15%

17. a.

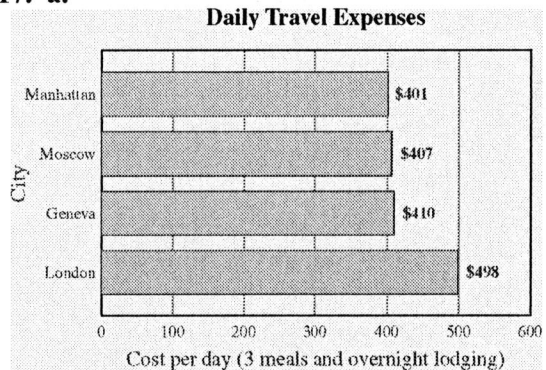

b. London is the most expensive city.
c. Manhattan is the least expensive city.
d. $498 - $401 = $97

19. a. The price was highest in years 3 and 4.
   b. The price was lowest in year 5.
   c. The price was about $1 a lb in year 5.

21. a. The boys rate of participation in 2001–2003 was about 47%.
   b. The girls rate of participation in 2001–2003 was about 33%.
   c. The difference was about $47 - 33 = 14\%$.
   d. The difference was about $42 - 22 = 20\%$.
   e. The difference was greater in 1980–1982.

23. a. Highest: 1982; lowest: 2001
   b. The percent difference is 64 − 36 = 28%.
   c. They felt safest in 2001, since the percent saying "no" is highest (or the percent saying "yes" is lowest) in this year.

25.

Average Amount of Money Spent on Apparel by Men Between 16 and 25

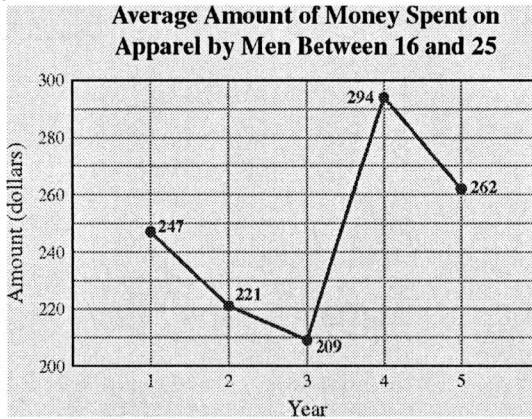

27.

Average Amount of Money Spent on Food by Persons Under 25

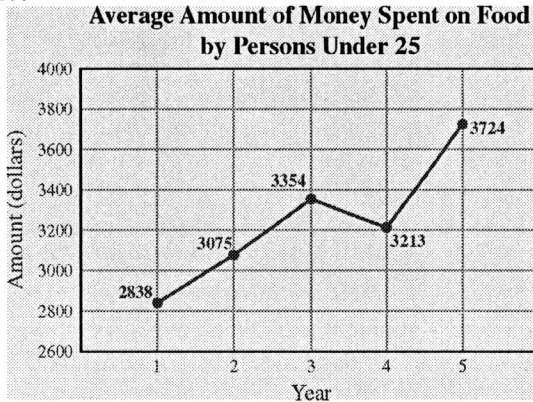

29.

Average Amount of Money Spent on Fresh Fruit by Persons Under 25

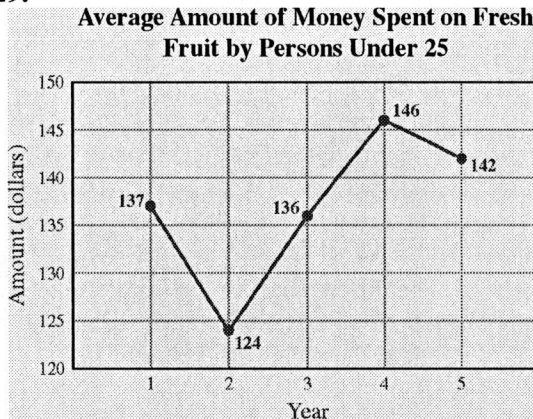

31.

Average Amount of Money Spent on Housing by Persons Under 25

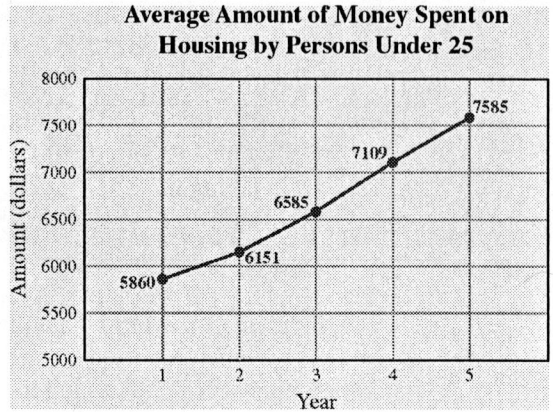

33.

Average Amount of Money Spent on Entertainment by Persons Under 25

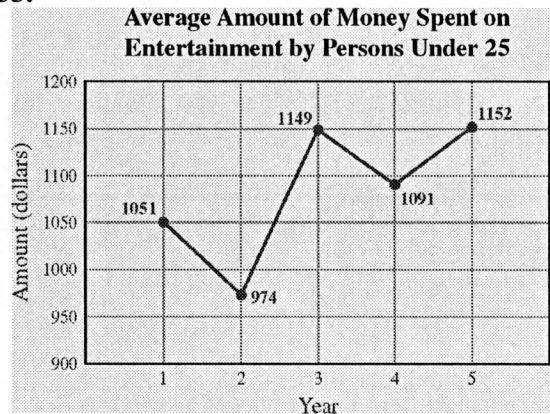

35.

Average Amount of Money Spent on Health Care by Persons Under 25

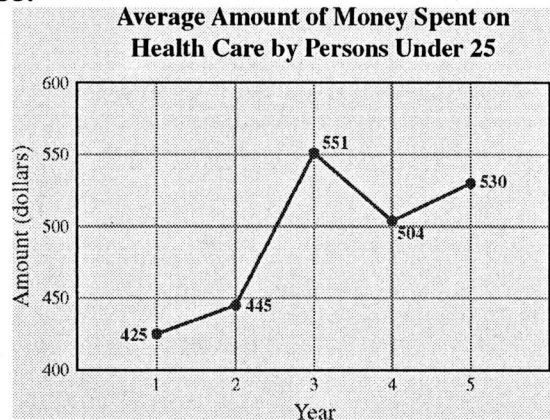

37.

Average Amount of Annual Wages-Salaries Earned by Persons Between 25 and 34

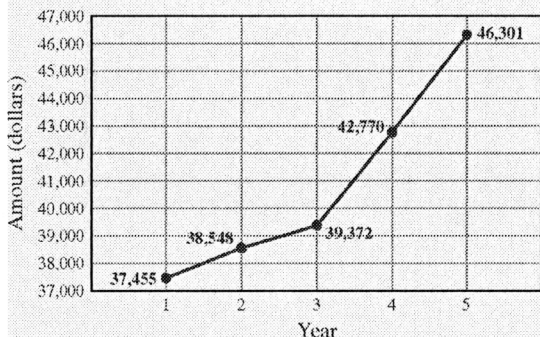

39.

Average Amount of Annual Federal Income Taxes Paid by Persons Between 25 and 34

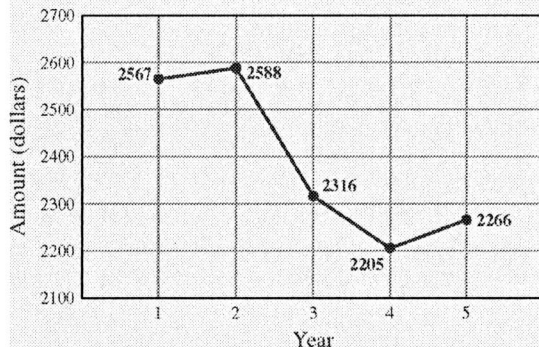

41. $22\% \times 500 = 0.22 \times 500 = 110$

43. $5\% \times 500 = 0.05 \times 500 = 25$

45. The increase was

$3,205,000 - 3,104,000 = 101,000$ so the percent of increase is

$$\frac{101,000}{3,104,000} = \frac{101}{3,104}$$

$$= 3104 \overline{)101.000} \approx 0.03 = 3\%$$
$$\begin{array}{r} 0.032 \\ \underline{93\,12} \\ 7880 \\ \underline{6208} \\ 1672 \end{array}$$

No, this is not even close to 200%!

47. Answers may vary.

Mastery Test 6.2

49.

Average Amount of Vehicle Insurance Paid by a Driver Less Than 25 Years Old

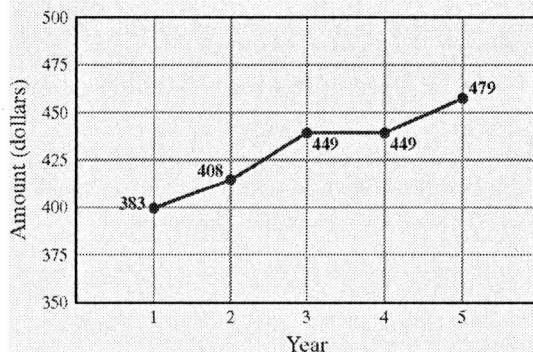

51.

Projected Percent of Age Ranges in the United States for the Year 2050

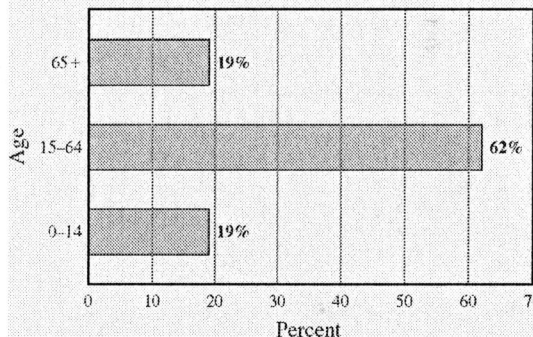

53. a. 47% answered better on October 6–8.
b. 57% answered worse on Jan. 10–14.
c. September 7–10

Section 6.3 – Circle Graphs (Pie Charts)

Problems

1. **a.** 38% of the people eat breakfast every day.
   b. The category *Some days* occupies the smallest sector, occupying 19% of the graph.
   c. $19\% \times 500 = 0.19 \times 500 = 95$ people

2. **a.** About 53% (52.66%) of the households contain people who are currently not on diets.
   b. The difference is about $53 - 44 = 9\%$.
   c. About $44\% \times 500 = 0.44 \times 500 = 220$ people

3. **a.** 11% of the commuters carpool.
   b. 4% of the commuters use other ways of commuting.
   c. 80% of the commuters drive alone, and $80\% \times 500 = 0.80 \times 500 = 400$. Thus 400 of the 500 commuters would be expected to drive alone.

4.

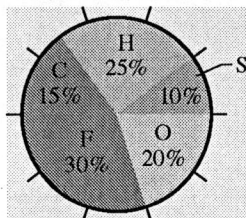

Exercises 6.3

1. **a.** The bus is the preferred mode of transportation (43%).
   b. The bike is the least preferred mode of transportation (5%).
   c. The percent difference is $91 - 20 = 71\%$.

3. **a.** Cheddar cheese was produced the most (36.0%).
   b. Swiss cheese was produced the least (2.8%).
   c. Mozzarella is the second most popular cheese (30.6%).

5. **a.** Water is used most in the toilet – 27%.
   b. $17\% \times 500 = 0.17 \times 500 = 85$ gal would be used for showering.
   c. The faucet uses more water.
   d. Set up a proportion using the ratio $\dfrac{\text{amount}}{\text{percent}}$. Let $f =$ amount used by

   faucets. $\dfrac{10}{5\%} = \dfrac{f}{15\%}$
   $$0.05f = 10 \times 0.15$$
   $$f = \frac{1.5}{0.05} = \frac{150}{5} = 30$$

   The faucet would use 30 gal of water.

7. **a.** Paper is the most prevalent item – 40%.
   b. Yard trimmings is the second most prevalent item – 18%.
   c. $40\% \times 50 = 0.40 \times 50 = 20$ lb of paper
   d. $18\% \times 50 = 0.18 \times 50 = 9$ lb of yard trimmings

9. **a.** Oil produces the most energy – 33%.
   b. Nuclear produces the least energy – 5%.
   c. Of these three energy sources, natural gas produces the least energy – 18%.

11. Food is the greatest expense since its percentage is highest.

13. Books is the smallest expense, since its percentage is the less than all other percentages.

15.

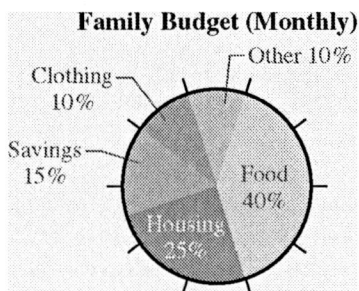

Family Budget (Monthly)

Other 10%
Clothing 10%
Savings 15%
Food 40%
Housing 25%

17.

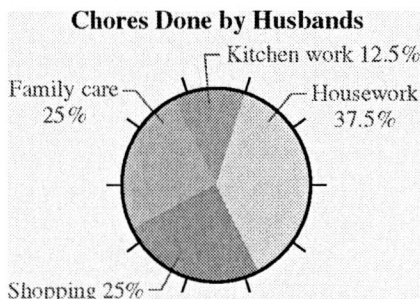

Chores Done by Husbands

Kitchen work 12.5%
Family care 25%
Housework 37.5%
Shopping 25%

19.

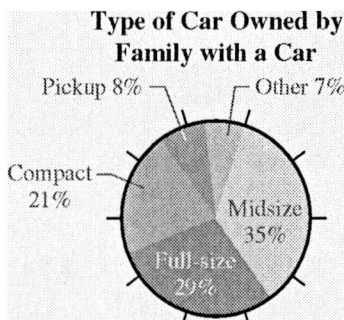

Type of Car Owned by Family with a Car

Pickup 8%
Other 7%
Compact 21%
Midsize 35%
Full-size 29%

21.
$$\begin{array}{r} 24 \\ \times\ 50 \\ \hline 1200 \end{array}$$

23.
$$\begin{array}{r} \overset{9\ 12}{\cancel{1.0}2} \\ -\ 0.84 \\ \hline 0.18 \end{array}$$

25. Transportation: $0.40 \times 360° = 144°$

27. Industry: $0.15 \times 360° = 54°$

29. Answers may vary.

Mastery Test 6.3

31. 34% of the workers stay late for one hour.

33. 13% of the workers stay late for three hours. $13\% \times 500 = 0.13 \times 500 = 65$. You would expect 65 of the 500 workers to say that they stay late for three hours.

Section 6.4 – Mean, Median, and Mode

Problems

1. The mean number of albums sold is
$$\frac{19+19+19+18+17}{5} = \frac{92}{5} = 18.4 \text{ million}$$
copies.

2. The average (mean) number of calories is:
$$\frac{511+590+550+607}{4} = \frac{2258}{4} = 564.5 \approx 565$$
calories

3. Add up the number of credit hours in column 4 to get 14 credit hours. The grade points are found by multiplying the number of credit hours in each course by the points earned from the grade in the corresponding course shown in the third column. Add the grade points to get 42 total grade points. Thus, the students

   GPA is $\dfrac{\text{Points earned}}{\text{Credit hours}} = \dfrac{42}{14} = 3.00$.

4. First write the amounts in ascending order:

        1170   1185   1256   1286

   Now take the average of two middle

   numbers: $\dfrac{1185 + 1256}{2} = \dfrac{2441}{2} = 1220.5$

   The median amount is \$1220.50.

5. The number that occurs most often is 4, so the mode is 4.

6. **a.** The mean salary is

       $\dfrac{29 + 15 + 5 + 5 + 5 + 3}{6} = \dfrac{62}{6} = 10.33$

       which is about \$10 million.

   b. The numbers are in descending order so the median is the average of the two

       middle numbers: $\dfrac{5 + 5}{2} = \dfrac{10}{2} = 5$. The

       median salary is \$5 million.

   c. The salary that occurs the most is \$5 million, so the mode is \$5 million.

   d. Because Shaquille O'Neal pulls up the mean salary.

Exercises 6.4

1. $\dfrac{1 + 5 + 9 + 13 + 17}{5} = \dfrac{45}{5} = 9$; The mean is 9.

3. $\dfrac{1 + 4 + 9 + 16 + 25 + 36}{6} = \dfrac{91}{6} = 15\dfrac{1}{6} \approx 15.2$;

   The mean is 15.2.

5. The numbers are in ascending order, and there are an odd number of values (5) so the median is the middle number. Thus, the median is 9.

7. First write the numbers in ascending order.

       1, 4, 9, 16, 25, 36

   Since there is an even number of values (6), the median is the average of the two middle numbers. Thus, the median is

   $\dfrac{9 + 16}{2} = \dfrac{25}{2} = 12.5$.

9. Both 5 and 52 occur the most number of times (twice), so there are two modes: 5 and 52.

11. The number 11 occurs the most often (three times) so the mode is 11.

13. **a.** The mean price is:

        $\dfrac{5.34 + 4.55 + 4.41 + 4.38 + 4.35}{5}$

        $= \dfrac{23.03}{5} = 4.606 = \$4.61$

    The prices are in descending order, and there are an odd number of prices given so the median price is \$4.41, the middle value. There is no mode.

    b. No.

    c. Answers will vary, but the median is the best answer.

    d. Deleting Hong Kong gives

        mean $= \dfrac{4.55 + 4.41 + 4.38 + 4.35}{4} = \$4.42$

        and median $= \dfrac{4.41 + 4.38}{2} = \$4.40$.

    Thus, both the mean and the median are most representative of the prices.

15. The points earned are as follows.

    A: $4 \times 3 = 12$
    B: $3 \times 3 = 9$
    D: $1 \times 3 = 3$ so GPA $= \dfrac{24}{13} = 1.85$
    F: $0 \times 4 = 0$
    $\overline{24}$

17. **a.** The sum of the salaries over the 5-year period is \$216,082. Thus the mean salary over the 5-year period is

        $\dfrac{216,082}{5} = \$43,216.40$.

    Arrange the salaries in ascending order:

39,276; 39,468; 40,827; 46,778; 49,733
Since there are an odd number of values
the median salary is $40,827, the
middle value. There is no mode.

b. Median; answers may vary.

19. a. The rounded values are given to the
right in the Salary column.

Player	Salary	
Jeter	$15,600,000	→ $16,000,000
Mondesi	$13,000,000	→ $13,000,000
Williams	$12,357,143	→ $12,000,000
Mussina	$12,000,000	→ $12,000,000
Pettitte	$11,500,000	→ $12,000,000
Giambi	$11,428,571	→ $11,000,000
Riveria	$10,500,000	→ $11,000,000
Posada	$8,000,000	→ $8,000,000
Clemens	$7,061,181	→ $7,000,000
Hitchcock	$6,000,000	→ $6,000,000

b. The mean salary is:

$$\frac{16+13+12+12+12+11+11+8+7+6}{10}$$

$$=\frac{108}{10}=\$10.8 \text{ million } (\$10,800,000)$$

The salaries are in descending order,
and there is an even number of salaries
given so the median salary is:

$$\frac{12+11}{2}=\$11.5 \text{ million } (\$11,500,000)$$

The mode is $12,000,000 since it is the
most frequent occurring salary.

c. All three are good measures because
they are close. (Answers will vary.)

21. $60 \cdot \dfrac{1}{12} = \dfrac{\overset{5}{\cancel{60}}}{1} \cdot \dfrac{1}{\cancel{12}_1} = 5$

23. $10,560 \cdot \dfrac{1}{5280} = \dfrac{10,56\cancel{0}}{1} \cdot \dfrac{1}{528\cancel{0}} = \dfrac{1056}{528} = 2$

25.
$$
\begin{array}{r}
{}^{1\ 1} \\
8.75 \\
\times\ \ 12 \\
\hline
1750 \\
875 \\
\hline
105.00 = 105
\end{array}
$$

27. This average is the mode, since there is
expected to be more workers at the
company than the other positions listed.

29. Answers may vary.

Mastery Test 6.4

31. The total of all spams received in July is
200 (found by adding up the values in the
July column). Since there are 9 types of
spams listed, divide 200 by 9:

$$\frac{200}{9} = 9\overline{)200.00} \quad \begin{array}{r} 22.\overline{2} \\ \end{array}$$

$$\begin{array}{r} 22.\overline{2} \\ 9\overline{)200.00} \\ \underline{18} \\ 20 \\ \underline{18} \\ 2\ 0 \end{array}$$

The mean number of
spams received in July is about 22.2.

33. The mode of the number of spams
received in July is 28, since this value
appears most often (twice).

35. Find the difference in the number of
spams from July to August for each
category using subtraction. The greatest
difference is for Internet spams, which is
$22 - 14 = 8$ more spams in August than in
July.

Review Exercises – Chapter 6

1. a. 42% of the market will be reached.
b. 39% of the market will be reached.
c. The potential revenue is $1514.44.
d. The potential revenue is $1878.16.
e. The highest percentage of the market
will be reached when the price per song
is $0.01.

2. a. McDonald's sold the most, since there
are more hamburgers shown.
b. Wendy's sold the second most
hamburgers.
c. 2 symbols × 100 burgers = 200 burgers
There were 200 Jack in the Box
hamburgers sold.

d. $5\frac{1}{2}$ symbols × 100 burgers = 550 burgers. There were 550 McDonald's hamburgers sold.

e. 1 symbol × 100 burgers = 100 burgers There were 100 Del Taco hamburgers sold.

3. a. Second highest percentage: Ages 18–24
b. Next-to-the-lowest percentage: Ages 25–34
c. 44% of the 18–24 age bracket downloaded music files.
d. The percent difference is 29% – 19% = 10%.
e. The percent difference between the age categories with the highest and lowest percent of downloads is 56% – 19% = 37%.

4.

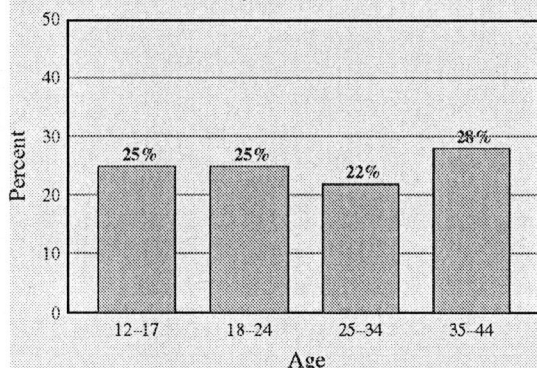

Consumers Who Have Downloaded Music and Said They Purchased More Music

5.

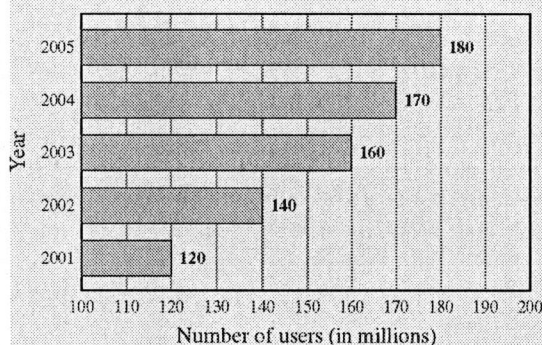

Number of Cell Phone Users by Year

6. a. 1960: About 175,000,000
b. 1970: About 200,000,000
c. 1990: About 250,000,000
d. 2010: About 300,000,000
e. 2020: About 325,000,000

7. a. 1960: About 12,500,000
b. 1980: About 25,000,000
c. 2020: About 50,000,000
d. 2040: About 75,000,000
e. 2050: About 80,000,000

8.

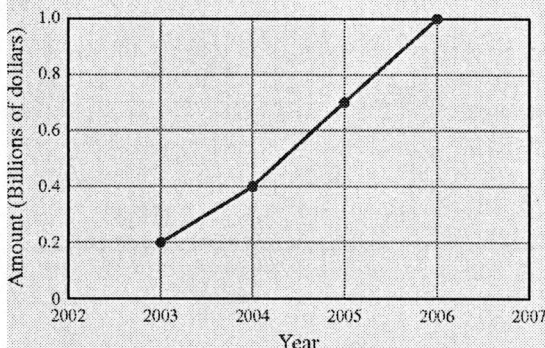

Online Spending Forecast for Kids

9. a. 13% of the people chose restaurant.
b. 42% of the people chose home.
c. 22% of the people chose public place.
d. Home
e. The percent difference is 42% – 22% = 20%.

10. a. $0.10 \times 1000 = \frac{1}{10} \times 1000 = 100$ people

b. $0.13 \times 1000 = \frac{13}{100} \times 1000 = 130$ people

c. $0.13 \times 1000 = \frac{13}{100} \times 1000 = 130$ people

d. $0.22 \times 1000 = \frac{22}{100} \times 1000 = 220$ people

e. $0.42 \times 1000 = \frac{42}{100} \times 1000 = 420$ people

11.

Ways We Commute to Work

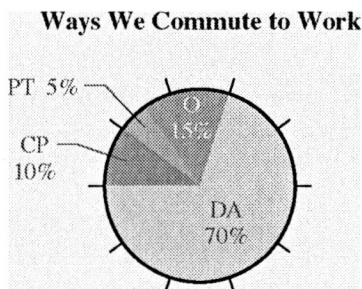

PT 5%

O 15%

CP 10%

DA 70%

12. a. Mean price: $\dfrac{79+99+99+99+89}{5}$

$= \dfrac{465}{5} = 93$ cents

b. Median price: 99 cents

c. Mode: 99 cents

d. Mean price for the first four:

$\dfrac{79+99+99+99}{4} = \dfrac{376}{4} = 94$ cents

e. Since we now have an even number of values, the median is found to be

$\dfrac{99+99}{2} = 99$ cents .

13. a. $\dfrac{17+13+18+14+12}{5} = \dfrac{74}{5} = 14.8$

The mean fat content for the five hamburgers is 15 grams.

b. Arranging the fat values in order gives 12, 13, 14, 17, and 18. Thus the median fat content for the five burgers is 14 grams.

c. The is no mode.

d. Mean fat content for the first four:

$\dfrac{17+13+18+14}{4} = \dfrac{62}{4} = 15.5 \approx 16$ grams

e. Arrange the first four values in order: 13, 14, 17, 18. Since we now have an even number of values, the median is found to be $\dfrac{14+17}{2} = \dfrac{31}{2} = 15.5 \approx 16$

grams.

14. Credit hours: $3+3+4+1+3 = 14$

Total points earned:

$3 \times 2 + 3 \times 3 + 4 \times 2 + 1 \times 4 + 3 \times 4$

$= 6 + 9 + 8 + 4 + 12$

$= 39$

The GPA is $\dfrac{39}{14} = 14\overline{)39.00} \approx 2.8$.

$$
\begin{array}{r}
2.78 \\
14\overline{)39.00} \\
\underline{28} \\
110 \\
\underline{98} \\
120 \\
\underline{112} \\
8
\end{array}
$$

15. Credit hours: $3+2+3+4+3 = 15$

Total points earned:

$3 \times 3 + 2 \times 1 + 3 \times 2 + 4 \times 4 + 3 \times 4$

$= 9 + 2 + 6 + 16 + 12$

$= 45$

The GPA is $\dfrac{45}{15} = 3.0$.

Cumulative Review Chapters 1–6

1. 6510

2. $4 \div 2 \cdot 2 + 9 - 6 = 2 \cdot 2 + 9 - 6$

$= 4 + 9 - 6$

$= 13 - 6$

$= 7$

3. $\dfrac{2}{9}$ is proper, since the numerator is less than the denominator.

4. $\dfrac{25}{8} = 3\dfrac{1}{8}$

5. $2\dfrac{1}{6} = \dfrac{6 \times 2 + 1}{6} = \dfrac{12 + 1}{6} = \dfrac{13}{6}$

6. $\left(\dfrac{6}{5}\right)^2 \cdot \dfrac{1}{36} = \left(\dfrac{6}{5} \cdot \dfrac{6}{5}\right) \cdot \dfrac{1}{36} = \dfrac{36}{25} \cdot \dfrac{1}{36} = \dfrac{1}{25}$

7. $\dfrac{15}{2} \div 4\dfrac{1}{6} = \dfrac{15}{2} \div \dfrac{25}{6} = \dfrac{\overset{3}{\cancel{15}}}{\underset{1}{\cancel{2}}} \cdot \dfrac{\overset{3}{\cancel{6}}}{\underset{5}{\cancel{25}}} = \dfrac{9}{5} = 1\dfrac{4}{5}$

8. $z - \dfrac{3}{4} = \dfrac{3}{8}$

$\quad z = \dfrac{3}{8} + \dfrac{3}{4}$

$\quad\quad = \dfrac{3}{8} + \dfrac{6}{8} = \dfrac{9}{8}$

9. Let n = the number.

$\dfrac{8}{9}n = 4\dfrac{1}{4}$

$\dfrac{8}{9}n = \dfrac{17}{4}$

$\dfrac{\cancel{\dfrac{8}{9}}n}{\cancel{\dfrac{8}{9}}} = \dfrac{\dfrac{17}{4}}{\dfrac{8}{9}}$

$n = \dfrac{17}{4} \cdot \dfrac{9}{8} = \dfrac{153}{32} = 4\dfrac{25}{32}$

10. 352.51: Three hundred fifty-two and fifty-one hundredths

11. $64.175 = 60 + 4 + \dfrac{1}{10} + \dfrac{7}{100} + \dfrac{5}{1000}$

12.
$$
\begin{array}{r}
\overset{\scriptstyle 3\;10\;14}{5\,\cancel{4}\,\cancel{1}.4\,2} \\
-\;\;1\,2.5\,0 \\
\hline
5\,2\,8.9\,2
\end{array}
$$

13.
$$
\begin{array}{r}
\overset{2\;2}{59.9} \\
\times\;0.013 \\
\hline
1797 \\
599 \\
\hline
0.7787
\end{array}
$$

14. $\dfrac{126}{0.21} = \dfrac{12{,}600}{21} = \dfrac{\cancel{3} \cdot \overset{600}{\cancel{4200}}}{\cancel{3} \cdot \underset{1}{\cancel{7}}} = 600$

15. $749.\underline{8}51 \rightarrow 749.9$

16.

$80 \div 0.13 = \dfrac{80}{0.13} = \dfrac{8000}{13} = $

$$
\begin{array}{r}
615.384 \\
13\overline{)8000.000} \\
\underline{78} \\
20 \\
\underline{13} \\
70 \\
\underline{65} \\
5\,0 \\
\underline{39} \\
110 \\
\underline{104} \\
60 \\
\underline{52} \\
8
\end{array}
$$

Thus $80 \div 0.13 = 615.38$.

17. $0.78 = \dfrac{78}{100} = \dfrac{\cancel{2} \cdot 39}{\cancel{2} \cdot 50} = \dfrac{39}{50}$

18. Let d = the decimal part.

$d \times 30 = 9$

$\quad d = \dfrac{9}{30} = \dfrac{3}{10} = 0.3$

19. $9.568 = 9.5680$

$9.56\overline{8} = 9.5688\cdots$

$9.5\overline{68} = 9.56868\cdots$

Considering the fourth decimal digit, we have $9.56\overline{8} > 9.5\overline{68} > 9.568$.

20. $0.53 = \dfrac{53}{100}$ and $\dfrac{19}{20} = \dfrac{95}{100}$. Since $53 < 95$,

we have $0.53 < \dfrac{19}{20}$.

21. $x + 3.7 = 7.9$

$\quad x = 7.9 - 3.7$

$\quad x = 4.2$

22. $4.5 = 0.5y$

$\dfrac{4.5}{0.5} = \dfrac{\cancel{0.5}y}{\cancel{0.5}}$

$\dfrac{45}{5} = y$

$\quad 9 = y$

23. $\quad 4 = \dfrac{z}{3.3}$

$\quad\quad 3.3 \times 4 = z$

$\quad\quad\quad 13.2 = z$

24. $\dfrac{28\not{0}}{100\not{0}} = \dfrac{28}{100} = \dfrac{\not{4} \cdot 7}{\not{4} \cdot 25} = \dfrac{7}{25}$

25. $\dfrac{4}{7} = \dfrac{28}{x}$

26. $\dfrac{10}{19} \overset{?}{=} \dfrac{40}{76}; \quad 10 \cdot 76 = 760$ and $19 \cdot 40 = 760$.

Since $10 \cdot 76 = 19 \cdot 40$, the proportion

$\dfrac{10}{19} = \dfrac{40}{76}$ is true. Hence yes, the flag is of

the correct ratio.

27. $\quad \dfrac{x}{2} = \dfrac{2}{8}$

$\quad 8x = 2 \cdot 2$

$\quad\quad x = \dfrac{4}{8} = \dfrac{1}{2}$

28. $\quad \dfrac{10}{p} = \dfrac{2}{3}$

$\quad 2p = 10 \cdot 3$

$\quad\quad p = \dfrac{30}{2} = 15$

29. $\dfrac{500 \text{ miles}}{19 \text{ gal}} = 19\overline{)500.0} \approx 26 \text{ mi/gal}$

$\quad\quad\quad\quad\quad\quad 26.3$

$\quad\quad\quad\quad\quad\quad \underline{38}$

$\quad\quad\quad\quad\quad\quad 120$

$\quad\quad\quad\quad\quad\quad \underline{114}$

$\quad\quad\quad\quad\quad\quad\quad 6\,0$

$\quad\quad\quad\quad\quad\quad\quad \underline{5\,7}$

$\quad\quad\quad\quad\quad\quad\quad\quad 3$

30. $\dfrac{\$2.49}{20 \text{ oz}} = 20\overline{)2.490} \approx 12\cancel{c}/\text{oz}$

$\quad\quad\quad\quad\quad\quad\quad 0.124$

$\quad\quad\quad\quad\quad\quad\quad 2\,0$

$\quad\quad\quad\quad\quad\quad\quad \underline{}$

$\quad\quad\quad\quad\quad\quad\quad\, 49$

$\quad\quad\quad\quad\quad\quad\quad\, \underline{40}$

$\quad\quad\quad\quad\quad\quad\quad\quad 90$

$\quad\quad\quad\quad\quad\quad\quad\quad \underline{80}$

$\quad\quad\quad\quad\quad\quad\quad\quad 10$

31. Let p = number of pounds needed.

$\quad \dfrac{1}{1000} = \dfrac{p}{4000}$

$\quad 1000p = 1 \cdot 4000$

$\quad\quad\quad p = \dfrac{4000}{1000} = 4$

You need 4 pounds of fertilizer.

32. Let n = number of ounces needed.

$\quad \dfrac{3}{4} = \dfrac{n}{52}$

$\quad 4n = 3 \cdot 52$

$\quad\quad n = \dfrac{3 \cdot \cancel{52}^{13}}{\cancel{4}} = 39$

You need 39 ounces.

33. $67\% = 0.67$

34. $3\dfrac{3}{4}\% = 3.75\% = 0.0375$

35. $0.09 = 9.\% = 9\%$

36. Let P = the number.

$\quad 80\% \times 60 = P$

$\quad 0.80 \times 60 = P$

$\quad\quad\quad\quad 48 = P$

Thus 80% of 60 is 48.

37. $66\dfrac{2}{3}\% \times 54 = \dfrac{2}{3} \times 54 = \dfrac{2}{\cancel{3}_1} \times \dfrac{\cancel{54}^{18}}{1} = 36$

Thus $66\dfrac{2}{3}\%$ of 54 is 36.

38. Let R = the percent.

$\quad R \times 32 = 16$

$\quad\quad\quad R = \dfrac{16}{32} = \dfrac{1}{2} = 0.5 = 50\%$

Thus 50% of 32 is 16.

39. Let B = the number.

$\quad\quad 12 = 40\% \times B$

$\quad\quad 12 = 0.40 \times B$

$\quad \dfrac{12}{0.40} = B \;$ so $\; B = \dfrac{120}{4} = 30$

Thus 12 is 40% of 30.

40. Total price $= 18 + 18 \times 4\%$
$$= 18 + 18 \times 0.04$$
$$= 18 + 0.72$$
$$= \$18.72$$

41. $I = P \times R \times T$
$$= 300 \times 0.045 \times 5$$
$$= \$67.50$$

42. 27% of families have incomes between $15,000 and $19,999.

43. About 26 inches

44. Average temperature in May $\approx 75°$
Average temperature in March $\approx 68°$
Thus the difference is about $75 - 68 = 7°$.

45. Percent of hours in P.E. is 30% and the percent of hours in art is 20%. Thus the number of hours in P.E. and art combined is $30 + 20 = 50\%$.

46. The mode is 2, since this value occurs most frequently (5 times).

47. First find the sum of the twelve numbers. The sum is 108. Thus the mean is
$$\frac{108}{12} = 9 \, .$$

48. First arrange the eleven numbers in order:
4, 4, 6, 7, 14, 18, 23, 24, 29, 29, 29.
The median is the middle number. Thus, the median is 18.

Chapter 7

Measurement and the Metric System

Section 7.1 – Linear (Length) Measures

Problems

1. 1 yd = 3 ft
   3 yd = 9 ft = 9·12 in. = 108 in.

2. **Method 1:**

   $40 \text{ in.} = 40 \cdot \dfrac{1}{12} \text{ ft} = \dfrac{40}{12} \text{ ft} = 3\dfrac{1}{3} \text{ ft}$

   Method 2:

   $40 \text{ in.} = 40 \text{ in.} \cdot \dfrac{1 \text{ ft}}{12 \text{ in.}} = \dfrac{40}{12} \text{ ft} = 3\dfrac{1}{3} \text{ ft}$

   Thus, 40 in. = $3\dfrac{1}{3}$ ft.

3. **Method 1:** $7 \text{ in.} = 7 \cdot \dfrac{1}{12} \text{ ft} = \dfrac{7}{12} \text{ ft}$

   Method 2: $7 \text{ in.} = 7 \text{ in.} \cdot \dfrac{1 \text{ ft}}{12 \text{ in.}} = \dfrac{7}{12} \text{ ft}$

   Thus, 7 in. = $\dfrac{7}{12}$ ft.

4. **Method 1:**

   $38 \text{ ft} = 38 \cdot \dfrac{1}{3} \text{ yd} = \dfrac{38}{3} \text{ yd} = 12\dfrac{2}{3} \text{ yd}$

   Method 2:

   $38 \text{ ft} = 38 \text{ ft} \cdot \dfrac{1 \text{ yd}}{3 \text{ ft}} = \dfrac{38}{3} \text{ yd} = 12\dfrac{2}{3} \text{ yd}$

   Thus, 38 ft = $12\dfrac{2}{3}$ yd.

5. **Method 1:**

   $10{,}560 \text{ ft} = 10{,}560 \cdot \dfrac{1}{5280} \text{ mi}$

   $= \dfrac{10{,}560}{5280} \text{ mi} = 2 \text{ mi}$

   Method 2:

   $10{,}560 \text{ ft} = 10{,}560 \text{ ft} \cdot \dfrac{1 \text{ mi}}{5280 \text{ ft}}$

   $= \dfrac{10{,}560}{5280} \text{ mi} = 2 \text{ mi}$

   Thus, 10,560 ft = 2 mi.

6. **Method 1:** 3 ft = 3·12 in. = 36 in.

   Method 2: $3 \text{ ft} = 3 \text{ ft} \cdot \dfrac{12 \text{ in.}}{1 \text{ ft}} = 36 \text{ in.}$

   Thus, 3 ft = 36 in.

7. $8.75 \text{ ft} = 8.75 \text{ ft} \cdot \dfrac{12 \text{ in.}}{1 \text{ ft}} = 105 \text{ in.}$

Exercises 7.1

1. 4 yd = 4·3 ft = 12 ft = 12·12 in. = 144 in.
   Thus, 4 yd = 144 in.

3. 2.5 yd = 2.5·3 ft
   = 7.5 ft = 7.5·12 in. = 90 in.
   Thus, 2.5 yd = 90 in.

5. $2\dfrac{1}{3} \text{ yd} = \dfrac{7}{3} \cdot 3 \text{ ft} = 7 \text{ ft} = 7 \cdot 12 \text{ in.} = 84 \text{ in.}$

   Thus, $2\dfrac{1}{3}$ yd = 84 in.

7. $50 \text{ in.} = 50 \text{ in.} \cdot \dfrac{1 \text{ ft}}{12 \text{ in.}} = \dfrac{50}{12} \text{ ft} = 4\dfrac{1}{6} \text{ ft}$

   Thus, 50 in. = $4\dfrac{1}{6}$ ft.

9. $84 \text{ in.} = 84 \text{ in.} \cdot \dfrac{1 \text{ ft}}{12 \text{ in.}} = \dfrac{84}{12} \text{ ft} = 7 \text{ ft}$

   Thus, 84 in. = 7 ft.

11. $3 \text{ in.} = 3 \text{ in.} \cdot \dfrac{1 \text{ ft}}{12 \text{ in.}} = \dfrac{3}{12} \text{ ft} = \dfrac{1}{4} \text{ ft}$

    Thus, 3 in. = $\dfrac{1}{4}$ ft.

13. $9 \text{ in.} = 9 \text{ in.} \cdot \dfrac{1 \text{ ft}}{12 \text{ in.}} = \dfrac{9}{12} \text{ ft} = \dfrac{3}{4} \text{ ft}$

Thus, $9 \text{ in.} = \dfrac{3}{4} \text{ ft.}$

15. $30 \text{ ft} = 30 \text{ ft} \cdot \dfrac{1 \text{ yd}}{3 \text{ ft}} = \dfrac{30}{3} \text{ yd} = 10 \text{ yd}$

Thus, $30 \text{ ft} = 10 \text{ yd.}$

17. $37 \text{ ft} = 37 \text{ ft} \cdot \dfrac{1 \text{ yd}}{3 \text{ ft}} = \dfrac{37}{3} \text{ yd} = 12\dfrac{1}{3} \text{ yd}$

Thus, $37 \text{ ft} = 12\dfrac{1}{3} \text{ yd.}$

19. $5280 \text{ ft} = 1 \text{ mi}$

21. $4 \text{ ft} = 4 \text{ ft} \cdot \dfrac{12 \text{ in.}}{1 \text{ ft}} = 4 \cdot 12 \text{ in.} = 48 \text{ in.}$

Thus, $4 \text{ ft} = 48 \text{ in.}$

23. $1 \text{ yd} = 3 \text{ ft}$

25. $1 \text{ mi} = 5280 \text{ ft}$

27. $1 \text{ mi} = 5280 \text{ ft}$

$\qquad = 5280 \text{ ft} \cdot \dfrac{1 \text{ yd}}{3 \text{ ft}}$

$\qquad = \dfrac{5280}{3} \text{ yd} = 1760 \text{ yd}$

Thus $1 \text{ mi} = 1760 \text{ yd.}$

29. $1760 \text{ yd} = 1760 \text{ yd} \cdot \dfrac{3 \text{ ft}}{1 \text{ yd}}$

$\qquad = 5280 \text{ ft}$

$\qquad = 5280 \text{ ft} \cdot \dfrac{1 \text{ mi}}{5280 \text{ ft}} = 1 \text{ mi}$

Thus, $1760 \text{ yd} = 1 \text{ mi.}$

31. a. $3 \text{ ft} = 3 \text{ ft} \cdot \dfrac{12 \text{ in.}}{1 \text{ ft}} = 36 \text{ in.}$

That is 36 inches.

b. Yes, it will fit.

33. a. $2\dfrac{1}{2} \text{ ft} = \dfrac{5}{2} \text{ ft} = \dfrac{5}{2} \cdot 12 \text{in.} = 30 \text{ in.}$

That is 30 inches.

b. Yes, it will fit.

35. $12{,}087 \text{ ft} = 12{,}087 \text{ ft} \cdot \dfrac{1 \text{ yd}}{3 \text{ ft}}$

$\qquad = \dfrac{12{,}087}{3} \text{ yd} = 4029 \text{ yd}$

That is 4029 yards.

37. $22 \text{ ft} = 22 \text{ ft} \cdot \dfrac{12 \text{ in.}}{1 \text{ ft}} = 264 \text{ in.}$ so

$22 \text{ ft } 1 \text{ in.} = 264 \text{ in.} + 1 \text{ in.} = 265 \text{ in.}$

That is 265 inches.

39. $8 \cdot 220 \text{ yd} = 1760 \text{ yd.}$ From exercise 29 we have $1760 \text{ yd} = 1 \text{ mi.}$ Therefore an 8-furlong race is 1 mile.

41. $11.75 \text{ ft} = 11.75 \cdot 12 \text{ in.} =$

$$\begin{array}{r} 11.75 \\ \times \quad 12 \\ \hline 2350 \\ 1175 \\ \hline 141.00 \text{ in.} \end{array}$$

That is 141 inches.

43. $5 \text{ ft} = 5 \cdot 12 \text{ in.} = 60 \text{ in.}$ so

$5 \text{ ft } 4 \text{ in.} = 60 \text{ in.} + 4 \text{ in.} = 64 \text{ in.}$

That is 64 inches.

45. $2 \text{ mi} = 2 \text{ mi} \cdot \dfrac{5280 \text{ ft}}{1 \text{ mi}} = 10{,}560 \text{ ft}$

That is 10,560 ft/hr.

47. a. $6 \text{ ft} = 6 \text{ ft} \cdot \dfrac{1 \text{ yd}}{3 \text{ ft}} = 2 \text{ yd}$

That is 2 yd/sec.

b. $2\dfrac{\text{yd}}{\text{sec}} = 2\dfrac{\text{yd}}{\text{sec}} \cdot \dfrac{60 \text{ sec}}{1 \text{ min}} = 120\dfrac{\text{yd}}{\text{min}}$

$- \text{ OR } -$

$1 \text{ sec} = 2 \text{ yd}$ so

$1 \text{ min} = 60 \text{ sec}$

$\qquad = 60 \cdot 1 \text{ sec} = 60 \cdot 2 \text{ yd} = 120 \text{ yd}$

An urbanite would go 120 yd in one minute.

49. $4 \text{ mi} = 4 \cdot 5280 \text{ ft} = 21{,}120 \text{ ft}$

$686 \text{ yd} = 686 \cdot 3 \text{ ft} = 2058 \text{ ft}$

Thus $4 \text{ mi } 686 \text{ ft} = 21{,}120 \text{ ft} + 2058 \text{ ft} = 23{,}178 \text{ ft.}$

4.55 mi = 4 mi + 0.55 mi
0.55 mi = 0.55 · 5280 ft = 2904 ft
Thus 4.55 mi = 21,120 ft + 2904 ft = 24,024 ft. Hence the 4.55 mile banana split was longer.

51. $4232 \div 1000 = \dfrac{4232}{1000} = 4.232$

53. $83.5 \cdot 100 = 8350$

55. $100 \cdot 0.465 = 46.5$

57. 36.75 mi = 36 mi + 0.75 mi
36 mi = 36 · 5280 ft = 190,080 ft
0.75 mi = 0.75 · 5280 ft = 3960 ft
Thus 36.75 mi = 190,080 ft + 3960 ft = 194,040 ft.

59. Answers may vary.

61. Answers may vary.

Mastery Test 7.1

63. $6.5 \text{ ft} = 6\dfrac{1}{2} \text{ ft} = \dfrac{13}{2} \text{ ft} \cdot \dfrac{\overset{6}{\cancel{12}} \text{ in.}}{1 \text{ ft}} = 78 \text{ in.}$
That is 78 inches.

65. $31,680 \text{ ft} = 31,680 \text{ ft} \cdot \dfrac{1 \text{ mi}}{5280 \text{ ft}}$
$= \dfrac{31,680}{5280} \text{ mi}$
$= 6 \text{ mi}$
Thus, 31,680 ft = 6 mi.

67. $25 \text{ ft} = 25 \text{ ft} \cdot \dfrac{1 \text{ yd}}{3 \text{ ft}} = \dfrac{25}{3} \text{ yd} = 8\dfrac{1}{3} \text{ yd}$
Thus, $25 \text{ ft} = 8\dfrac{1}{3} \text{ yd}.$

69. $75 \text{ in.} = 75 \text{ in.} \cdot \dfrac{1 \text{ ft}}{12 \text{ in.}}$
$= \dfrac{75}{12} \text{ ft} = \dfrac{25}{4} \text{ ft} = 6\dfrac{1}{4} \text{ ft}$
Thus, $75 \text{ in.} = 6\dfrac{1}{4} \text{ ft.}$

Section 7.2 – The Metric System

Problems

1. Method 1: 4 km = 4 · 1000 m = 4000 m
Method 2: $4 \text{ km} = 4 \text{ km} \cdot \dfrac{1000 \text{ m}}{1 \text{ km}}$
$= 4000 \text{ m}$
Thus, 4 km = 4000 m.

2. Method 1:
$3 \text{ km} = 3 \cdot \dfrac{1}{10} \text{ dam} = \dfrac{3}{10} \text{ dam} = 0.3 \text{ dam}$
Method 2:
$3 \text{ m} = 3 \text{ m} \cdot \dfrac{1 \text{ dam}}{10 \text{ m}} = \dfrac{3}{10} \text{ dam} = 0.3 \text{ dam}$
Thus, 3 m = 0.3 m.

3. Method 1: 4 dam = 4 · 10 m = 40 m
Method 2: $4 \text{ dam} = 4 \text{ dam} \cdot \dfrac{10 \text{ m}}{1 \text{ dam}}$
$= 40 \text{ m}$
Thus, 4 dam = 40 m.

4. Method 1:
215 m = 215 · 100 cm = 21,500 cm
Method 2:
$215 \text{ m} = 215 \text{ m} \cdot \dfrac{100 \text{ cm}}{1 \text{ m}} = 21,500 \text{ cm}$
Thus, 215 m = 21,500 cm.

5. Method 1:

$581 \text{ cm} = 581 \cdot \dfrac{1}{100} \text{ m} = 5.81 \text{ m}$

Method 2:

$581 \text{ cm} = 581 \cancel{\text{cm}} \cdot \dfrac{1 \text{ m}}{100 \cancel{\text{cm}}} = 5.81 \text{ m}$

Thus, $581 \text{ cm} = 5.81 \text{ m}$.

Exercises 7.2

1. $5 \text{ km} = \underline{5000} \text{ m}$

3. $1877 \text{ m} = \underline{1.877} \text{ km}$

5. $4 \text{ dm} = \underline{0.4} \text{ m}$

7. $49 \text{ m} = \underline{490} \text{ dm}$

9. $182 \text{ cm} = \underline{1.82} \text{ m}$

11. $22 \text{ m} = \underline{2200} \text{ cm}$

13. $3 \text{ m} = \underline{3000} \text{ mm}$

15. $2358 \text{ mm} = \underline{2.358} \text{ m}$

17. $30 \text{ cm} = \underline{300} \text{ mm}$

19. $67 \text{ mm} = \underline{6.7} \text{ cm}$

21. For comparison purposes, $1 \text{ in.} \approx 2.5 \text{ cm}$. Consider that most basketball players are at least $6 \text{ ft} = 6 \cdot 12 \text{ in.} = 72 \text{ in.}$ tall. Converting, $72 \text{ in.} \approx 72 \cdot 2.5 \text{ cm} = 180$ cm.
a) $200 \text{ mm} = 20 \text{ cm}$ and
b) $200 \text{ m} = 20{,}000 \text{ cm}$. Thus, the most nearly correct would be c) 200 cm.

23. Using $1 \text{ in.} \approx 2.5 \text{ cm}$, we have b) $1 \text{ mm} = 0.1 \text{ cm} \approx 0.1 \cdot \dfrac{1}{2.5} \text{ in.} = 0.04 \text{ in.}$ and c)

$1 \text{ m} = 100 \text{ cm} \approx 100 \cdot \dfrac{1}{2.5} \text{ in.} = 40 \text{ in.}$

Neither of these are reasonable so the most nearly correct answer would be
a) $1 \text{ cm} \approx 1 \div 2.5 \text{ in.} = 0.4 \text{ in.}$

25. a) $19 \text{ mm} = 1.9 \text{ cm} \approx 1.9 \cdot \dfrac{1}{2.5} \text{ in.}$
    $= 0.76 \text{ in.}$

    b) $19 \text{ cm} \approx 19 \cdot \dfrac{1}{2.5} \text{ in.} = 7.6 \text{ in.}$

    c) $19 \text{ m} = 1900 \text{ cm} \approx 1900 \cdot \dfrac{1}{2.5} \text{ in.}$
    $= 760 \text{ in.}$

The most nearly correct answer would be b).

27. $6 \text{ mm} = .6 \text{ cm} = 0.6 \text{ cm}$

29. $1.6 \text{ m} = 160. \text{ cm} = 160 \text{ cm}$

31.
$$\begin{array}{r} 2.54 \\ \times\ \ 72 \\ \hline 508 \\ 1778\ \ \\ \hline 182.88 \end{array}$$

33.
$$\begin{array}{r} 1.6 \\ \times\ 50 \\ \hline 80.0 = 80 \end{array}$$

35. $10 \cdot 0.4 = 4$

37. **i.** d
ii. e
iii. b
iv. c
v. a

39. Answers may vary.

Mastery Test 7.2

41. $92.4 \text{ m} = \underline{9240} \text{ cm}$

43. $3 \text{ m} = \underline{0.3} \text{ dam}$

45. $2300 \text{ m} = \underline{2.3} \text{ km}$

Section 7.3 – Converting Between American and Metric Units

Problems

1. **Method 1:**
   7 ft = 7·12 in.
   = 84 in. = 84·2.54 cm = 213.36 cm
   Method 2:
   7 ft = 7·12 in.
   $= 84 \text{ in.} = 84 \text{ in.} \cdot \dfrac{2.54 \text{ cm}}{1 \text{ in.}}$
   = 213.36 cm
   Thus, 7 ft = 213.36 cm.

2. **Method 1:**
   300 yd = 300·0.914 m = 274.2 m
   Method 2:
   $300 \text{ yd} = 300 \text{ yd} \cdot \dfrac{0.914 \text{ m}}{1 \text{ yd}} = 274.2 \text{ m}$
   Thus, 300 yd = 274.2 m.

3. **Method 1:** 70 mi = 70·1.6 km = 112 km
   Method 2: $70 \text{ mi} = 70 \text{ mi} \cdot \dfrac{1.6 \text{ km}}{1 \text{ mi}}$
   = 112 km
   Thus, 70 mi = 112 km.

4. **Method 1:** 200 cm = 200·0.4 in. = 80 in.
   Method 2: $200 \text{ cm} = 200 \text{ cm} \cdot \dfrac{0.4 \text{ in.}}{1 \text{ cm}}$
   = 80 in.
   Thus, 200 cm = 80 in.

5. **Method 1:** 300 m = 300·1.1 yd = 330 yd
   Method 2: $300 \text{ m} = 300 \text{ m} \cdot \dfrac{1.1 \text{ yd}}{1 \text{ m}}$
   = 330 yd
   Thus, 300 m = 330 yd.

6. **Method 1:** 70 km = 70·0.62 mi
   = 43.4 mi
   Method 2: $70 \text{ km} = 70 \text{ km} \cdot \dfrac{0.62 \text{ mi}}{1 \text{ km}}$
   = 43.4 mi
   Thus, 70 km = 43.4 mi.

7. **Method 1:**
   5 ft = 5·12 in.
   = 60 in. = 60·2.54 cm = 152.4 cm
   Method 2:
   5 ft = 5·12 in.
   $= 60 \text{ in.} = 60 \text{ in.} \cdot \dfrac{2.54 \text{ cm}}{1 \text{ in.}}$
   = 152.4 cm

8. **Method 1:** 50 yd = 50·0.914 m = 45.7 m
   Method 2: $50 \text{ yd} = 50 \text{ yd} \cdot \dfrac{0.914 \text{ m}}{1 \text{ yd}}$
   = 45.7 m

9. **Method 1:** 30 mi = 30·1.6 km = 48 km
   Method 2: $30 \text{ mi} = 30 \text{ mi} \cdot \dfrac{1.6 \text{ km}}{1 \text{ mi}}$
   = 48 km
   Thus, 30 mi/hr is equivalent to 48 km/hr.

10. **Method 1:** 200 m = 200·1.1 yd = 220 yd
    Method 2: $200 \text{ m} = 200 \text{ m} \cdot \dfrac{1.1 \text{ yd}}{1 \text{ m}}$
    = 220 yd

11. **Method 1:** 80 km = 80·0.62 mi
    = 49.6 mi
    Method 2: $80 \text{ km} = 80 \text{ km} \cdot \dfrac{0.62 \text{ mi}}{1 \text{ km}}$
    = 49.6 mi
    The car is traveling 49.6 mi/hr.

12. **Method 1:** 30 cm = 30·0.4 in. = 12 in.
    Method 2: $30 \text{ cm} = 30 \text{ cm} \cdot \dfrac{0.4 \text{ in.}}{1 \text{ cm}}$
    = 12 in.

Exercises 7.3

1. 4 ft = 4·12 in.
   = 48 in = 48·2.54 cm = <u>121.92</u> cm

3. 30 yd = 30·0.914 m = <u>27.42</u> m

5. 90 mi = 90·1.6 km = <u>144</u> km

7. 20 cm = 20·0.4 in. = <u>8</u> in.

9. 600 m = 600·1.1 yd = <u>660</u> yd

11. 10 km = 10·0.62 mi = <u>6.2</u> mi

13. 98 yd = 98·0.914 m = 89.572 m
That is about 89.6 meters.

15. 8848 m = 8848·1.1 yd = 9732.8 yd
That is about 9700 yards.

17. 100 km = 100·0.62 mi = 62 mi
That is 62 mi/hr.

19. 90 cm = 90·0.4 in. = 36 in.
60 cm = 60·0.4 in. = 24 in.
In inches this would be 36-24-36.

21. $\frac{1}{2} \cdot 15 \cdot 20 = \frac{1}{\cancel{2}} \cdot \frac{15}{1} \cdot \frac{\cancel{20}^{10}}{1} = 15 \cdot 10 = 150$

23. $\frac{1}{2} \cdot 3 \cdot 5 = \frac{1}{2} \cdot \frac{3}{1} \cdot \frac{5}{1} = \frac{15}{2} = 7\frac{1}{2}$

25. 1900 m = 1.9 km = 1.9·0.62 mi = 1.178 mi

27. 15 mi = 15·1.6 km = 24 km

29. Answers may vary.

Mastery Test 7.3

31. 400 km = 40·0.62 mi = 24.8 mi
That is 24.8 mi/hr.

33. 60 mi = 60·1.6 km = 96 km
That is 96 km/hr.

35. 7.15 ft = 7.15·12 in.
= 85.8 in.
= 85.8·2.54 cm
= 217.932 cm
= 2.17932 m
That is 2.179 meters.

37. 800 yd = 800·0.914 m = <u>731.2</u> m

39. 600 cm = 600·0.4 in. = <u>240</u> in.

41. 20 km = 20·0.62 mi = <u>12.4</u> mi

Section 7.4 – Converting Units of Area

Problems

1. $5 \text{ m}^2 = 5 \cdot (100 \text{ cm})^2$
$= 5 \cdot (100 \text{ cm}) \cdot (100 \text{ cm})$
$= \underline{50{,}000} \text{ cm}^2$

2. $2 \text{ km}^2 = 2 \cdot (1000 \text{ m})^2$
$= 2 \cdot (1000 \text{ m}) \cdot (1000 \text{ m})$
$= \underline{2{,}000{,}000} \text{ m}^2$

3. $5 \text{ ft}^2 = 5 \cdot (12 \text{ in.})^2$
$= 5 \cdot (12 \text{ in.}) \cdot (12 \text{ in.}) = \underline{720} \text{ in.}^2$

4. $432 \text{ in.}^2 = 432 \cdot \left(\frac{1}{144} \text{ ft}^2 \right)$
$= 432 \cdot \frac{1}{144} \text{ ft}^2 = \underline{3} \text{ ft}^2$

5. $36 \text{ ft}^2 = 36 \cdot \left(\frac{1}{9} \text{ yd}^2 \right) = 36 \cdot \frac{1}{9} \text{ yd}^2 = \underline{4} \text{ yd}^2$

6. $20 \text{ acres} = 20 \cdot 4840 \text{ yd}^2 = 96{,}800 \text{ yd}^2$

7. $12 \text{ hectares} = 12 \cdot 10{,}000 \text{ m}^2$
$= \underline{120{,}000} \text{ m}^2$

8. $6 \text{ hectares} = 6 \cdot 2.47 \text{ acres} = \underline{14.82} \text{ acres}$

Exercises 7.4

1. $3 \text{ km}^2 = 3 \cdot (1000 \text{ m})^2$
$= 3 \cdot (1000 \text{ m}) \cdot (1000 \text{ m})$
$= \underline{3,000,000} \text{ m}^2$

3. $2 \text{ ft}^2 = 2 \cdot (12 \text{ in.})^2$
$= 2 \cdot (12 \text{ in.}) \cdot (12 \text{ in.}) = \underline{288} \text{ in.}^2$

5. $432 \text{ in.}^2 = 432 \cdot \left(\dfrac{1}{144} \text{ ft}^2 \right)$
$= 432 \cdot \dfrac{1}{144} \text{ ft}^2 = \underline{3} \text{ ft}^2$

7. $54 \text{ ft}^2 = 54 \cdot \left(\dfrac{1}{9} \text{ yd}^2 \right) = 54 \cdot \dfrac{1}{9} \text{ yd}^2 = \underline{6} \text{ yd}^2$

9. $2 \text{ acres} = 2 \cdot 4840 \text{ yd}^2 = \underline{9680} \text{ yd}^2$

11. $3 \text{ hectares} = 3 \cdot 2.47 \text{ acres} = \underline{7.41} \text{ acres}$

13. $2 \text{ hectares} = 2 \cdot 10,000 \text{ m}^2$
$= \underline{20,000} \text{ m}^2$

15. $154,250 \text{ ft}^2 = 154,250 \cdot \left(\dfrac{1}{9} \text{ yd}^2 \right)$
$= 154,250 \cdot \dfrac{1}{9} \text{ yd}^2$
$= 17,138.\overline{8} \text{ yd}^2$
That is about 17,139 square yards.

17. $1730 \text{ acres} = 1730 \cdot 4840 \text{ yd}^2$
$= 8,373,200 \text{ yd}^2$

19. a. $50 \text{ ft} \times 30 \text{ ft} = 1500 \text{ ft}^2$
b. $1500 \text{ ft}^2 = 1500 \cdot \left(\dfrac{1}{9} \text{ yd}^2 \right) = 166\dfrac{2}{3} \text{ yd}^2$

21. $1 \text{ acre} = 4840 \text{ yd}^2$
$= 4840 \cdot (9 \text{ ft}^2)$
$= 43,560 \text{ ft}^2 \Rightarrow \dfrac{1}{43,560} \text{ acre} = 1 \text{ ft}^2$

$125,000 \text{ ft}^2 = 125,000 \cdot \dfrac{1}{43,560} \text{ acre}$
$\approx 2.869605 \text{ acres}$
That is about 2.87 acres.

23. From exercise 21 we have
$1 \text{ acre} = 43,560 \text{ ft}^2 \Rightarrow \dfrac{1}{43,560} \text{ acre} = 1 \text{ ft}^2$.

$129,000 \text{ ft}^2 = 129,000 \cdot \dfrac{1}{43,560} \text{ acre}$
$\approx 2.9614325 \text{ acres}$
That is about 2.96 acres.

25. $5 \cdot 100 = 500$

27. Area $= 12 \text{ ft} \times 11 \text{ ft} = 132 \text{ ft}^2$
$132 \text{ ft}^2 = 132 \cdot \dfrac{1}{9} \text{ yd}^2 = 14\dfrac{2}{3} \text{ yd}^2$
We need $14\dfrac{2}{3}$ square yards of carpet.

29. Area $= 50 \text{ yd} \times 20 \text{ yd} = 1000 \text{ yd}^2$;
$1000 \text{ yd}^2 = 1000 \cdot (3 \text{ ft})^2$
$= 1000 \cdot 9\text{ft}^2 = 9000 \text{ ft}^2$
We need 9000 squares of sod.

31. Area $= 12 \text{ ft} \times 9 \text{ ft} = 108 \text{ ft}^2$; $1 \text{ roll} = 36 \text{ ft}^2$
$108 \cancel{\text{ft}^2} \cdot \dfrac{1 \text{ roll}}{36 \cancel{\text{ft}^2}} = \dfrac{108}{36} \text{ rolls} = 3 \text{ rolls}$

33. One square meter; answers will vary.

Mastery Test 7.4

35. $54 \text{ ft}^2 = 54 \cdot \left(\dfrac{1}{9} \text{ yd}^2 \right) = 54 \cdot \dfrac{1}{9} \text{ yd}^2 = \underline{6} \text{ yd}^2$

37. $3 \text{ ft}^2 = 3 \cdot (12 \text{ in.})^2$
$= 3 \cdot (12 \text{ in.}) \cdot (12 \text{ in.}) = \underline{432} \text{ in.}^2$

39. $4 \text{ hectares} = 4 \cdot 2.47 \text{ acres} = \underline{9.88} \text{ acres}$

41. $10 \text{ acres} = 10 \cdot 4840 \text{ yd}^2 = 48,400 \text{ yd}^2$

Section 7.5 – Capacity

Problems

1. a. $\frac{1}{4}$ gal $= \frac{1}{4}(4$ qt$) = 1$ qt $\approx \underline{1}$ L

    b. 16 oz $= 1$ pt $= \frac{1}{2}$ qt $\approx \underline{\frac{1}{2}}$ L

    c. 10 gal $= 10 (4$ qt$) = 40$ qt $\approx \underline{40}$ L

2. 9 kL $= \underline{9000}$ L

3. 247 mL $= \underline{0.247}$ L

4. a. $\frac{1}{4}$ cup $= \frac{1}{4}(240$ mL$) = 60$ mL

    You need 60 mL of ammonia.

    b. $1\frac{1}{2}$ cup $= \frac{3}{2}(240$ mL$)$

$$= \frac{3}{\cancel{2}} \cdot \frac{\overset{120}{\cancel{240}}}{1} \text{ mL} = 360 \text{ mL}$$

    You need 360 mL of warm water.

5. 240 mL $= 1$ cup $\Rightarrow$ 1 mL $= \frac{1}{240}$ cup

$$400 \text{ mL} = 400 \cdot \left(\frac{1}{240} \text{ cup}\right)$$

$$= \frac{40\cancel{0}}{24\cancel{0}} \text{ cup} = \frac{5}{3} \text{ cup} = 1\frac{2}{3} \text{ cups}$$

That is $1\frac{2}{3}$ cups.

6. a. 30 fl. oz $= 30 (30$ mL$) = 900$ mL

    b. 900 mL $= 900 (1$ cc$) = 900$ cc

Exercises 7.5

1. $\frac{3}{4}$ gal $= \frac{3}{4}(4$ qt$) = 3$ qt $\approx \underline{3}$ L

3. 2 L $\approx \underline{2}$ qt

5. 8 L ≈ 8 qt $= 8 \left(\frac{1}{4}$ gal$\right) = \underline{2}$ gal

7. 5 gal $= 5 (4$ qt$) = 20$ qt $\approx \underline{20}$ L

9. 8 qt $\approx \underline{8}$ L

11. 177 mL $= \underline{0.177}$ L

13. 3847 mL $= \underline{3.847}$ L

15. 205 cL $= \underline{2.05}$ L

17. 55 dL $= \underline{5.5}$ L

19. 6 daL $= \underline{60}$ L

21. 7 hL $= \underline{700}$ L

23. 6 kL $= \underline{6000}$ L

25. 5 kL $= \underline{500}$ daL

27. 4 daL $= \underline{4000}$ cL

29. 9 daL $= \underline{900}$ L

31. a. 200 mL $= 0.2$ L
    b. 1800 mL $= 1.8$ L

33. 1500 mL $= 1.5$ L

35. 1 teaspoon $= 5$ mL
    There are 5 mL of borax in the cleaner.

37. 2 teaspoons $= 2 (5$ mL$) = 10$ mL
    There are 10 mL of vinegar in the cleaner.

39. 2 cups $= 2 (240$ mL$) = 480$ mL
    There are 480 mL of hot water in the cleaner.

41. 15 fl. oz $= 15 (30$ mL$) = 450$ mL

43. 0.5 fl. oz $= 0.5 (30$ mL$) = 15$ mL

45. 240 mL $= 240 \left(\frac{1}{30}$ fl. oz$\right)$

$$= \frac{240}{30} \text{ fl. oz} = 8 \text{ fl. oz}$$

47. $80 \cdot \dfrac{1}{16} = \dfrac{\overset{5}{\cancel{80}}}{1} \cdot \dfrac{1}{\cancel{16}_1} = 5$

49.
$$
\begin{array}{r}
\overset{3}{0.45} \\
\times \quad 6 \\
\hline
2.70 = 2.7
\end{array}
$$

51. a. $2 \text{ eyes} \times 2 \text{ drops} \times 4 \text{ times}$
   $= 16 \text{ drops/day}$

   b. $16 \cdot 0.2 \text{ mL} = 3.2 \text{ mL}$

   c. $\dfrac{15 \text{ } \cancel{mL}}{3.2 \text{ } \cancel{mL}/\text{day}} = \dfrac{150}{32} \text{ days} = 4.6875 \text{ days}$

53. Answers may vary.

Mastery Test 7.5

55. a. 25 fl. oz = 25 (30 mL) = 750 mL

   b. 750 mL = 750 (1 cc) = 750 cc

57. $\dfrac{1}{4}$ teaspoon $= \dfrac{1}{4}$ (5 mL)

   $= \dfrac{5}{4}$ mL = 1.25 mL

   and 1 cup = 240 mL. Hence you need
   1.25 mL of dishwashing liquid and 240
   mL of water.

59. 10 kL = <u>10,000</u> L

Section 7.6 – Weight and Temperature

Problems

1. 4 lb = 4 · (16 oz) = <u>64</u> oz

2. 32 oz = $32 \cdot \left(\dfrac{1}{16} \text{ lb} \right)$ = <u>2</u> lb

3. 2 tons = 2 · (2000 lb) = <u>4000</u> lb

4. 7000 lb = $7000 \cdot \left(\dfrac{1}{2000} \text{ ton} \right) = 3\dfrac{1}{2}$ tons

5. 3 dag = <u>300</u> dg

6. 103 mg = <u>0.103</u> g

7. 2 kg = 2 · (2.2 lb) = <u>4.4</u> lb

8. 4 lb = 4 · (0.45 kg) = <u>1.80</u> kg

9. $C = \dfrac{5(F-32)}{9} = \dfrac{5(50-32)}{9} = \dfrac{5 \cdot \cancel{18}^{2}}{\cancel{9}} = 10$
   Thus 50°F = 10°C.

10. $F = \dfrac{9C}{5} + 32 = \dfrac{9 \cdot \cancel{20}^{4}}{\cancel{5}} + 32 = 36 + 32 = 68$
   Thus 20°C = 68°F.

Exercises 7.6

1. 3 lb = 3 · (16 oz) = <u>48</u> oz

3. 4.5 lb = 4.5 · (16 oz) = <u>72</u> oz

5. 64 oz = $64 \cdot \left(\dfrac{1}{16} \text{ lb} \right)$ = <u>4</u> lb

7. 72 oz = $72 \cdot \left(\dfrac{1}{16} \text{ lb} \right) = \dfrac{9}{2}$ lb = <u>4.5</u> lb

9. 4 tons = 4 · (2000 lb) = <u>8000</u> lb

11. $2\dfrac{1}{2}$ tons = $\dfrac{5}{2} \cdot$ (2000 lb) = <u>5,000</u> lb

13. 3000 lb = $3000 \cdot \left(\dfrac{1}{2000} \text{ ton} \right)$ = <u>1.5</u> tons

15. 4000 lb = $4000 \cdot \left(\dfrac{1}{2000} \text{ ton} \right)$ = <u>2</u> tons

17. 2 kg = <u>200,000</u> cg

19. 2 dag = <u>200</u> dg

21. 2 kg = <u>200</u> dag

23. 5 dag = <u>50</u> g

25. 899 mg = <u>0.899</u> g

27. 30 mg = <u>0.030</u> g

29. 57 cg = <u>0.57</u> g

31. 4 kg = 4 · (2.2 lb) = <u>8.8</u> lb

33. 10 kg = 10 · (2.2 lb) = <u>22</u> lb

35. 2 lb = 2 · (0.45 kg) = <u>0.9</u> kg

37. 10 kg = <u>100</u> hg

39. 16 oz = 1 lb = <u>0.45</u> kg

41. $C = \dfrac{5(F-32)}{9} = \dfrac{5(59-32)}{9} = \dfrac{5 \cdot \cancel{27}^{3}}{\cancel{9}} = 15$

Thus, 59°F = 15°C.

43. $C = \dfrac{5(F-32)}{9} = \dfrac{5(77-32)}{9} = \dfrac{5 \cdot \cancel{45}^{5}}{\cancel{9}} = 25$

Thus, 77°F = 25°C.

45. $C = \dfrac{5(F-32)}{9}$

$= \dfrac{5(113-32)}{9} = \dfrac{5 \cdot \cancel{81}^{9}}{\cancel{9}} = 45$

Thus, 113°F = 45°C.

47. $F = \dfrac{9C}{5} + 32 = \dfrac{9 \cdot \cancel{10}^{2}}{\cancel{5}} + 32 = 18 + 32 = 50$

Thus, 10°C = 50°F.

49. $F = \dfrac{9C}{5} + 32 = \dfrac{9 \cdot \cancel{35}^{7}}{\cancel{5}} + 32 = 63 + 32 = 95$

Thus, 35°C = 95°F.

51. $F = \dfrac{9C}{5} + 32$

$= \dfrac{9 \cdot \cancel{1000}^{200}}{\cancel{5}} + 32 = 1800 + 32 = 1832$

That is 1832°F.

53. $C = \dfrac{5(F-32)}{9}$

$= \dfrac{5(104-32)}{9} = \dfrac{5 \cdot \cancel{72}^{8}}{\cancel{9}} = 40$

That is 40°C.

55. 160 lb = 160 · (0.45 kg) = 72 kg

57. $F = \dfrac{9C}{5} + 32$

$= \dfrac{9 \cdot \cancel{70}^{14}}{\cancel{5}} + 32 = 126 + 32 = 158$

That is 158°F.

59. 52 kg = 52 · (2.2 lb) = 114.4 lb

That is about 114 pounds.

61. $\dfrac{4}{5}$ and $\dfrac{6}{7}$; cross-multiplying we get

$4 \cdot 7 = 28 < 5 \cdot 6 = 30$ so $\dfrac{4}{5} < \dfrac{6}{7}$.

63. Get the denominators to be the same:

$\dfrac{9}{11} = \dfrac{54}{66}$ and $\dfrac{5}{6} = \dfrac{55}{66}$. Since $\dfrac{54}{66} < \dfrac{55}{66}$ we

have $\dfrac{9}{11} < \dfrac{6}{7}$.

65. 130 lb = 130 · (0.45 kg) = 58.5 kg; <u>59</u> kg

67. 158 lb = 158 · (0.45 kg) = 71.1 kg; <u>71</u> kg

69. 184 lb = 184 · (0.45 kg) = 82.8 kg; <u>83</u> kg

71. 135 lb = 135 · (0.45 kg) = 60.75 kg; <u>61</u> kg

73. 155 lb = 155 · (0.45 kg) = 69.75 kg; <u>70</u> kg

75. Answers may vary.

Mastery Test 7.6

77. $F = \dfrac{9C}{5} + 32 = \dfrac{9 \cdot \cancel{40}^{8}}{\cancel{5}} + 32 = 72 + 32 = 104$

Thus 40°C = 104°F.

79. $8 \text{ kg} = 8 \cdot (2.2 \text{ lb}) = \underline{17.6} \text{ lb}$

81. $384 \text{ mg} = \underline{0.384} \text{ g}$

83. $12{,}000 \text{ lb} = 12{,}000 \cdot \left(\dfrac{1}{2000} \text{ ton} \right) = \underline{6} \text{ tons}$

85. $160 \text{ oz} = 160 \cdot \left(\dfrac{1}{16} \text{ lb} \right) = \underline{10} \text{ lb}$

Review Exercises – Chapter 7

1. **a.** $2 \text{ yd} = 2 \cdot (3 \text{ ft}) = 2 \cdot 3 \cdot (12 \text{ in.}) = 72 \text{ in.}$
 b. $3 \text{ yd} = 3 \cdot (3 \text{ ft}) = 3 \cdot 3 \cdot (12 \text{ in.}) = 108 \text{ in.}$
 c. $4 \text{ yd} = 4 \cdot (3 \text{ ft}) = 4 \cdot 3 \cdot (12 \text{ in.}) = 144 \text{ in.}$
 d. $6 \text{ yd} = 6 \cdot (3 \text{ ft}) = 6 \cdot 3 \cdot (12 \text{ in.}) = 216 \text{ in.}$
 e. $7 \text{ yd} = 7 \cdot (3 \text{ ft}) = 7 \cdot 3 \cdot (12 \text{ in.}) = 252 \text{ in.}$

2. **a.** $12 \text{ in.} = 1 \text{ ft}$
 b. $24 \text{ in.} = 24 \text{ in.} \cdot \dfrac{1 \text{ ft}}{12 \text{ in.}} = 2 \text{ ft}$
 c. $36 \text{ in.} = 36 \text{ in.} \cdot \dfrac{1 \text{ ft}}{12 \text{ in.}} = 3 \text{ ft}$
 d. $40 \text{ in.} = 40 \text{ in.} \cdot \dfrac{1 \text{ ft}}{12 \text{ in.}} = 3\dfrac{1}{3} \text{ ft}$
 e. $65 \text{ in.} = 65 \text{ in.} \cdot \dfrac{1 \text{ ft}}{12 \text{ in.}} = 5\dfrac{5}{12} \text{ ft}$

3. **a.** $2 \text{ in.} = 2 \text{ in.} \cdot \dfrac{1 \text{ ft}}{12 \text{ in.}} = \dfrac{1}{6} \text{ ft}$
 b. $3 \text{ in.} = 3 \text{ in.} \cdot \dfrac{1 \text{ ft}}{12 \text{ in.}} = \dfrac{1}{4} \text{ ft}$
 c. $5 \text{ in.} = 5 \text{ in.} \cdot \dfrac{1 \text{ ft}}{12 \text{ in.}} = \dfrac{5}{12} \text{ ft}$
 d. $7 \text{ in.} = 7 \text{ in.} \cdot \dfrac{1 \text{ ft}}{12 \text{ in.}} = \dfrac{7}{12} \text{ ft}$
 e. $14 \text{ in.} = 14 \text{ in.} \cdot \dfrac{1 \text{ ft}}{12 \text{ in.}} = \dfrac{7}{6} \text{ ft} = 1\dfrac{1}{6} \text{ ft}$

4. **a.** $6 \text{ ft} = 6 \text{ ft} \cdot \dfrac{1 \text{ yd}}{3 \text{ ft}} = 2 \text{ yd}$
 b. $12 \text{ ft} = 12 \text{ ft} \cdot \dfrac{1 \text{ yd}}{3 \text{ ft}} = 4 \text{ yd}$
 c. $18 \text{ ft} = 18 \text{ ft} \cdot \dfrac{1 \text{ yd}}{3 \text{ ft}} = 6 \text{ yd}$
 d. $20 \text{ ft} = 20 \text{ ft} \cdot \dfrac{1 \text{ yd}}{3 \text{ ft}} = 6\dfrac{2}{3} \text{ yd}$
 e. $29 \text{ ft} = 29 \text{ ft} \cdot \dfrac{1 \text{ yd}}{3 \text{ ft}} = 9\dfrac{2}{3} \text{ yd}$

5. **a.** $5280 \text{ ft} = 1 \text{ mi}$
 b. $10{,}560 \text{ ft} = 10{,}560 \cdot \dfrac{1}{5280} \text{ mi} = 2 \text{ mi}$
 c. $26{,}400 \text{ ft} = 26{,}400 \cdot \dfrac{1}{5280} \text{ mi} = 5 \text{ mi}$
 d. $21{,}120 \text{ ft} = 21{,}120 \cdot \dfrac{1}{5280} \text{ mi} = 4 \text{ mi}$
 e. $15{,}840 \text{ ft} = 15{,}840 \cdot \dfrac{1}{5280} \text{ mi} = 3 \text{ mi}$

6. **a.** $2 \text{ km} = 2000 \text{ m}$
 b. $7 \text{ km} = 7000 \text{ m}$
 c. $4.6 \text{ km} = 4600 \text{ m}$
 d. $0.45 \text{ km} = 450 \text{ m}$
 e. $45 \text{ km} = 45{,}000 \text{ m}$

7. **a.** $2 \text{ dam} = 20 \text{ m}$
 b. $3 \text{ dam} = 30 \text{ m}$
 c. $7 \text{ dam} = 70 \text{ m}$
 d. $9 \text{ dam} = 90 \text{ m}$
 e. $10 \text{ dam} = 100 \text{ m}$

8. **a.** $100 \text{ m} = 1000 \text{ dm}$
 b. $300 \text{ m} = 3000 \text{ dm}$
 c. $350 \text{ m} = 3500 \text{ dm}$
 d. $450 \text{ m} = 4500 \text{ dm}$
 e. $600 \text{ m} = 6000 \text{ dm}$

9. **a.** $200 \text{ cm} = 2 \text{ m}$
 b. $395 \text{ cm} = 3.95 \text{ m}$
 c. $405 \text{ cm} = 4.05 \text{ m}$
 d. $234 \text{ cm} = 2.34 \text{ m}$
 e. $499 \text{ cm} = 4.99 \text{ m}$

10. a. 30 in. $= 30 \cdot (2.54 \text{ cm}) = 76.2$ cm

    b. 40 in. $= 40 \cdot (2.54 \text{ cm}) = 101.6$ cm

    c. 50 ft $= 50 \cdot (12 \text{ in.})$

         $= 600$ in.

         $= 600 \cdot (2.54 \text{ cm}) = 1524$ cm

    d. 60 ft $= 60 \cdot (12 \text{ in.})$

         $= 720$ in.

         $= 720 \cdot (2.54 \text{ cm}) = 1828.8$ cm

    e. 70 ft $= 70 \cdot (12 \text{ in.})$

         $= 840$ in.

         $= 840 \cdot (2.54 \text{ cm}) = 2133.6$ cm

11. a. 100 yd $= 100 \cdot (0.914 \text{ m}) = 91.4$ m

    b. 200 yd $= 200 \cdot (0.914 \text{ m}) = 182.8$ m

    c. 350 yd $= 350 \cdot (0.914 \text{ m}) = 319.9$ m

    d. 450 yd $= 450 \cdot (0.914 \text{ m}) = 411.3$ m

    e. 500 yd $= 500 \cdot (0.914 \text{ m}) = 457$ m

12. a. 30 mi $= 30 \cdot (1.6 \text{ km}) = 48$ km

    b. 50 mi $= 50 \cdot (1.6 \text{ km}) = 80$ km

    c. 90 mi $= 90 \cdot (1.6 \text{ km}) = 144$ km

    d. 100 mi $= 100 \cdot (1.6 \text{ km}) = 160$ km

    e. 250 mi $= 250 \cdot (1.6 \text{ km}) = 400$ km

13. a. 40 km $= 40 \cdot (0.62 \text{ mi}) = 24.8$ mi

    b. 50 km $= 50 \cdot (0.62 \text{ mi}) = 31$ mi

    c. 60 km $= 60 \cdot (0.62 \text{ mi}) = 37.2$ mi

    d. 70 km $= 70 \cdot (0.62 \text{ mi}) = 43.4$ mi

    e. 80 km $= 80 \cdot (0.62 \text{ mi}) = 49.6$ mi

14. a. $2 \text{ m}^2 = 2 \cdot (100 \text{ cm})^2$

         $= 2 \cdot (100 \text{ cm}) \cdot (100 \text{ cm})$

         $= 2 \cdot 10,000 \text{ cm}^2 = 20,000 \text{ cm}^2$

    b. $3 \text{ m}^2 = 3 \cdot (100 \text{ cm})^2$

         $= 3 \cdot (100 \text{ cm}) \cdot (100 \text{ cm})$

         $= 3 \cdot 10,000 \text{ cm}^2 = 30,000 \text{ cm}^2$

    c. $4 \text{ m}^2 = 4 \cdot (100 \text{ cm})^2$

         $= 4 \cdot (100 \text{ cm}) \cdot (100 \text{ cm})$

         $= 4 \cdot 10,000 \text{ cm}^2 = 40,000 \text{ cm}^2$

    d. $5 \text{ m}^2 = 5 \cdot (100 \text{ cm})^2$

         $= 5 \cdot (100 \text{ cm}) \cdot (100 \text{ cm})$

         $= 5 \cdot 10,000 \text{ cm}^2 = 50,000 \text{ cm}^2$

    e. $6 \text{ m}^2 = 6 \cdot (100 \text{ cm})^2$

         $= 6 \cdot (100 \text{ cm}) \cdot (100 \text{ cm})$

         $= 6 \cdot 10,000 \text{ cm}^2 = 60,000 \text{ cm}^2$

15. a. $2 \text{ km}^2 = 2 \cdot (1000 \text{ m})^2$

         $= 2 \cdot (1000 \text{ m}) \cdot (1000 \text{ m})$

         $= 2 \cdot 1,000,000 \text{ m}^2$

         $= 2,000,000 \text{ m}^2$

    b. $3 \text{ km}^2 = 3 \cdot (1000 \text{ m})^2$

         $= 3 \cdot (1000 \text{ m}) \cdot (1000 \text{ m})$

         $= 3 \cdot 1,000,000 \text{ m}^2$

         $= 3,000,000 \text{ m}^2$

    c. $4 \text{ km}^2 = 4 \cdot (1000 \text{ m})^2$

         $= 4 \cdot (1000 \text{ m}) \cdot (1000 \text{ m})$

         $= 4 \cdot 1,000,000 \text{ m}^2$

         $= 4,000,000 \text{ m}^2$

    d. $5 \text{ km}^2 = 5 \cdot (1000 \text{ m})^2$

         $= 5 \cdot (1000 \text{ m}) \cdot (1000 \text{ m})$

         $= 5 \cdot 1,000,000 \text{ m}^2$

         $= 5,000,000 \text{ m}^2$

    e. $6 \text{ km}^2 = 6 \cdot (1000 \text{ m})^2$

         $= 6 \cdot (1000 \text{ m}) \cdot (1000 \text{ m})$

         $= 6 \cdot 1,000,000 \text{ m}^2$

         $= 6,000,000 \text{ m}^2$

16. a. $2 \text{ ft}^2 = 2 \cdot (12 \text{ in.})^2$

         $= 2 \cdot (12 \text{ in.}) \cdot (12 \text{ in.}) = 288 \text{ in.}^2$

    b. $3 \text{ ft}^2 = 3 \cdot (12 \text{ in.})^2$

         $= 3 \cdot (12 \text{ in.}) \cdot (12 \text{ in.}) = 432 \text{ in.}^2$

    c. $4 \text{ ft}^2 = 4 \cdot (12 \text{ in.})^2$

         $= 4 \cdot (12 \text{ in.}) \cdot (12 \text{ in.}) = 576 \text{ in.}^2$

    d. $5 \text{ ft}^2 = 5 \cdot (12 \text{ in.})^2$

         $= 5 \cdot (12 \text{ in.}) \cdot (12 \text{ in.}) = 720 \text{ in.}^2$

    e. $6 \text{ ft}^2 = 6 \cdot (12 \text{ in.})^2$

         $= 6 \cdot (12 \text{ in.}) \cdot (12 \text{ in.}) = 864 \text{ in.}^2$

17. a. $144 \text{ in.}^2 = 144 \cdot \left(\dfrac{1}{12} \text{ ft} \right)^2$

         $= 144 \cdot \dfrac{1}{144} \text{ ft}^2 = 1 \text{ ft}^2$

    b. $288 \text{ in.}^2 = 288 \cdot \left(\dfrac{1}{12} \text{ ft} \right)^2$

         $= 288 \cdot \dfrac{1}{144} \text{ ft}^2 = 2 \text{ ft}^2$

c. $360 \text{ in.}^2 = 360 \cdot \left(\frac{1}{12} \text{ ft}\right)^2$

$= 350 \cdot \frac{1}{144} \text{ ft}^2 = 2.5 \text{ ft}^2$

d. $432 \text{ in.}^2 = 432 \cdot \left(\frac{1}{12} \text{ ft}\right)^2$

$= 432 \cdot \frac{1}{144} \text{ ft}^2 = 3 \text{ ft}^2$

e. $504 \text{ in.}^2 = 504 \cdot \left(\frac{1}{12} \text{ ft}\right)^2$

$= 504 \cdot \frac{1}{144} \text{ ft}^2 = 3.5 \text{ ft}^2$

18. a. $1 \text{ acre} = 4840 \text{ yd}^2$

b. $3 \text{ acres} = 3 \cdot 4840 \text{ yd}^2 = 14{,}520 \text{ yd}^2$

c. $2 \text{ acres} = 2 \cdot 4840 \text{ yd}^2 = 9680 \text{ yd}^2$

d. $1.5 \text{ acres} = 1.5 \cdot 4840 \text{ yd}^2 = 7260 \text{ yd}^2$

e. $4 \text{ acres} = 4 \cdot 4840 \text{ yd}^2 = 19{,}360 \text{ yd}^2$

19. a. $5 \text{ hectares} = 5 \cdot (2.47 \text{ acres})$
$= 12.35 \text{ acres}$

b. $4 \text{ hectares} = 4 \cdot (2.47 \text{ acres})$
$= 9.88 \text{ acres}$

c. $1 \text{ hectares} = 2.47 \text{ acres}$

d. $3 \text{ hectares} = 3 \cdot (2.47 \text{ acres})$
$= 7.41 \text{ acres}$

e. $2 \text{ hectares} = 2 \cdot (2.47 \text{ acres})$
$= 4.94 \text{ acres}$

20. a. $7 \text{ qt} \approx 7 \cdot (1 \text{ L}) = 7 \text{ L}$

b. $9 \text{ qt} \approx 9 \cdot (1 \text{ L}) = 9 \text{ L}$

c. $10 \text{ qt} \approx 10 \cdot (1 \text{ L}) = 10 \text{ L}$

d. $13 \text{ qt} \approx 13 \cdot (1 \text{ L}) = 13 \text{ L}$

e. $2.2 \text{ qt} \approx 2.2 \cdot (1 \text{ L}) = 2.2 \text{ L}$

21. a. $4 \text{ kL} = 4000 \text{ L}$

b. $7 \text{ kL} = 7000 \text{ L}$

c. $9 \text{ kL} = 9000 \text{ L}$

d. $2.3 \text{ kL} = 2300 \text{ L}$

e. $5.97 \text{ kL} = 5970 \text{ L}$

22. a. $452 \text{ mL} = 0.452 \text{ L}$

b. $48 \text{ mL} = 0.048 \text{ L}$

c. $3 \text{ mL} = 0.003 \text{ L}$

d. $1657 \text{ mL} = 1.657 \text{ L}$

e. $456 \text{ mL} = 0.456 \text{ L}$

23. a. $4 \text{ cups} = 4 \cdot (240 \text{ mL}) = 960 \text{ mL}$

b. $5 \text{ cups} = 5 \cdot (240 \text{ mL}) = 1200 \text{ mL}$

c. $6 \text{ cups} = 6 \cdot (240 \text{ mL}) = 1440 \text{ mL}$

d. $7 \text{ cups} = 7 \cdot (240 \text{ mL}) = 1680 \text{ mL}$

e. $9 \text{ cups} = 9 \cdot (240 \text{ mL}) = 2160 \text{ mL}$

24. a. $120 \text{ mL} = 120 \cdot \left(\frac{1}{240} \text{ cups}\right) = \frac{1}{2} \text{ cup}$

b. $360 \text{ mL} = 360 \cdot \left(\frac{1}{240} \text{ cups}\right) = 1\frac{1}{2} \text{ cups}$

c. $480 \text{ mL} = 480 \cdot \left(\frac{1}{240} \text{ cups}\right) = 2 \text{ cups}$

d. $600 \text{ mL} = 600 \cdot \left(\frac{1}{240} \text{ cups}\right) = 2\frac{1}{2} \text{ cups}$

e. $720 \text{ mL} = 720 \cdot \left(\frac{1}{240} \text{ cups}\right) = 3 \text{ cups}$

25. a. $120 \text{ fl. oz} = 120 \cdot (30 \text{ mL}) = 3600 \text{ mL}$

b. $180 \text{ fl. oz} = 180 \cdot (30 \text{ mL}) = 5400 \text{ mL}$

c. $240 \text{ fl. oz} = 240 \cdot (30 \text{ mL}) = 7200 \text{ mL}$

d. $300 \text{ fl. oz} = 300 \cdot (30 \text{ mL}) = 9000 \text{ mL}$

e. $360 \text{ fl. oz} = 360 \cdot (30 \text{ mL}) = 10{,}800 \text{ mL}$

26. a. $7 \text{ tablespoons} = 7 \cdot (15 \text{ mL}) = 105 \text{ mL}$

b. $8 \text{ tablespoons} = 8 \cdot (15 \text{ mL}) = 120 \text{ mL}$

c. $10 \text{ teaspoons} = 10 \cdot (5 \text{ mL}) = 50 \text{ mL}$

d. $11 \text{ teaspoons} = 11 \cdot (5 \text{ mL}) = 55 \text{ mL}$

e. $13 \text{ teaspoons} = 13 \cdot (5 \text{ mL}) = 65 \text{ mL}$

27. a. $3 \text{ lb} = 3 \cdot (16 \text{ oz}) = 48 \text{ oz}$

b. $4 \text{ lb} = 4 \cdot (16 \text{ oz}) = 64 \text{ oz}$

c. $5 \text{ lb} = 5 \cdot (16 \text{ oz}) = 80 \text{ oz}$

d. $6 \text{ lb} = 6 \cdot (16 \text{ oz}) = 96 \text{ oz}$

e. $7 \text{ lb} = 7 \cdot (16 \text{ oz}) = 112 \text{ oz}$

28. a. $16 \text{ oz} = 1 \text{ lb}$

b. $24 \text{ oz} = 24 \text{ o\!\!\!/z} \cdot \frac{1 \text{ lb}}{16 \text{ o\!\!\!/z}} = 1\frac{1}{2} \text{ lb}$

c. $32 \text{ oz} = 32 \text{ o\!\!\!/z} \cdot \frac{1 \text{ lb}}{16 \text{ o\!\!\!/z}} = 2 \text{ lb}$

d. $40 \text{ oz} = 40 \text{ o\!\!\!/z} \cdot \frac{1 \text{ lb}}{16 \text{ o\!\!\!/z}} = 2\frac{1}{2} \text{ lb}$

e. $48 \text{ oz} = 48 \text{ o\!\!\!/z} \cdot \frac{1 \text{ lb}}{16 \text{ o\!\!\!/z}} = 3 \text{ lb}$

29. a. 2 tons = $2 \cdot (2000 \text{ lb}) = 4000$ lb

    b. 3 tons = $3 \cdot (2000 \text{ lb}) = 6000$ lb

    c. 4 tons = $4 \cdot (2000 \text{ lb}) = 8000$ lb

    d. 5 tons = $5 \cdot (2000 \text{ lb}) = 10,000$ lb

    e. 6 tons = $6 \cdot (2000 \text{ lb}) = 12,000$ lb

30. a. $3000 \text{ lb} = 3000 \cdot \left(\dfrac{1}{2000} \text{ ton}\right) = 1.5$ tons

    b. $5000 \text{ lb} = 5000 \cdot \left(\dfrac{1}{2000} \text{ ton}\right) = 2.5$ tons

    c. $7000 \text{ lb} = 7000 \cdot \left(\dfrac{1}{2000} \text{ ton}\right) = 3.5$ tons

    d. $9000 \text{ lb} = 9000 \cdot \left(\dfrac{1}{2000} \text{ ton}\right) = 4.5$ tons

    e. $18,000 \text{ lb} = 18,000 \cdot \left(\dfrac{1}{2000} \text{ ton}\right)$
                 $= 9$ tons

31. a. 1 dag = 100 dg

    b. 3 dag = 300 dg

    c. 5 dag = 500 dg

    d. 4 dag = 400 dg

    e. 2 dag = 200 dg

32. a. 307 mg = 0.307 g

    b. 40 mg = 0.04 g

    c. 3245 mg = 3.245 g

    d. 2 mg = 0.002 g

    e. 10,342 mg = 10.342

33. a. 1 kg = 2.2 lb

    b. $7 \text{ kg} = 7 \cdot (2.2 \text{ lb}) = 15.4$ lb

    c. $6 \text{ kg} = 6 \cdot (2.2 \text{ lb}) = 13.2$ lb

    d. $4 \text{ kg} = 4 \cdot (2.2 \text{ lb}) = 8.8$ lb

    e. $8 \text{ kg} = 8 \cdot (2.2 \text{ lb}) = 17.6$ lb

34. a. 1 lb = 0.45 kg

    b. $3 \text{ lb} = 3 \cdot (0.45 \text{ kg}) = 1.35$ kg

    c. $6 \text{ lb} = 6 \cdot (0.45 \text{ kg}) = 2.7$ kg

    d. $4 \text{ lb} = 4 \cdot (0.45 \text{ kg}) = 1.8$ kg

    e. $10 \text{ lb} = 10 \cdot (0.45 \text{ kg}) = 4.5$ kg

35. a. $C = \dfrac{5(F-32)}{9} = \dfrac{5(32-32)}{9} = \dfrac{5 \cdot 0}{9} = 0$

    Thus 32°F = 0°C.

    b. $C = \dfrac{5(F-32)}{9} = \dfrac{5(41-32)}{9} = \dfrac{5 \cdot \cancel{9}}{\cancel{9}} = 5$

    Thus 41°F = 5°C.

    c. $C = \dfrac{5(F-32)}{9}$

        $= \dfrac{5(50-32)}{9} = \dfrac{5 \cdot \cancel{18}^{2}}{\cancel{9}} = 10$

    Thus 50°F = 10°C.

    d. $C = \dfrac{5(F-32)}{9}$

        $= \dfrac{5(59-32)}{9} = \dfrac{5 \cdot \cancel{27}^{3}}{\cancel{9}} = 15$

    Thus 59°F = 15°C.

    e. $C = \dfrac{5(F-32)}{9}$

        $= \dfrac{5(212-32)}{9} = \dfrac{5 \cdot 180}{9} = 100$

    Thus 212°F = 100°C.

36. a. $F = \dfrac{9C}{5} + 32$

        $= \dfrac{9 \cdot \cancel{10}^{2}}{\cancel{5}} + 32 = 18 + 32 = 50$

    Thus 10°C = 50°F.

    b. $F = \dfrac{9C}{5} + 32$

        $= \dfrac{9 \cdot \cancel{15}^{3}}{\cancel{5}} + 32 = 27 + 32 = 59$

    Thus 15°C = 59°F.

    c. $F = \dfrac{9C}{5} + 32$

        $= \dfrac{9 \cdot \cancel{20}^{4}}{\cancel{5}} + 32 = 36 + 32 = 68$

    Thus 20°C = 68°F.

    d. $F = \dfrac{9C}{5} + 32$

        $= \dfrac{9 \cdot \cancel{25}^{5}}{\cancel{5}} + 32 = 45 + 32 = 77$

    Thus 25°C = 77°F.

e. $F = \dfrac{9C}{5} + 32$

$= \dfrac{9 \cdot \cancel{30}^{6}}{\cancel{5}} + 32 = 54 + 32 = 86$

Thus $30°C = 86°F$.

Cumulative Review Chapters 1–7

1. 2910

2. $4 \div 2 \cdot 2 + 9 - 7 = 2 \cdot 2 + 9 - 7$
$= 4 + 9 - 7$
$= 13 - 7$
$= 6$

3. $\dfrac{31}{6} = 5\dfrac{1}{6}$

4. $4\dfrac{1}{3} = \dfrac{3 \times 4 + 1}{3} = \dfrac{13}{3}$

5. $\begin{array}{r} \overset{3\ \ 14\ 14}{7\cancel{4}\cancel{5}.42} \\ -\ 17.50 \\ \hline 727.92 \end{array}$

6. $\begin{array}{r} 0.503 \\ \times\ \ 0.16 \\ \hline 3018 \\ 503\ \ \\ \hline 0.08048 \end{array}$

7. $549.\underline{8}51 \rightarrow 549.9$

8. $50 \div 0.13 = \dfrac{50}{0.13} = \dfrac{5000}{13} =$

$\begin{array}{r} 384.615 \\ 13\overline{)5000.000} \approx 384.62 \\ \underline{39}\ \ \ \ \ \ \ \ \\ 110\ \ \ \ \ \ \\ \underline{104}\ \ \ \ \ \ \\ 60\ \ \ \ \\ \underline{52}\ \ \ \ \\ 8\ 0\ \ \\ \underline{7\ 8}\ \ \\ 20\ \\ \underline{13}\ \\ 70 \\ \underline{65} \\ 5 \end{array}$

9. Let d = the decimal part.
$d \times 12 = 3$
$d = \dfrac{3}{12} = \dfrac{1}{4} = 0.25$

10. $1.6 = 0.4y$
$\dfrac{1.6}{0.4} = \dfrac{0.4y}{0.4}$
$\dfrac{16}{4} = y$
$4 = y$

11. $9 = \dfrac{z}{3.9}$
$3.9 \cdot 9 = z$
$35.1 = z$

12. $\dfrac{10}{19} \overset{?}{=} \dfrac{50}{97}$; $10 \cdot 97 = 970 \neq 19 \cdot 50 = 950$.

Therefore no, the flag is not of the correct ratio.

13. $\dfrac{s}{3} = \dfrac{3}{27}$
$27s = 3 \cdot 3$
$s = \dfrac{9}{27} = \dfrac{1}{3}$

14. Let n = number of ounces.
$\dfrac{3}{4} = \dfrac{n}{60}$
$4n = 3 \cdot 60$
$n = \dfrac{180}{4} = 45$

You need 45 ounces of the product.

15. $12\% = .12 = 0.12$

16. $7\dfrac{1}{4}\% = 7.25\% = 0.0725$

17. Let P = the percentage.
$40\% \times 50 = P$
$0.40 \times 50 = P$
$20 = P$
Thus 40% of 50 is 20.

18. Let P = the percentage.

$$P = 33\frac{1}{3}\% \times 6 = \frac{1}{3} \times 6 = 2$$

Thus $33\frac{1}{3}\%$ of 6 is 2.

19. Let R = the percent.

$$R \times 28 = 14$$
$$R = \frac{14}{28} = \frac{1}{2} = 0.5 = 50\%$$

Thus 50% of 28 is 14.

20. Let B = the number.

$$6 = 30\% \times B$$
$$6 = 0.30 \times B$$
$$\frac{6}{0.30} = B \text{ so } B = \frac{60}{3} = 20$$

Thus 6 is 30% of 20.

21. $I = P \times R \times T = 500 \times 0.085 \times 5 = 212.5$

The interest is $212.50.

22. Buses and trucks

23.

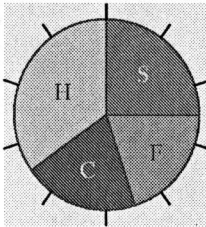

24. 43% of the families have incomes between $20,000 and $24,999

25. The rainfall for 2005 is about 34 inches.

26. The average temperature is $80 - 75 = 5$ degrees higher in July than in May.

27. The percent of these hours in Math and English combined is $13\% + 17\% = 30\%$.

28. The mode is 6, since it occurs most often.

29. The sum of the 11 numbers is 88. Thus the mean is $\frac{88}{11} = 8$.

30. Arrange the numbers in order: 6, 12, 13, 16, 25, 27, 27, 27, 27. Since there are an odd number of values, the median is the middle term. Thus the median is 25.

31. 12 yd = $12 \cdot (3 \text{ ft}) = 12 \cdot 3 \cdot (12 \text{ in.}) = 432$ in.

32. 17 in. = $17 \text{ in.} \cdot \dfrac{1 \text{ ft}}{12 \text{ in.}} = \dfrac{17}{12}$ ft $= 1\dfrac{5}{12}$ ft

33. 47 ft = $47 \text{ ft} \cdot \dfrac{1 \text{ yd}}{3 \text{ ft}} = \dfrac{47}{3}$ yd $= 15\dfrac{2}{3}$ yd

34. 26,400 ft = $26,400 \text{ ft} \cdot \dfrac{1 \text{ mi}}{5280 \text{ ft}} = 5$ mi

35. 8 ft = $8 \cdot (12 \text{ in.}) = 96$ in.

36. 2 km = 2000 m

37. 5 dam = 50 m

38. 150 m = 1500 dm

39. 50 yd = $50 \cdot (0.914 \text{ m}) = 45.7$ m

40. 66 mi = $66 \cdot (1.6 \text{ km}) = 105.6$ km

41. 100 km = $100 \cdot (0.62 \text{ mi}) = 62$ mi

42. 4 acres = $4 \cdot (4840 \text{ yd}^2) = 19,360 \text{ yd}^2$

43. 6 hectares = $6 \cdot (2.47 \text{ acres}) = 14.82$ acres

Chapter 8

Geometry

Section 8.1 – Finding Perimeters

Problems

1. The perimeter is
   $(2+4+2+4+1)$ yd $= 13$ yd.

2. The perimeter is
   $(5+2+5+2)$ cm $= 14$ cm.

3. $P = (2 \cdot 6.3 + 2 \cdot 3.4)$ in.
   $= (12.6 + 6.8)$ in.
   $= 19.4$ in.

4. The perimeter is $(3+3+3+3)$ m $= 12$ m.

5. $P = 4 \cdot 2\frac{3}{4}$ cm
   $= 4 \cdot \frac{11}{4}$ cm
   $= 11$ cm

6. $P = 5 \cdot 21$ m $= 105$ m

7. $d = 2 \cdot r = 2 \cdot 5$ ft $= 10$ ft. Thus the
   circumference is $C = \pi \cdot d$
   $\qquad = 3.14 \cdot 10$ ft
   $\qquad = 31.4$ ft

8. The distance traveled is
   $C = \pi \cdot d$
   $\quad = 3.14 \cdot 200$ million km
   $\quad = 628$ million km

9. $P = 2(400$ ft$) + 2(300$ ft$)$
   $= 800$ ft $+ 600$ ft
   $= 1400$ ft
   The security cable would be 1400 ft.

Exercises 8.1

1. $P = (2 \cdot 6 + 2 \cdot 4)$ ft
   $= (12 + 8)$ ft
   $= 20$ ft

3. $P = 2(3.1$ cm$) + 2(1.2$ cm$)$
   $= 6.2$ cm $+ 2.4$ cm
   $= 8.6$ cm

5. $P = \left(2 \cdot 5\frac{1}{8} + 2 \cdot 3\frac{1}{4}\right)$ in.
   $= \left(2 \cdot \frac{41}{8} + 2 \cdot \frac{13}{4}\right)$ in.
   $= \left(\frac{41}{4} + \frac{26}{4}\right)$ in.
   $= \frac{67}{4}$ in.
   $= 16\frac{3}{4}$ in.

7. $P = 4 \cdot (5.25$ cm$) = 21$ cm

9. $P = 4 \cdot \left(4\frac{1}{3}$ yd$\right)$
   $= 4 \cdot \frac{13}{3}$ yd
   $= \frac{52}{3}$ yd
   $= 17\frac{1}{3}$ yd

11. $P = (2 + 4 + 3)$ ft
    $= 11$ ft

13. $P = \left(2 + 3\frac{1}{3} + 2\frac{1}{4} + 4\frac{1}{3}\right)$ yd
    $= \left(2 + \frac{10}{3} + \frac{9}{4} + \frac{13}{3}\right)$ yd
    $= \left(\frac{24}{12} + \frac{40}{12} + \frac{27}{12} + \frac{52}{12}\right)$ yd
    $= \frac{143}{12}$ yd
    $= 11\frac{11}{12}$ yd

15. $P = \left(2 \cdot 1\frac{3}{8} + 2 \cdot 2\frac{1}{4}\right)$ ft

$\quad = \left(2 \cdot \frac{11}{8} + 2 \cdot \frac{9}{4}\right)$ ft

$\quad = \left(\frac{11}{4} + \frac{18}{4}\right)$ ft

$\quad = \frac{29}{4}$ ft

$\quad = 7\frac{1}{4}$ ft

17. $P = 5 \cdot (13.1 \text{ m})$

$\quad = 78.6 \text{ m}$

19. $P = (17 + 14.1 + 25.2 + 42.3)$ km $= 98.6$ km

21. $C = \pi \cdot d$

$\quad = 3.14 \cdot (3 \text{ cm})$

$\quad = 9.42 \text{ cm}$

23. $C = \pi \cdot d$

$\quad = 3.14 \cdot (3.5 \text{ ft})$

$\quad = 10.99 \text{ ft}$

25. $d = 2 \cdot r = 2(2.1 \text{ m}) = 4.2 \text{ m}$ so

$\quad C = \pi \cdot d$

$\quad = 3.14 \cdot (4.2 \text{ m})$

$\quad = 13.188 \text{ m}$

27. $C = \pi \cdot d$

$\quad = 3.14 \cdot (2.4 \text{ mi})$

$\quad = 7.536 \text{ mi}$

29. $P = 2 \cdot 480 + 2 \cdot 75$

$\quad = 960 + 150$

$\quad = 1110$

The perimeter of this pool is 1110 m.

31. $P = 2(1772.4) + 2(430)$

$\quad = 3544.8 + 860$

$\quad = 4404.8$

You have to walk 4404.8 ft.

33. $P = 2(30.5) + 2(20.9)$

$\quad = 61 + 41.8$

$\quad = 102.8$

There were 102.8 in. of frame needed.

35. $C = \pi \cdot d$

$\quad = 3.14 \cdot 60$

$\quad = 188.4$

The circumference of the tire is 188.4 cm.

37. $\frac{1}{\cancel{2}} \cdot 15 \cdot \cancel{20}^{\,10} = 150$

39. $\frac{1}{2} \cdot 3 \cdot 5 = \frac{15}{2}$

41. The distance is the perimeter of the (approximate) triangular region made by these three cities. Thus the distance is $P = 154 + 155 + 180 = 489$ mi.

43. The distance is the perimeter of the triangular region made by these three cities. Thus the distance is $P = 190 + 151 + 223 = 564$ mi.

45. Answers may vary.

Mastery Test 8.1

47. The distance traveled is

$\quad C = \pi \cdot d$

$\quad = 3.14 \cdot 800$ million km

$\quad = 2512$ million km

49. $P = 4 \cdot \left(5\frac{1}{4} \text{ cm}\right)$

$\quad = 4 \cdot \frac{21}{4} \text{ cm}$

$\quad = 21 \text{ cm}$

51. $P = 2(3.4 \text{ cm}) + 2(1.2 \text{ cm})$

$\quad = 6.8 \text{ cm} + 2.4 \text{ cm}$

$\quad = 9.2 \text{ cm}$

53. $P = (3 + 6 + 4 + 3 + 5)$ ft

$\quad = 21 \text{ ft}$

Section 8.2 – Finding Areas

Problems

1. $A = L \cdot W = (8 \text{ yd}) \cdot (5 \text{ yd}) = 40 \text{ yd}^2$

2. $A = S^2 = S \cdot S = (8 \text{ cm}) \cdot (5 \text{ cm}) = 64 \text{ cm}^2$

3. $A = \dfrac{1}{2} \cdot b \cdot h$

   $= \dfrac{1}{\cancel{2}} \cdot (\cancel{10}^{\,5} \text{ in.}) \cdot (12 \text{ in.}) = 60 \text{ in.}^2$

4. $A = b \cdot h = (4 \text{ in.}) \cdot (2 \text{ in.}) = 8 \text{ in.}^2$

5. $A = \dfrac{1}{2} \cdot h \cdot (a + b)$

   $= \dfrac{1}{\cancel{2}} \cdot (\cancel{4}^{\,2} \text{ cm}) \cdot (3 \text{ cm} + 5 \text{ cm})$

   $= 2 \text{ cm} \cdot 8 \text{ cm}$

   $= 16 \text{ cm}^2$

6. $A = \pi r^2$

   $= (3.14) \cdot (2 \text{ in.})^2$

   $= 3.14 \cdot (2 \text{ in.}) \cdot (2 \text{ in.})$

   $= 3.14 \cdot (4 \text{ in.}^2)$

   $= 12.56 \text{ in.}^2$

7. $r = \dfrac{d}{2} = \dfrac{140}{2} = 70 \text{ ft}$ so the area is

   $A = \pi r^2$

   $= (3.14) \cdot (70 \text{ ft})^2$

   $= 3.14 \cdot (70 \text{ ft}) \cdot (70 \text{ ft})$

   $= 3.14 \cdot (4900 \text{ ft}^2)$

   $= 15,386 \text{ ft}^2$

Exercises 8.2

1. $A = L \cdot W = (15 \text{ ft}) \cdot (10 \text{ ft}) = 150 \text{ ft}^2$

3. $A = L \cdot W = (3 \text{ in.}) \cdot (2 \text{ in.}) = 6 \text{ in.}^2$

5. $A = L \cdot W = (9 \text{ cm}) \cdot (8 \text{ cm}) = 72 \text{ cm}^2$

7. $A = S^2 = (9 \text{ in.})^2 = 81 \text{ in.}^2$

9. $A = S^2 = (9 \text{ yd})^2 = 81 \text{ yd}^2$

11. $A = \dfrac{1}{2} \cdot b \cdot h$

    $= \dfrac{1}{\cancel{2}} \cdot (\cancel{8}^{\,4} \text{ cm}) \cdot (7 \text{ cm}) = 28 \text{ cm}^2$

13. $A = \dfrac{1}{2} \cdot b \cdot h$

    $= \dfrac{1}{\cancel{2}} \cdot (\cancel{50}^{\,25} \text{ mm}) \cdot (7 \text{ mm}) = 175 \text{ mm}^2$

15. $A = \dfrac{1}{2} \cdot b \cdot h$

    $= \dfrac{1}{\cancel{2}} \cdot (\cancel{4}^{\,2} \text{ ft}) \cdot (5 \text{ ft}) = 10 \text{ ft}^2$

17. $A = \dfrac{1}{2} \cdot b \cdot h$

    $= \dfrac{1}{\cancel{2}} \cdot (\cancel{12}^{\,6} \text{ in.}) \cdot (5 \text{ in.}) = 30 \text{ in.}^2$

19. $A = \dfrac{1}{2} \cdot b \cdot h$

    $= \dfrac{1}{\cancel{2}} \cdot (\cancel{4}^{\,2} \text{ km}) \cdot (10 \text{ km}) = 20 \text{ km}^2$

21. $A = b \cdot h = (5 \text{ in.}) \cdot (3 \text{ in.}) = 15 \text{ in.}^2$

23. $A = b \cdot h = \left(5\dfrac{1}{2} \text{ cm}\right) \cdot \left(2\dfrac{1}{2} \text{ cm}\right)$

    $= \left(\dfrac{11}{2} \text{ cm}\right) \cdot \left(\dfrac{5}{2} \text{ cm}\right)$

    $= \dfrac{55}{4} \text{ cm}^2 = 13\dfrac{3}{4} \text{ cm}^2$

25. $A = b \cdot h = (10 \text{ m}) \cdot (6 \text{ m}) = 60 \text{ m}^2$

27. $A = b \cdot h = (9 \text{ yd}) \cdot (5 \text{ yd}) = 45 \text{ yd}^2$

29. $A = \frac{1}{2} \cdot h \cdot (a+b)$

$= \frac{1}{\cancel{2}} \cdot (\cancel{2}^{1} \text{ in.}) \cdot \left(4 \text{ in.} + 7\frac{1}{2} \text{ in.}\right)$

$= (1 \text{ in.}) \cdot \left(11\frac{1}{2} \text{ in.}\right)$

$= (1 \text{ in.}) \cdot \left(\frac{23}{2} \text{ in.}\right)$

$= 11\frac{1}{2} \text{ in.}^2$

31. $A = \frac{1}{2} \cdot h \cdot (a+b)$

$= \frac{1}{2} \cdot (45 \text{ ft}) \cdot (120 \text{ ft} + 130 \text{ ft})$

$= \frac{1}{\cancel{2}} \cdot (45 \text{ ft}) \cdot \left(\cancel{250}^{125} \text{ ft}\right)$

$= 5625 \text{ ft}^2$

33. $A = \frac{1}{2} \cdot h \cdot (a+b)$

$= \frac{1}{\cancel{2}} \cdot (\cancel{10}^{5} \text{ ft}) \cdot (18 \text{ ft} + 21 \text{ ft})$

$= 5 \text{ ft} \cdot 39 \text{ ft}$

$= 195 \text{ ft}^2$

35. $A = \pi r^2 = 3.14 \cdot (4 \text{ in.})^2$

$= 3.14 \cdot (4 \text{ in.}) \cdot (4 \text{ in.}) = 50.24 \text{ in.}^2$

37. $A = \pi r^2 = 3.14 \cdot (7 \text{ cm})^2$

$= 3.14 \cdot (7 \text{ cm}) \cdot (7 \text{ cm}) = 153.86 \text{ cm}^2$

39. $A = \pi r^2 = 3.14 \cdot (1 \text{ m})^2$

$= 3.14 \cdot (1 \text{ m}) \cdot (1 \text{ m}) = 3.14 \text{ m}^2$

41. $A = S^2 = (90 \text{ ft})^2$

$= (90 \text{ ft}) \cdot (90 \text{ ft}) = 8100 \text{ ft}^2$

43. $A = \pi r^2 = 3.14 \cdot (40 \text{ ft})^2$

$= 3.14 \cdot (40 \text{ ft}) \cdot (40 \text{ ft}) = 5024 \text{ ft}^2$

45. $r = \frac{d}{2} = \frac{5200}{2} = 2600 \text{ ft}$ so the area is

$A = \pi r^2$

$= (3.14) \cdot (2600 \text{ ft})^2$

$= 3.14 \cdot (2600 \text{ ft}) \cdot (2600 \text{ ft})$

$= 21,226,400 \text{ ft}^2$

47. $r = \frac{d}{2} = \frac{12}{2} = 6 \text{ m}$ so the area is

$A = \pi r^2$

$= (3.14) \cdot (6 \text{ m})^2$

$= 3.14 \cdot (6 \text{ m}) \cdot (6 \text{ m})$

$= 113.04 \text{ m}^2$

49. $r = \frac{d}{2} = \frac{20}{2} = 10 \text{ in.}$ so the area is

$A = \pi r^2$

$= (3.14) \cdot (10 \text{ in.})^2$

$= 3.14 \cdot (10 \text{ in.}) \cdot (10 \text{ in.})$

$= 314.00 \text{ in.}^2$

51. $A = \pi r^2$

$= (3.14) \cdot (20 \text{ ft})^2$

$= 3.14 \cdot (20 \text{ ft}) \cdot (20 \text{ ft})$

$= 1256.00 \text{ ft}^2$

53. $A = \frac{1}{2} \cdot \pi r^2$

$= \frac{1}{2} \cdot (3.14) \cdot (3 \text{ ft})^2$

$= \frac{1}{2} \cdot (3.14) \cdot (3 \text{ ft}) \cdot (3 \text{ ft})$

$= 14.13 \text{ ft}^2$

55. $A = \pi r^2$

$= (3.14) \cdot (27 \text{ ft})^2$

$= 3.14 \cdot (27 \text{ ft}) \cdot (27 \text{ ft})$

$= 2289.06 \text{ ft}^2$

57. $\frac{10}{3} \cdot \cancel{2} \cdot \frac{1}{\cancel{2}} = \frac{10}{3}$

59. $(0.5)(22)(20) = \left(\frac{1}{\cancel{2}}\right)(\cancel{22}^{11})(20) = 220$

61. $A_{\text{rectangle}} = L \cdot W = (5 \text{ in.}) \cdot (3 \text{ in.}) = 15 \text{ in.}^2$

63. Total area $= A_{\text{rectangle}} + 2 \cdot A_{\text{half-circle}}$

$= 15 \text{ in.}^2 + 2 \cdot (6.28 \text{ in.}^2)$

$= 15 \text{ in.}^2 + 12.56 \text{ in.}^2$

$= 27.56 \text{ in.}^2$

65. $A_{\text{rectangle}} = L \cdot W = (3 \text{ in.}) \cdot (1 \text{ in.}) = 3 \text{ in.}^2$

67. *Small pizza:* $r = \frac{11}{2} = 5.5$ in. so

$A = 3.14 \cdot (5.5)^2 = 94.985$ in.^{2}. Two small pizzas would give

$2 \cdot (94.985) = 189.97$ in.^{2} of pizza or

$\frac{189.97 \text{ in.}^2}{\$16} = 11.87$ in.^{2}/dollar.

Large pizza: $r = \frac{15}{2} = 7.5$ in. so the area is

$A = 3.14 \cdot (7.5)^2 = 176.625$ in.^{2} yielding

$\frac{176.625 \text{ in.}^2}{\$15} = 11.78$ in.^{2}/dollar. Thus the two small pizzas is the better buy.

Mastery Test 8.2

69. $r = \frac{d}{2} = \frac{120}{2} = 60$ ft so the area is

$A = \pi r^2 = (3.14) \cdot (60 \text{ ft})^2$
$= 3.14 \cdot (60 \text{ ft}) \cdot (60 \text{ ft}) = 11{,}304.00 \text{ ft}^2$

71. $A = \frac{1}{2} \cdot h \cdot (a + b)$

$= \frac{1}{\cancel{2}} \cdot (\cancel{4}^{\,2} \text{ cm}) \cdot (3 \text{ cm} + 6 \text{ cm})$

$= 2 \text{ cm} \cdot 9 \text{ cm} = 18 \text{ cm}^2$

73. $A = \frac{1}{2} \cdot b \cdot h$

$= \frac{1}{\cancel{2}} \cdot (\cancel{20}^{\,10} \text{ cm}) \cdot (10 \text{ cm}) = 100 \text{ cm}^2$

75. $A = L \cdot W = (8 \text{ m}) \cdot (3 \text{ m}) = 24 \text{ m}^2$

Section 8.3 – Volume of Solids

Problems

1. $V = L \cdot W \cdot H = 6 \text{ m} \cdot 2 \text{ m} \cdot 4 \text{ m} = 48 \text{ m}^3$

2. 18 ft = 6 yd, 4 ft = $3\frac{1}{3}$ yd, and

9 in. = $\frac{3}{4}$ ft = $\frac{1}{4}$ yd so the volume is

$V = L \cdot W \cdot H$

$= 6 \text{ yd} \cdot 1\frac{1}{3} \text{ yd} \cdot \frac{1}{4} \text{ yd}$

$= \cancel{6}^{\,2} \cdot \frac{\cancel{4}}{\cancel{3}} \cdot \frac{1}{\cancel{4}} \text{ yd}^3 = 2 \text{ yd}^3$

3. $V = \pi \cdot r^2 \cdot h$

$= 3.14 \cdot (1 \text{ in.})^2 \cdot (6 \text{ in.}) \approx 18.84 \text{ in.}^3$

4. $r = \frac{4.3}{2} = 2.15$ in. so the volume of the

ball is $V = \frac{4}{3} \cdot \pi \cdot r^3$

$= \frac{4}{3} \cdot 3.14 \cdot (2.15 \text{ in.})^3$

$= \frac{4(3.14)(9.938375)}{3} \text{ in.}^3$

$\approx 41.61 \text{ in.}^3$

5. $r = \frac{4}{2} = 2$ in. so the volume of the hat is

$V = \frac{1}{3} \cdot \pi \cdot r^2 \cdot h$

$= \frac{1}{3} \cdot 3.14 \cdot (2 \text{ in.})^2 \cdot (14 \text{ in.}) \approx 58.61 \text{ in.}^3$

6. 120 ft = 40 yd, 6 ft = 2 yd, and

3 ft = 1 yd so the volume is

$V = L \cdot W \cdot H$

$= 40 \text{ yd} \cdot 2 \text{ yd} \cdot 1 \text{ yd} = 80 \text{ yd}^3$

Thus 80 yd^3 of soil was removed from that trench.

7. $V_{\text{cylinder}} = \pi \cdot r^2 \cdot h$

$= 3.14 \cdot (1 \text{ mm})^2 \cdot (10 \text{ mm})$

$= 31.4 \text{ mm}^3$

$V_{\text{sphere}} = \frac{4}{3} \cdot \pi \cdot r^3$

$= \frac{4}{3} \cdot 3.14 \cdot (1 \text{ mm})^3$

$= \frac{4(3.14)(1)}{3} \text{ mm}^3 \approx 4.19 \text{ mm}^3$

Total volume $= V_{\text{cylinder}} + V_{\text{sphere}}$
$$= 31.4 \text{ mm}^3 + 4.19 \text{ mm}^3$$
$$= 35.59 \text{ mm}^3$$
Each capsule is 35.59 mm^3.

Exercises 8.3

1. $V = 8 \text{ cm} \cdot 5 \text{ cm} \cdot 6 \text{ cm} = 240 \text{ cm}^3$

3. $V = 9\frac{1}{2} \text{ in.} \cdot 3 \text{ in.} \cdot 4\frac{1}{2} \text{ in.}$
$$= \frac{19}{2} \text{ in.} \cdot 3 \text{ in.} \cdot \frac{9}{2} \text{ in.}$$
$$= \frac{513}{4} \text{ in.}^3 = 128\frac{1}{4} \text{ in.}^3$$

5. $V = 26 \text{ in.} \cdot 16 \text{ in.} \cdot 19 \text{ in.} = 7904 \text{ in.}^3$

7. $V = 11 \text{ in.} \cdot 8 \text{ in.} \cdot 5 \text{ in.} = 440 \text{ in.}^3$

9. 30 ft = 10 yd, 18 ft = 6 yd, and
6 ft = 2 yd so the volume is
$$V = L \cdot W \cdot H$$
$$= 10 \text{ yd} \cdot 6 \text{ yd} \cdot 2 \text{ yd} = 120 \text{ yd}^3$$
Thus 120 yd^3 of dirt must be removed.

11. $V = L \cdot W \cdot H$
$$= 30 \text{ ft} \cdot 3 \text{ ft} \cdot 2 \text{ ft} = 180 \text{ ft}^3 \text{ and so}$$
$$180 \text{ ft}^3 = 180 \cdot (7.5 \text{ gal}) = 1350 \text{ gal}$$

13. 18 in. $= 1\frac{1}{2}$ ft, 8 in. $= \frac{2}{3}$ ft, and
12 in. = 1 ft so the volume is
$$V = L \cdot W \cdot H$$
$$= 1\frac{1}{2} \text{ ft} \cdot \frac{2}{3} \text{ ft} \cdot 1 \text{ ft}$$
$$= \frac{3}{2} \text{ ft} \cdot \frac{2}{3} \text{ ft} \cdot 1 \text{ ft} = 1 \text{ ft}^3$$
and $1 \text{ ft}^3 = 7.5 \text{ gal} \approx 8 \text{ gal}$ of water.

15. $V = \pi \cdot r^2 \cdot h$
$$= 3.14 \cdot (10 \text{ in.})^2 \cdot (8 \text{ in.}) \approx 2512.0 \text{ in.}^3$$

17. $V = \pi \cdot r^2 \cdot h$
$$= 3.14 \cdot (10 \text{ cm})^2 \cdot (20 \text{ cm}) \approx 6280.0 \text{ cm}^3$$

19. $V = \pi \cdot r^2 \cdot h$
$$= 3.14 \cdot (1.5 \text{ m})^2 \cdot (4.5 \text{ m}) \approx 31.8 \text{ m}^3$$

21. $r = \frac{1 \text{ ft}}{2} = \frac{1}{2}$ ft so
$$V = \pi \cdot r^2 \cdot h$$
$$= 3.14 \cdot \left(\frac{1}{2} \text{ ft}\right)^2 \cdot (4 \text{ ft})$$
$$= 3.14 \cdot \frac{1}{\cancel{4}} \cdot \cancel{4} \text{ ft}^3 = 3.14 \text{ ft}^3$$

23. $r = \frac{3}{2}$ in. so
$$V = \pi \cdot r^2 \cdot h$$
$$= 3.14 \cdot \left(\frac{3}{2} \text{ in.}\right)^2 \cdot (5 \text{ in.})$$
$$= 3.14 \cdot \frac{9}{4} \cdot 5 \text{ in.}^3 \approx 35.33 \text{ in.}^3$$

25. $r = \frac{20}{2} = 10$ ft so
$$V = \pi \cdot r^2 \cdot h$$
$$= 3.14 \cdot (10 \text{ ft})^2 \cdot (40 \text{ ft})$$
$$= 3.14 \cdot 100 \cdot 40 \text{ ft}^3 \approx 12,560 \text{ ft}^3$$

27. $r = \frac{3}{2} = 1.5$ in. so
$$V = \pi \cdot r^2 \cdot h$$
$$= 3.14 \cdot (1.5 \text{ in.})^2 \cdot (4 \text{ in.})$$
$$= 3.14 \cdot (2.25) \cdot 4 \text{ in.}^3 = 28.26 \text{ in.}^3;$$
$$28.26 \text{ in.}^3 = 28.26 \cdot (0.6 \text{ fl oz}) \approx 17 \text{ fl oz}$$

29. $V = \frac{4}{3} \cdot \pi \cdot r^3$
$$= \frac{4}{3} \cdot 3.14 \cdot (24 \text{ ft})^3$$
$$= \frac{4(3.14)(13,824)}{3} \text{ ft}^3 \approx 57,876.48 \text{ ft}^3;$$
$$57,876.48 \text{ ft}^3 = 57,876.48 \cdot (7.5 \text{ gal})$$
$$= 434,073.6 \text{ gal of water}$$

31. $r = \dfrac{120}{2} = 60$ ft so

$$V = \dfrac{4}{3} \cdot \pi \cdot r^3$$

$$= \dfrac{4}{3} \cdot 3.14 \cdot (60 \text{ ft})^3$$

$$= \dfrac{4(3.14)(216,000)}{3} \text{ ft}^3 = 904,320 \text{ ft}^3;$$

$$904,320 \text{ ft}^3 = 904,320 \cdot (7.5 \text{ gal})$$
$$= 6,782,400.0 \text{ gal of fuel}$$

33. $V = \dfrac{1}{3} \cdot \pi \cdot r^2 \cdot h$

$$= \dfrac{1}{3} \cdot 3.14 \cdot (10 \text{ in.})^2 \cdot (6 \text{ in.}) \approx 628.0 \text{ in.}^3$$

35. $V = \dfrac{1}{3} \cdot \pi \cdot r^2 \cdot h$

$$= \dfrac{1}{3} \cdot 3.14 \cdot (50 \text{ ft})^2 \cdot (20 \text{ ft})$$

$$\approx 52,333.3 \text{ ft}^3$$

37. $V = \dfrac{1}{3} \cdot \pi \cdot r^2 \cdot h$

$$= \dfrac{1}{3} \cdot 3.14 \cdot (0.6 \text{ m})^2 \cdot (1.2 \text{ m}) \approx 0.5 \text{ m}^3$$

39. $r = \dfrac{7}{2} = 3.5$ cm so

$$V = \dfrac{1}{3} \cdot \pi \cdot r^2 \cdot h$$

$$= \dfrac{1}{3} \cdot 3.14 \cdot (3.5 \text{ cm})^2 \cdot (6 \text{ cm})$$

$$= \dfrac{1}{3} \cdot 3.14 \cdot (12.25) \cdot (6) \text{ cm}^3 \approx 76.9 \text{ cm}^3$$

41. a. $V = 22 \cdot 7 \cdot 8 = 1232 \text{ ft}^3$

   b. $V = 3 \cdot 7 \cdot 3 = 63 \text{ ft}^3$

   c. The total volume of the truck is approximately $1232 + 63 = 1295 \text{ ft}^3$, which is less than the 1300 ft^3 that you have to be moved. Therefore no, your stuff does not fit in the truck.

43. a. $r = \dfrac{34}{2} = 17$ in.

$$V = \dfrac{4}{3} \cdot \pi \cdot r^3$$

$$= \dfrac{4}{3} \cdot (3.14) \cdot (17)^3$$

$$\approx 20,569.1 \text{ in.}^3 \text{ and}$$

$$20,569.1 \text{ in.}^3 = 20,569.1 \cdot \dfrac{1}{1728} \text{ ft}^3$$
$$= 11.9 \text{ ft}^3$$

The volume is $20,569 \text{ in.}^3 \approx 11.9 \text{ ft}^3$.

   b. $V = \dfrac{1}{2} \cdot 20,569.1 \text{ in.}^3 = 10,284.6 \text{ in.}^3$

and $V = \dfrac{1}{2} \cdot 11.9 \text{ ft}^3 = 5.95 \text{ ft}^3 \approx 6.0 \text{ ft}^3$

This volume is about $10,284.6 \text{ in.}^3 \approx 6.0 \text{ ft}^3$.

45. a. $V = L \cdot W \cdot H$
$$= 36 \cdot 24 \cdot 20 = 17,280 \text{ in.}^3 \text{ and}$$

$$17,280 \text{ in.}^3 \cdot \dfrac{1 \text{ ft}^3}{1728 \text{ in.}^3} = 10 \text{ ft}^3$$

The volume is $17,280 \text{ in.}^3 = 10 \text{ ft}^3$.

   b. $V = L \cdot W \cdot H$
$$= 24 \cdot 24 \cdot 20 = 11,520 \text{ in.}^3 \text{ and}$$

$$11,520 \text{ in.}^3 \cdot \dfrac{1 \text{ ft}^3}{1728 \text{ in.}^3} \approx 6.7 \text{ ft}^3$$

The volume of the effective area for the box is $11,520 \text{ in.}^3 \approx 6.7 \text{ ft}^3$.

47. $r = \dfrac{300}{2} = 150$ ft so

$$V = \dfrac{1}{3} \cdot \pi \cdot r^2 \cdot h$$

$$= \dfrac{1}{3} \cdot 3.14 \cdot (150 \text{ ft})^2 \cdot (600 \text{ ft})$$

$$= \dfrac{3.14(22,5000)(600)}{3} \text{ ft}^3$$

$$= 14,130,000 \text{ ft}^3$$

49. $r = \dfrac{1.5}{2} = 0.75$ ft so

$$V_{\text{hemisphere}} = \dfrac{1}{2} \cdot \dfrac{4}{3} \cdot \pi \cdot r^3$$

$$= \dfrac{2}{3} \cdot 3.14 \cdot (0.75)^3 \approx 0.9 \text{ ft}^3$$

51. 3 cans on 4 floors = 3 · 4 = 12 trash cans. From exercise 50 the total volume of one trash can is 5.3 ft³. Thus the volume of the trash cans on the three floors is

$$V = 12 \cdot (5.3 \text{ ft}^3) = 63.6 \text{ ft}^3.$$

53. From exercise 52, the truck can carry about 94.3 cans of garbage; thus we have 1 truck = 94.3 cans. Using the fact that 1 building = 12 cans, we have:

$$1 \text{ truck} = 94.3 \text{ cans}$$

$$= 94.3 \cancel{\text{ cans}} \cdot \frac{1 \text{ building}}{12 \cancel{\text{ cans}}}$$

$$= \frac{94.3}{12} \text{ buildings}$$

$$\approx 7.9 \text{ buildings}$$

55. $x + 45 + 30 = 180$

   $x + 75 = 180$

   $x = 180 - 75 = 105$

57. $r = \dfrac{4}{2} = 4$ in. so

$$V = \pi \cdot r^2 \cdot h$$

$$= 3.14 \cdot (4 \text{ in.})^2 \cdot (2 \text{ in.})$$

$$= 3.14 \cdot 16 \cdot 2 \text{ in.}^3 \approx 100.5 \text{ in.}^3$$

59. After performing the cut and "stretch", the length would be the circumference of the loaf, which is $\pi \cdot d = 3.14 \cdot 8 \approx 25.1$ in.

61. Answers may vary.

Mastery Test 8.3

63. $V = \dfrac{1}{3} \cdot \pi \cdot r^2 \cdot h$

$$= \frac{1}{3} \cdot 3.14 \cdot (6 \text{ cm})^2 \cdot (10 \text{ cm})$$

$$= \frac{3.14 \cdot 36 \cdot 10}{3} \text{ cm}^3$$

$$\approx 376.8 \text{ cm}^3$$

65. $V = \pi \cdot r^2 \cdot h$

$$= 3.14 \cdot (1 \text{ in.})^2 \cdot (8 \text{ in.})$$

$$= 3.14 \cdot 1 \cdot 8 \text{ in.}^3 \approx 25.1 \text{ in.}^3$$

67. $V = L \cdot W \cdot H = (3 \text{ m}) \cdot (1 \text{ m}) \cdot (2 \text{ m}) = 6 \text{ m}^2$

Section 8.4 – Angles and Triangles

Problems

1. $\angle \gamma$, $\angle C$, or $\angle BCA$ (or $\angle ACB$)

2. α is exactly 90° it is a right angle.
β is between 0° and 90° so it is an acute angle.
δ is between 90° and 180° so it is an obtuse angle.
γ is exactly 180° so it is a straight angle.

3. Both γ and β are less than 90° so they are acute angles. δ is exactly 90° so it is a right angle.

4. $\angle ❶$ and $\angle ❷$; $\angle ❷$ and $\angle ❸$; $\angle ❸$ and $\angle ❹$; $\angle ❶$ and $\angle ❹$

5. The measure of the complement of a 30° angle is 90° − 30° = 60°.

6. $m\angle ❶ = 180° - m\angle ❷$

$$= 180° - 100° = 80°$$

7. a. Equilateral equiangular triangle
   b. Isosceles right triangle
   c. Obtuse scalene triangle

8. The measure of the third angle is
180° − 62° − 63° = 55°.

Exercises 8.4

1. $\angle \delta$, $\angle P$, or $\angle RPQ$ (or $\angle QRP$)

3. $\angle CAB$ or $\angle BAC$

5. $\angle EAF$ or $\angle FAE$

7. Acute

9. Right

11. Straight

13. $\angle DAF$, $\angle DAB$

15. $\angle FAB$

17. $\angle DAE$

19. $\angle CAD$

21. $\angle EAB$

23. $\angle \alpha$ and $\angle CAD$ are complementary
    angles so $m\angle CAD = 90° - m\angle \alpha$
    $$= 90° - 15°$$
    $$= 75°$$

25. $\angle \beta$ and $\angle DAE$ are complementary
    angles so $m\angle \beta = 90° - m\angle DAE$
    $$= 90° - 35°$$
    $$= 55°$$

27. The triangle has a right angle with no
    equal sides; it is a *scalene right triangle*.

29. The triangle has no equal sides and acute
    angles; it is a *scalene acute triangle*.

31. The triangle has two equal sides and all
    acute angles; it is an *isosceles acute
    triangle*.

33. The triangle has an obtuse angle and no
    equal sides; it is a *scalene obtuse triangle*.

35. $180° - 97° - 28° = 55°$
    The measure of the missing angle is 55°.

37. $180° - 45° - 120° = 15°$
    The measure of the missing angle is 15°.

39. $180° - 110° - 45° = 25°$
    The measure of the missing angle is 25°.

41. $m\angle A + m\angle B + m\angle C = 180°$
    $37° + m\angle B + 53° = 180°$
    $$m\angle B = 180° - 37° - 53°$$
    $$= 90°$$

43. Acute

45. Acute

47. Obtuse

49. Acute

51. $m\angle \alpha = 180° - 43° - 92° = 45°$

53. Obtuse

55. Right

57. The left and right side of the body is
    assumed to be equal and all angles are
    acute so it is an isosceles acute triangle.

59. $5^2 + 3^2 = 5 \cdot 5 + 3 \cdot 3 = 25 + 9 = 34$

61. $(4.28 - 4)^2 + (4.10 - 4)^2 = (0.28)^2 + (0.1)^2$
    $$= 0.0784 + 0.01$$
    $$= 0.0884$$

63. Rap

65. Three sections show acute angles (Rock &
    roll, Country, and Classical).

67. $25\% \times 360° = 0.25 \times 360° = 90°$

69. $10\% \times 360° = 0.10 \times 360 = 36$
    Thus $m\angle \beta = 36°$.

71. Answers may vary (you don't want an
    obtuse angle!).

73. Answers may vary.

Mastery Test 8.4

75. **a.** The triangle has an obtuse angle and no equal sides; it is an *obtuse scalene* triangle.
    b. The triangle has a right angle and two equal sides; it is a *right isosceles* triangle.
    c. All three angles are acute (and equal) and all three sides are equal; it is an *acute equilateral* triangle.

   d. All three angles are acute and two sides are equal; it is an *acute isosceles* triangle.

77. The measure of the supplement of the 35° angle is $180° - 35° = 145°$.

79. $\angle \theta$, $\angle C$, or $\angle EDC$ (or $\angle DCE$)

Section 8.5 – Square Roots and the Pythagorean Theorem

Problems

1. **a.** We need a number whose square is 49. The number is 7. Thus, $\sqrt{49} = 7$.
   b. We need a number whose square is 81. The number is 9. Thus, $\sqrt{81} = 9$.

2. $\sqrt{7} \approx 2.6457513 \approx 2.646$

3. $a^2 + b^2 = c^2$
   $8^2 + 6^2 = c^2$
   $64 + 36 = c^2$
   $100 = c^2$
   $c = \sqrt{100} = 10$
   The length of the hypotenuse is 10.

4. $a^2 + b^2 = c^2$
   $a^2 + 9^2 = 10^2$
   $a^2 + 81 = 100$
   $a^2 = 19$
   $a = \sqrt{19} = 4.3588989$
   The exact length of a is $\sqrt{19}$. The approximate length of a is 4.359.

5. The diagonal is the hypotenuse c.
   $a^2 + b^2 = c^2$
   $12^2 + 16^2 = c^2$
   $144 + 256 = c^2$
   $400 = c^2$
   $c = \sqrt{400} = 20$
   The length of the diagonal is 20 inches.

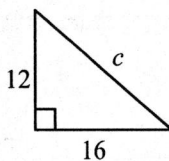

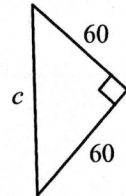

6. The distance from home plate to second base is the hypotenuse c.
   $a^2 + b^2 = c^2$
   $60^2 + 60^2 = c^2$
   $3600 + 3600 = c^2$
   $7200 = c^2$
   $c = \sqrt{7200}$
   ≈ 85.853
   It is about 85 ft. from home plate to second base.

Exercises 8.5

1. We need a number whose square is 100. The number is 10. Thus $\sqrt{100} = 10$.

3. We need a number whose square is 196. The number is 14. Thus $\sqrt{196} = 14$.

5. We need a number whose square is 361. The number is 19. Thus $\sqrt{361} = 19$.

7. We need a number whose square is 400. The number is 20. Thus $\sqrt{400} = 20$.

9. We need a number whose square is 169. The number is 13. Thus $\sqrt{169} = 13$.

11. $\sqrt{8} \approx 2.8284271 \approx 2.828$

13. $\sqrt{11} \approx 3.3166248 \approx 3.317$

15. $\sqrt{23} \approx 4.7958315 \approx 4.796$

17. $\sqrt{29} \approx 5.3851648 \approx 5.385$

19. $\sqrt{108} \approx 10.3923048 \approx 10.392$

21. $a^2 + b^2 = c^2$
$6^2 + 10^2 = c^2$
$36 + 100 = c^2$
$136 = c^2$
$c = \sqrt{136} = 11.6619038$
The length of the hypotenuse is
$\sqrt{136} \approx 11.662$.

23. $a^2 + b^2 = c^2$
$9^2 + 12^2 = c^2$
$81 + 144 = c^2$
$225 = c^2$
$c = \sqrt{225} = 15$
The length of the hypotenuse is 15.

25. $a^2 + b^2 = c^2$
$12^2 + 16^2 = c^2$
$144 + 256 = c^2$
$400 = c^2$
$c = \sqrt{400} = 20$
The length of the hypotenuse is 20.

27. $a^2 + b^2 = c^2$
$25^2 + 40^2 = c^2$
$625 + 1600 = c^2$
$2225 = c^2$
$c = \sqrt{2225} = 47.1699057$
The length of the hypotenuse is
$\sqrt{2225} \approx 47.170$.

29. $a^2 + b^2 = c^2$
$10^2 + 44^2 = c^2$
$100 + 1936 = c^2$
$2036 = c^2$
$c = \sqrt{2036} = 45.1220567$

The length of the hypotenuse is
$\sqrt{2036} \approx 45.122$.

31. Let b = the missing side.
$a^2 + b^2 = c^2$
$9^2 + b^2 = 15^2$
$81 + b^2 = 225$
$b^2 = 144$
$b = \sqrt{144} = 12$
The missing side is 12 in.

33. Let a = the missing side.
$a^2 + b^2 = c^2$
$a^2 + 16^2 = 20^2$
$a^2 + 256 = 400$
$a^2 = 144$
$a = \sqrt{144} = 12$
The missing side is 12 in.

35. Let a = the missing side.
$a^2 + b^2 = c^2$
$a^2 + 2^2 = 3^2$
$a^2 + 4 = 9$
$a^2 = 5$
$a = \sqrt{5} \approx 2.236068$
The missing side is $\sqrt{5}$ ft ≈ 2.236 ft.

37. Let b = the missing side.
$a^2 + b^2 = c^2$
$4^2 + b^2 = 6^2$
$16 + b^2 = 36$
$b^2 = 20$
$b = \sqrt{20} \approx 4.472136$
The missing side is $\sqrt{20}$ m ≈ 4.472 m.

39. The distance from the hip to the knee is
the hypotenuse c.
$a^2 + b^2 = c^2$
$(0.44)^2 + (0.30)^2 = c^2$
$0.1936 + 0.09 = c^2$
$0.2836 = c^2$
$c = \sqrt{0.2836} \approx 0.5325411$
The distance from the hip to the ankle is
about 0.533 m.

41. Let a = the height.
$$a^2 + b^2 = c^2$$
$$a^2 + 150^2 = 289^2$$
$$a^2 + 22,500 = 83,521$$
$$a^2 = 61,021$$
$$a = \sqrt{61,021}$$
$$\approx 247.02429$$
The height was about 247.024 ft.

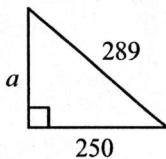

43. Let c = the diagonal measurement.
$$a^2 + b^2 = c^2$$
$$70^2 + 96^2 = c^2$$
$$4900 + 9216 = c^2$$
$$14,116 = c^2$$
$$c = \sqrt{14,116}$$
$$\approx 118.81077$$
The diagonal measurement is about 118.811 ft.

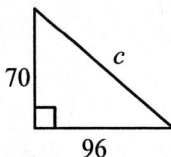

45. Let c = the diagonal length.
$$a^2 + b^2 = c^2$$
$$11^2 + 14^2 = c^2$$
$$121 + 196 = c^2$$
$$317 = c^2$$
$$c = \sqrt{317}$$
$$\approx 17.804494$$
The diagonal length is about 17.804 cm.

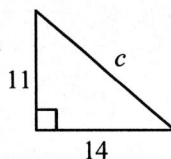

47. $38 - 23 = 15$
check: $15 + 23 = 38$ is true.

49. $53 - 27 = 26$
check: $26 + 27 = 53$ is true.

51. a. $\sqrt{25} = 5$
$\sqrt{26} \approx 5 + ?$
$\sqrt{36} = 6$
Now $26 - 25 = 1$ and $36 - 25 = 11$.
Thus $\sqrt{26} \approx 5\frac{1}{11}$.

b. $\sqrt{25} = 5$
$\sqrt{28} \approx 5 + ?$
$\sqrt{36} = 6$
$28 - 25 = 3$ and $36 - 25 = 11$ so
$\sqrt{28} \approx 5\frac{3}{11}$.

c. $\sqrt{25} = 5$
$\sqrt{30} \approx 5 + ?$
$\sqrt{36} = 6$
$30 - 25 = 5$ and $36 - 25 = 11$ so
$\sqrt{28} \approx 5\frac{5}{11}$.

53. Answers may vary.

Mastery Test 8.5

55. $a^2 + b^2 = c^2$
$$a^2 + 5^2 = 10^2$$
$$a^2 + 25 = 100$$
$$a^2 = 75$$
$$a = \sqrt{75} = 8.660254$$
The length of a is $\sqrt{75} \approx 8.660$.

57. $a^2 + b^2 = c^2$
$$6^2 + 5^2 = c^2$$
$$36 + 25 = c^2$$
$$61 = c^2$$
$$c = \sqrt{61} \approx 7.81025$$
The hypotenuse is $\sqrt{61} \approx 7.810$.

59. $\sqrt{21} \approx 4.5825757 \approx 4.583$

Review Exercises – Chapter 8

1. a. $P = 2 \cdot (5.1) + 2 \cdot (3.2)$
$$= 10.2 + 6.4$$
$$= 16.6 \text{ m}$$
b. $P = 2 \cdot (4.1) + 2 \cdot (3.2)$
$$= 8.2 + 6.4$$
$$= 14.6 \text{ cm}$$
c. $P = 2 \cdot (3.5) + 2 \cdot (4.1)$
$$= 7 + 8.2$$
$$= 15.2 \text{ in.}$$
d. $P = 2 \cdot (5.2) + 2 \cdot (4.1)$
$$= 10.4 + 8.2$$
$$= 18.6 \text{ yd}$$
e. $P = 2 \cdot (4.1) + 2 \cdot (6.2)$
$$= 8.2 + 12.4$$
$$= 20.6 \text{ ft}$$

2. a. $P = 3 + 3 + 3 + 6 = 15$ m
  b. $P = 3 + 5 + 2 + 2 + 7 = 19$ yd
  c. $P = 2 + 5 + 2 + 2 + 4 = 15$ ft
  d. $P = 3 + 5 + 2 + 2 + 5 = 17$ cm
  e. $P = 6 \cdot (2 \text{ in.}) = 12$ in.

3. a. $d = 2 \cdot r = 2 \cdot 6 = 12$ cm;
  $C = \pi \cdot d = 3.14 \cdot 12 = 37.68$ cm
  b. $d = 2 \cdot r = 2 \cdot 8 = 16$ cm;
  $C = \pi \cdot d = 3.14 \cdot 16 = 50.24$ cm
  c. $d = 2 \cdot r = 2 \cdot 10 = 20$ in.;
  $C = \pi \cdot d = 3.14 \cdot 20 = 62.8$ in.
  d. $d = 2 \cdot r = 2 \cdot 12 = 24$ in.;
  $C = \pi \cdot d = 3.14 \cdot 24 = 75.36$ in.
  e. $d = 2 \cdot r = 2 \cdot 14 = 28$ ft;
  $C = \pi \cdot d = 3.14 \cdot 28 = 87.92$ ft

4. a. $A = L \cdot W = (5 \text{ m}) \cdot (6 \text{ m}) = 30$ m^2
  b. $A = L \cdot W = (7 \text{ m}) \cdot (8 \text{ m}) = 56$ m^2
  c. $A = L \cdot W = (4 \text{ in.}) \cdot (6 \text{ in.}) = 24$ in.^{2}
  d. $A = L \cdot W = (3 \text{ in.}) \cdot (7 \text{ in.}) = 21$ in.^{2}
  e. $A = L \cdot W = (9 \text{ in.}) \cdot (12 \text{ in.}) = 108$ in.^{2}

5. a. $A = \dfrac{1}{2} \cdot b \cdot h$
  $= \dfrac{1}{\cancel{2}} \cdot (\overset{5}{\cancel{10}} \text{ in.}) \cdot (12 \text{ in.}) = 60$ in.^{2}
  b. $A = \dfrac{1}{2} \cdot b \cdot h$
  $= \dfrac{1}{\cancel{2}} \cdot (\overset{4}{\cancel{8}} \text{ in.}) \cdot (10 \text{ in.}) = 40$ in.^{2}
  c. $A = \dfrac{1}{2} \cdot b \cdot h$
  $= \dfrac{1}{\cancel{2}} \cdot (\overset{3}{\cancel{6}} \text{ cm}) \cdot (8 \text{ cm}) = 24$ cm^2
  d. $A = \dfrac{1}{2} \cdot b \cdot h$
  $= \dfrac{1}{\cancel{2}} \cdot (\overset{2}{\cancel{4}} \text{ m}) \cdot (6 \text{ m}) = 12$ m^2
  e. $A = \dfrac{1}{2} \cdot b \cdot h$
  $= \dfrac{1}{\cancel{2}} \cdot (\overset{1}{\cancel{2}} \text{ m}) \cdot (4 \text{ m}) = 4$ m^2

6. a. $A = b \cdot h = (3 \text{ in.}) \cdot (2 \text{ in.}) = 6$ in.^{2}
  b. $A = b \cdot h = (5 \text{ cm}) \cdot (1.5 \text{ cm}) = 7.5$ cm^2
  c. $A = b \cdot h = (4 \text{ m}) \cdot (2 \text{ m}) = 8$ m^2
  d. $A = b \cdot h = (10 \text{ ft}) \cdot (3.5 \text{ ft}) = 35$ ft^2
  e. $A = b \cdot h = (16 \text{ m}) \cdot (4 \text{ m}) = 64$ m^2

7. a. $A = \dfrac{1}{2} \cdot h \cdot (a + b)$
  $= \dfrac{1}{\cancel{2}} \cdot (\overset{1}{\cancel{2}} \text{ in.}) \cdot (3 \text{ in.} + 5 \text{ in.})$
  $= (1 \text{ in.}) \cdot (8 \text{ in.}) = 8$ in.^{2}
  b. $A = \dfrac{1}{2} \cdot h \cdot (a + b)$
  $= \dfrac{1}{\cancel{2}} \cdot (\overset{3}{\cancel{6}} \text{ ft}) \cdot (8 \text{ ft} + 12 \text{ ft})$
  $= (3 \text{ ft}) \cdot (20 \text{ ft}) = 60$ ft^2
  c. $A = \dfrac{1}{2} \cdot h \cdot (a + b)$
  $= \dfrac{1}{2} \cdot (3 \text{ m}) \cdot (3 \text{ m} + 5 \text{ m})$
  $= \dfrac{1}{\cancel{2}} \cdot (3 \text{ m}) \cdot (\overset{4}{\cancel{8}} \text{ m}) = 12$ m^2
  d. $A = \dfrac{1}{2} \cdot h \cdot (a + b)$
  $= \dfrac{1}{2} \cdot (2.5 \text{cm}) \cdot (5 \text{ cm} + 7 \text{ cm})$
  $= \dfrac{1}{\cancel{2}} \cdot (2.5 \text{ cm}) \cdot (\overset{6}{\cancel{12}} \text{ cm}) = 15$ cm^2
  e. $A = \dfrac{1}{2} \cdot h \cdot (a + b)$
  $= \dfrac{1}{\cancel{2}} \cdot (\overset{2}{\cancel{4}} \text{ m}) \cdot (4 \text{ m} + 10 \text{ m})$
  $= (2 \text{ m}) \cdot (14 \text{ m}) = 28$ m^2

8. a. $A = \pi \cdot r^2$
  $= 3.14 \cdot (7 \text{ in.})^2$
  $= 3.14 \cdot 49 \text{ in.}^2 = 153.86$ in.^{2}
  b. $A = \pi \cdot r^2$
  $= 3.14 \cdot (1 \text{ cm})^2$
  $= 3.14 \cdot 1 \text{ cm}^2 = 3.14$ cm^2
  c. $A = \pi \cdot r^2$
  $= 3.14 \cdot (3 \text{ in.})^2$
  $= 3.14 \cdot 9 \text{ in.}^2 = 28.26$ in.^{2}

d. $A = \pi \cdot r^2$
$= 3.14 \cdot (2 \text{ ft})^2$
$= 3.14 \cdot 4 \text{ ft}^2 = 12.56 \text{ ft}^2$

e. $A = \pi \cdot r^2$
$= 3.14 \cdot (5 \text{ yd})^2$
$= 3.14 \cdot 25 \text{ yd}^2 = 78.5 \text{ yd}^2$

9. a. $V = L \cdot W \cdot H$
$= (6 \text{ cm}) \cdot (2 \text{ cm}) \cdot (7 \text{ cm}) = 84 \text{ cm}^3$

b. $V = L \cdot W \cdot H$
$= (4 \text{ cm}) \cdot (3 \text{ cm}) \cdot (5 \text{ cm}) = 60 \text{ cm}^3$

c. $V = L \cdot W \cdot H$
$= (5 \text{ cm}) \cdot (2 \text{ cm}) \cdot (4 \text{ cm}) = 40 \text{ cm}^3$

d. $V = L \cdot W \cdot H$
$= (6 \text{ cm}) \cdot (3 \text{ cm}) \cdot (5 \text{ cm}) = 90 \text{ cm}^3$

e. $V = L \cdot W \cdot H$
$= (7 \text{ cm}) \cdot (2 \text{ cm}) \cdot (5 \text{ cm}) = 70 \text{ cm}^3$

10. a. $V = \pi \cdot r^2 \cdot h$
$= 3.14 \cdot (1 \text{ in.})^2 \cdot (6 \text{ in.})$
$= 3.14 \cdot 1 \cdot 6 \text{ in.}^3 = 18.84 \text{ in.}^3$

b. $V = \pi \cdot r^2 \cdot h$
$= 3.14 \cdot (1 \text{ in.})^2 \cdot (7 \text{ in.})$
$= 3.14 \cdot 1 \cdot 7 \text{ in.}^3 = 21.98 \text{ in.}^3$

c. $V = \pi \cdot r^2 \cdot h$
$= 3.14 \cdot (1 \text{ in.})^2 \cdot (8 \text{ in.})$
$= 3.14 \cdot 1 \cdot 6 \text{ in.}^3 = 25.12 \text{ in.}^3$

d. $V = \pi \cdot r^2 \cdot h$
$= 3.14 \cdot (2 \text{ in.})^2 \cdot (9 \text{ in.})$
$= 3.14 \cdot 4 \cdot 9 \text{ in.}^3 = 113.04 \text{ in.}^3$

e. $V = \pi \cdot r^2 \cdot h$
$= 3.14 \cdot (2 \text{ in.})^2 \cdot (10 \text{ in.})$
$= 3.14 \cdot 4 \cdot 10 \text{ in.}^3 = 125.6 \text{ in.}^3$

11. a. $V = \dfrac{4}{3} \cdot \pi \cdot r^3$
$= \dfrac{4}{3} \cdot 3.14 \cdot (6 \text{ in.})^3$
$= \dfrac{4 \cdot 3.14 \cdot 216}{3} \text{ in.}^3 = 904.32 \text{ in.}^3$

b. $V = \dfrac{4}{3} \cdot \pi \cdot r^3$
$= \dfrac{4}{3} \cdot 3.14 \cdot (7 \text{ in.})^3$
$= \dfrac{4 \cdot 3.14 \cdot 343}{3} \text{ in.}^3 = 1436.03 \text{ in.}^3$

c. $V = \dfrac{4}{3} \cdot \pi \cdot r^3$
$= \dfrac{4}{3} \cdot 3.14 \cdot (8 \text{ in.})^3$
$= \dfrac{4 \cdot 3.14 \cdot 512}{3} \text{ in.}^3 = 2143.57 \text{ in.}^3$

d. $V = \dfrac{4}{3} \cdot \pi \cdot r^3$
$= \dfrac{4}{3} \cdot 3.14 \cdot (9 \text{ in.})^3$
$= \dfrac{4 \cdot 3.14 \cdot 729}{3} \text{ in.}^3 = 3052.08 \text{ in.}^3$

e. $V = \dfrac{4}{3} \cdot \pi \cdot r^3$
$= \dfrac{4}{3} \cdot 3.14 \cdot (10 \text{ in.})^3$
$= \dfrac{4 \cdot 3.14 \cdot 1000}{3} \text{ in.}^3 = 4186.67 \text{ in.}^3$

12. a. $V = \dfrac{1}{3} \cdot \pi \cdot r^2 \cdot h$
$= \dfrac{1}{\cancel{3}} \cdot 3.14 \cdot (10 \text{ in.})^2 \cdot (\overset{5}{\cancel{15}} \text{ in.})$
$= 3.14 \cdot 100 \cdot 5 \text{ in.}^3$
$= 1570 \text{ in.}^3$

b. $V = \dfrac{1}{3} \cdot \pi \cdot r^2 \cdot h$
$= \dfrac{1}{\cancel{3}} \cdot 3.14 \cdot (8 \text{ in.})^2 \cdot (\overset{4}{\cancel{12}} \text{ in.})$
$= 3.14 \cdot 64 \cdot 6 \text{ in.}^3$
$= 803.84 \text{ in.}^3$

c. $V = \dfrac{1}{3} \cdot \pi \cdot r^2 \cdot h$
$= \dfrac{1}{\cancel{3}} \cdot 3.14 \cdot (6 \text{ in.})^2 \cdot (\overset{3}{\cancel{9}} \text{ in.})$
$= 3.14 \cdot 36 \cdot 3 \text{ in.}^3$
$= 339.12 \text{ in.}^3$

d. $V = \dfrac{1}{3} \cdot \pi \cdot r^2 \cdot h$
$= \dfrac{1}{\cancel{3}} \cdot 3.14 \cdot (4 \text{ in.})^2 \cdot (\overset{2}{\cancel{6}} \text{ in.})$
$= 3.14 \cdot 16 \cdot 2 \text{ in.}^3$
$= 100.48 \text{ in.}^3$

e. $V = \dfrac{1}{3} \cdot \pi \cdot r^2 \cdot h$
$= \dfrac{1}{\cancel{3}} \cdot 3.14 \cdot (2 \text{ in.})^2 \cdot (\overset{1}{\cancel{3}} \text{ in.})$
$= 3.14 \cdot 4 \cdot 1 \text{ in.}^3$
$= 12.56 \text{ in.}^3$

13. a. $\angle a$, $\angle A$, $\angle CAB$ (or $\angle BAC$)
  b. $\angle d$, $\angle D$, $\angle EDF$ (or $\angle FDE$)
  c. $\angle g$, $\angle G$, $\angle HGI$ (or $\angle IGH$)
  d. $\angle j$, $\angle J$, $\angle KJL$ (or $\angle LJK$)
  e. $\angle m$, $\angle M$, $\angle NMO$ (or $\angle OMN$)

14. a. Acute
  b. Acute
  c. Right
  d. Straight
  e. Obtuse

15. a. The complement of 10° is
  $90° - 10° = 80°$.
  b. The complement of 15° is
  $90° - 15° = 75°$.
  c. The complement of 20° is
  $90° - 20° = 70°$.
  d. The complement of 30° is
  $90° - 30° = 60°$.
  e. The complement of 80° is
  $90° - 80° = 10°$.

16. a. The supplement of 10° is
  $180° - 10° = 170°$.
  b. The supplement of 15° is
  $180° - 15° = 165°$.
  c. The supplement of 20° is
  $180° - 20° = 160°$.
  d. The supplement of 30° is
  $180° - 30° = 150°$.
  e. The supplement of 80° is
  $180° - 80° = 100°$.

17. a. Acute equilateral
  b. Acute isosceles
  c. Right isosceles
  d. Obtuse scalene
  e. Right scalene

18. a. $m\angle C = 180° - m\angle A - m\angle B$
  $= 180° - 30° - 40°$
  $= 110°$
  b. $m\angle C = 180° - m\angle A - m\angle B$
  $= 180° - 40° - 50°$
  $= 90°$

c. $m\angle C = 180° - m\angle A - m\angle B$
  $= 180° - 50° - 60°$
  $= 70°$
  d. $m\angle C = 180° - m\angle A - m\angle B$
  $= 180° - 60° - 70°$
  $= 50°$
  e. $m\angle C = 180° - m\angle A - m\angle B$
  $= 180° - 70° - 80°$
  $= 30°$

19. a. $\sqrt{25} = 5$
  b. $\sqrt{9} = 3$
  c. $\sqrt{121} = 11$
  d. $\sqrt{225} = 15$
  e. $\sqrt{169} = 13$

20. a. $\sqrt{2} \approx 1.4142136 \approx 1.414$
  b. $\sqrt{8} \approx 2.8284271 \approx 2.828$
  c. $\sqrt{12} \approx 3.4641016 \approx 3.464$
  d. $\sqrt{22} \approx 4.6904158 \approx 4.690$
  e. $\sqrt{17} \approx 4.1231056 \approx 4.123$

21. a. $a^2 + b^2 = c^2$
  $5^2 + 12^2 = c^2$
  $25 + 144 = c^2$
  $169 = c^2$
  $c = \sqrt{169} = 13$
  The hypotenuse is 13 cm.
  b. $a^2 + b^2 = c^2$
  $4^2 + 3^2 = c^2$
  $16 + 9 = c^2$
  $25 = c^2$
  $c = \sqrt{25} = 5$
  The hypotenuse is 5 cm.
  c. $a^2 + b^2 = c^2$
  $9^2 + 12^2 = c^2$
  $81 + 144 = c^2$
  $225 = c^2$
  $c = \sqrt{225} = 15$
  The hypotenuse is 15 cm.

d. $a^2 + b^2 = c^2$
$12^2 + 16^2 = c^2$
$144 + 256 = c^2$
$400 = c^2$
$c = \sqrt{400} = 20$
The hypotenuse is 20 cm.

e. $a^2 + b^2 = c^2$
$15^2 + 20^2 = c^2$
$225 + 400 = c^2$
$625 = c^2$
$c = \sqrt{625} = 25$
The hypotenuse is 25 cm.

22. a. $a^2 + b^2 = c^2$
$2^2 + 3^2 = c^2$
$4 + 9 = c^2$
$13 = c^2$
$c = \sqrt{13} \approx 3.6055513$
The hypotenuse is about 3.606 cm.

b. $a^2 + b^2 = c^2$
$4^2 + 2^2 = c^2$
$16 + 4 = c^2$
$20 = c^2$
$c = \sqrt{20} \approx 4.472136$
The hypotenuse is about 4.472 cm.

c. $a^2 + b^2 = c^2$
$5^2 + 4^2 = c^2$
$25 + 16 = c^2$
$41 = c^2$
$c = \sqrt{41} \approx 6.4031242$
The hypotenuse is about 6.403 cm.

d. $a^2 + b^2 = c^2$
$3^2 + 6^2 = c^2$
$9 + 36 = c^2$
$45 = c^2$
$c = \sqrt{45} \approx 6.7082039$
The hypotenuse is about 6.708 cm.

e. $a^2 + b^2 = c^2$
$7^2 + 1^2 = c^2$
$49 + 1 = c^2$
$50 = c^2$
$c = \sqrt{50} \approx 7.0710678$
The hypotenuse is about 7.071 cm.

23. a. $a^2 + b^2 = c^2$
$a^2 + 3^2 = 5^2$
$a^2 + 9 = 25$
$a^2 = 16$
$a = \sqrt{16} = 4$
The length of side a is 4.

b. $a^2 + b^2 = c^2$
$a^2 + 4^2 = 8^2$
$a^2 + 16 = 64$
$a^2 = 48$
$a = \sqrt{48} \approx 6.9282032$
The length of side a is about 6.928.

c. $a^2 + b^2 = c^2$
$a^2 + 5^2 = 9^2$
$a^2 + 25 = 81$
$a^2 = 56$
$a = \sqrt{56} \approx 7.4833148$
The length of side a is about 7.483.

d. $a^2 + b^2 = c^2$
$a^2 + 6^2 = 10^2$
$a^2 + 36 = 100$
$a^2 = 64$
$a = \sqrt{64} = 8$
The length of side a is 8.

e. $a^2 + b^2 = c^2$
$a^2 + 8^2 = 12^2$
$a^2 + 64 = 144$
$a^2 = 80$
$a = \sqrt{80} \approx 8.9442719$
The length of side a is about 8.944.

Cumulative Review Chapters 1–8

1. $9 \div 3 \cdot 3 + 5 - 4 = 3 \cdot 3 + 5 - 4$
$= 9 + 5 - 4$
$= 14 - 4$
$= 10$

2.
$$\begin{array}{r} \overset{3\ \,14\ 14}{7\cancel{4}\cancel{5}.42} \\ -\ \ 17.50 \\ \hline 727.92 \end{array}$$

3. $7\underline{4}9.851 \to 750$

4. $40 \div 0.13 = \dfrac{40}{0.13} = \dfrac{4000}{13} =$

$$
\begin{array}{r}
307.692 \\
13\overline{)4000.000} \approx 307.69 \\
\underline{39} \\
100 \\
\underline{91} \\
9\,0 \\
7\,8 \\
\overline{1\,2}0 \\
117 \\
\overline{3}0 \\
\underline{26} \\
4
\end{array}
$$

5. $3.6 = 0.6y$

$\dfrac{3.6}{0.6} = y$

$\dfrac{36}{6} = y$

$6 = y$

6. $\quad 4 = \dfrac{z}{4.4}$

$4.4 \times 4 = z$

$17.6 = z$

7. $\quad \dfrac{j}{5} = \dfrac{5}{125}$

$125j = 5 \cdot 5$

$j = \dfrac{25}{125} = \dfrac{1}{5}$

8. $89\% = 0.89$

9. Let P = the percentage.

$50\% \times 90 = P$

$0.50 \times 90 = P$

$\qquad 45 = P$

Thus 50% of 90 is 45.

10. Let R = the percent.

$R \times 24 = 6$

$R = \dfrac{6}{24} = \dfrac{1}{4} = 0.25 = 25\%$

Thus 25% of 24 is 6.

11. Let B = the number.

$18 = 30\% \times B$

$18 = 0.30 \times B$

$\dfrac{18}{0.30} = B \quad$ so $\quad B = \dfrac{180}{3} = 60$

Thus 18 is 30% of 60.

12. $I = P \times R \times T$

$\quad = 600 \times 0.065 \times 2 = \$78$

13. Automobiles

14.

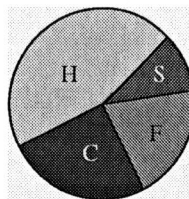

15. 10% of families in Portland have incomes between \$10,000 and \$14,999.

16. The average temperature is $80 - 75 = 5$ degrees higher in July than May.

17. The percent of these hours in math and English combined is $17\% + 23\% = 40\%$.

18. The mode is 18, since it occurs most often.

19. The sum of the 11 numbers is 220. Thus the mean is $\dfrac{220}{11} = 20$.

20. Arrange the numbers in order:
4, 4, 11, 15, 16, 21, 22, 23, 23, 23, 25.
Since there are an odd number of values, the median is the middle term. Thus the median is 21.

21. $11 \text{ yd} = 11 \cdot (3 \text{ ft}) = 33 \cdot (12 \text{ in.}) = 396 \text{ in.}$

22. $9 \text{ in.} = 9 \,\cancel{\text{in.}} \cdot \dfrac{1 \text{ ft}}{12 \,\cancel{\text{in.}}} = \dfrac{9}{12} \text{ ft} = \dfrac{3}{4} \text{ ft}$

23. $7 \text{ ft} = 7 \cancel{\text{ft}} \cdot \dfrac{1 \text{ yd}}{3 \cancel{\text{ft}}} = \dfrac{7}{3} \text{ yd} = 2\dfrac{1}{3} \text{ yd}$

24. $5 \text{ ft} = 5 \cdot (12 \text{ ft}) = 60 \text{ in.}$

25. $4 \text{ km} = 4000 \text{ m}$

26. $2 \text{ dam} = 20 \text{ m}$

27. $140 \text{ m} = 1400 \text{ dm}$

28. $10 \text{ acres} = 10 \cdot 4840 \text{ yd}^2 = 48{,}400 \text{ yd}^2$

29. $P = 2 \cdot L + 2 \cdot W$
$= 2 \cdot (5.1 \text{ in.}) + 2 \cdot (2.8 \text{ in.})$
$= 10.2 \text{ in.} + 5.6 \text{ in.}$
$= 15.8 \text{ in.}$

30. $d = 2 \cdot r = 2 \cdot (6 \text{ cm}) = 12 \text{ cm};$
$C = \pi \cdot d = 3.14 \cdot 12 = 37.68 \text{ cm}$

31. $A = \dfrac{1}{2} \cdot b \cdot h$
$= \dfrac{1}{\cancel{2}} \cdot (\overset{10}{\cancel{20}} \text{ cm}) \cdot (12 \text{ cm}) = 120 \text{ cm}^2$

32. $A = \pi \cdot r^2$
$= 3.14 \cdot (8 \text{ cm})^2$
$= 3.14 \cdot 64 \text{ cm}^2 = 200.96 \text{ cm}^2$

33. $A = \dfrac{1}{2} \cdot h \cdot (a + b)$
$= \dfrac{1}{2} \cdot (7 \text{ in.}) \cdot (18 \text{ in.} + 30 \text{ in.})$
$= \dfrac{1}{\cancel{2}} \cdot (7 \text{ in.}) \cdot (\overset{24}{\cancel{48}} \text{ in.})$
$= 168 \text{ in.}^2$

34. $V = L \cdot W \cdot H$
$= (2 \text{ cm}) \cdot (4 \text{ cm}) \cdot (9 \text{ cm}) = 72 \text{ cm}^3$

35. $V = \dfrac{4}{3} \cdot r^3$
$= \dfrac{4}{3} \cdot (6 \text{ in.})^3 = \dfrac{4}{\cancel{3}} \cdot \overset{72}{\cancel{216}} \text{ in.}^3 = 288 \text{ in.}^3$

36. The angle is between 0° and 90° so the angle is *acute.*

37. $m\angle A + m\angle B + m\angle C = 180°$
$20° + 35° + m\angle C = 180°$
$55° + m\angle C = 180°$
$m\angle C = 180° - 55°$
$= 125°$

38. We need a number whose square is 3600. The number is 60. Thus $\sqrt{3600} = 60$.

39. $a^2 + b^2 = c^2$
$4^2 + 3^2 = c^2$
$16 + 9 = c^2$
$25 = c^2$
$c = \sqrt{25} = 5$
The hypotenuse is 5.

Chapter 9

The Real Numbers

Section 9.1 – Addition and Subtraction of Integers

Problems

1. **a.** The additive inverse of 17 is –17.
   b. The additive inverse of –17 is 17.

2. **a.** $|-18| = 18$ since –18 is 18 units from 0.
   b. $|8| = 8$ since 8 is 8 units from 0.
   c. $|-0| = 0$ since $-0 = 0$ is 0 units from 0.

3.
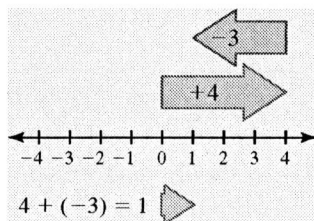

4.
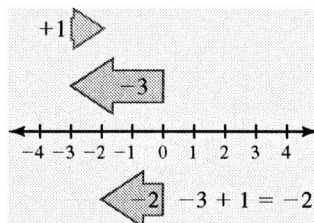

5.
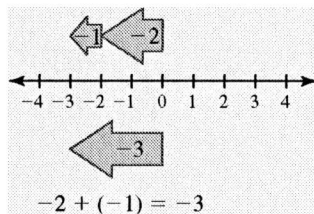

6. **a.** $-9 + (-4) = -(9 + 4) = -13$
   b. $-17 + 8 = -(17 - 8) = -9$
   c. $17 + (-8) = 17 - 8 = 9$

7. **a.** $13 - 7 = 6$
   b. $-18 - 4 = -18 + (-4) = -(18 + 4) = -22$
   c. $-3 - (-9) = -3 + 9 = 6$
   d. $7 - 15 = 7 + (-15) = -(15 - 7) = -8$

8. The score –14 corresponds to the verbal scores for test takers whose parents earned an associates degree. This means that the verbal scores of the test takers were 14 points below the verbal average of 506, that is, $506 - 14 = 492$.

Exercises 9.1

1. The additive inverse of 0 is 0.

3. The additive inverse of –17 is 17.

5. $|2| = 2$

7. $|-11| = 11$

9. $|-30| = 30$

11. $2 + 1 = 3$

13. $-4 + 3 = -1$

15. $5 + (-1) = 4$

17. $-3 + (-3) = -6$

19. $-4 + 4 = 0$

21. $-3 + 5 = +(5 - 3) = 2$

23. $8 + (-1) = 8 - 1 = 7$

25. $-8 + 13 = +(13 - 8) = 5$

27. $-17 + 4 = -(17 - 4) = -11$

29. $-4 + (+8) = -4 + 8 = +(8 - 4) = 4$

31. $-4 - 7 = -(4 + 7) = -11$

33. $-9 - 11 = -(9 + 11) = -20$

35. $8 - 12 = 8 + (-12) = -(12 - 8) = -4$

37. $8 - (-4) = 8 + 4 = 12$

39. $0 - (-4) = 0 + 4 = 4$

41. $23,615 - 7615 = 16,000$
The ocean floor is 16,000 ft below sea level.

43. $53° - (-20°) = 53° + 20° = 73°$
The difference in temperature is 73°C.

45. $5000 - 1500 = 3500$
The difference in temperature is 3500°C.

47. $99 - (-46) = 99 + 46 = 145$
The difference between these extremes is 145°F.

49. $10 + 3 - 2 + 5 - 7 = 13 - 2 + 5 - 7$
$$= 11 + 5 - 7$$
$$= 16 - 7$$
$$= 9$$
Joe is now on the ninth floor.

51. The integer +19 corresponds to the verbal scores of test takers whose parents earned a bachelor's degree. Their score was $506 + 19 = 525$.

53. The integer +53 corresponds to the verbal scores of test takers whose parents earned a bachelor's degree. Their score was $506 + 53 = 559$.

55. $-34 - (-38) = -34 + 38 = 4$

57. $19 - 19 = 0$

59. The highest net farm income was about $27,000 and the corresponding age bracket was 36-40.

61. The farms break even at the 46-50 age bracket.

63. About −$12,000

65. $5^2 = 5 \cdot 5 = 25$

67. $2^3 \cdot 3^2 = (2 \cdot 2 \cdot 2) \cdot (3 \cdot 3) = 8 \cdot 9 = 72$

69. Considering B.C. dates to represent negative numbers and A.D. dates positive numbers, we have
$476 - (-323) = 476 + 323 = 799$ years
elapsed between the fall of the Roman Empire and the death of Alexander the Great.

71. $1776 - 1492 = 284$ years

73. $1939 - (-323) = 1939 + 323 = 2262$ years

75. Answer may vary.

77. Answers may vary.

Mastery Test 9.1

79. a. $\left|-8\right| = 8$ since −8 is 8 units from 0.
b. $\left|4\right| = 4$ since 4 is 4 units from 0.
c. $\left|0\right| = 0$ since 0 is 0 units from 0.

81. $-4 + 2 = -2$

83. a. $-7 + (-5) = -7 - 5 = -(7 + 5) = -12$
b. $-9 + 5 = -(9 - 5) = -4$
c. $9 + (-5) = 9 - 5 = 4$

85. The percent change was lowest in Q1 2003 when it was −2%.

Section 9.2 – Multiplication and Division of Integers

Problems

1. a. $6 \cdot 7 = 42$ **b.** $(-4) \cdot 2 = -8$
c. $-3 \cdot (-8) = 24$ **d.** $(-2) \cdot (8) = -16$

2. a. $42 \div 6 = 7$ **b.** $\dfrac{-18}{9} = -2$
c. $49 \div (-7) = -7$ **d.** $\dfrac{-15}{-3} = 5$

3. a. $(-4)^2 = (-4) \cdot (-4) = 16$
b. $-4^2 = -(4 \cdot 4) = -16$

4. a. $(-3)^3 = (-3) \cdot (-3) \cdot (-3) = 9 \cdot (-3) = -27$
b. $-3^3 = -(3 \cdot 3 \cdot 3) = -27$

Exercises 9.2

1. $16 \cdot 2 = 32$

3. $-7 \cdot 8 = -56$

5. $2 \cdot (-5) = -10$

7. $-4 \cdot (-5) = 20$

9. $-7 \cdot (-10) = 70$

11. $10 \div 2 = 5$

13. $\dfrac{-20}{5} = -4$

15. $-40 \div 8 = -5$

17. $150 \div (-15) = -10$

19. $\dfrac{140}{-7} = -20$

21. $-98 \div (-14) = 7$

23. $\dfrac{-98}{-7} = 14$

25. $\dfrac{0}{-10} = 0$

27. $-0 \div 8 = 0$

29. $\dfrac{-8}{0} = $ Not defined

31. $(-4)^2 = (-4) \cdot (-4) = 16$

33. $-5^2 = -(5 \cdot 5) = -25$

35. $-6^3 = -(6 \cdot 6 \cdot 6) = -216$

37. $(-3)^4 = (-3) \cdot (-3) \cdot (-3) \cdot (-3)$
$= 9 \cdot (-3) \cdot (-3)$
$= -27 \cdot (-3)$
$= 81$

39. $(-5)^3 = (-5) \cdot (-5) \cdot (-5) = -125$

41. $\dfrac{6400}{3200} = 2.$ Each stockholder lost \$2.

43. $2 \cdot (45 \text{ cal}) + 5 \cdot (-15 \text{ cal}) = 90 \text{ cal} - 75 \text{ cal}$
$= 15 \text{ cal}$
There is a 15-calorie gain.

45. $2 \cdot (45) + 2 \cdot (65) + 8 \cdot (-15) = 90 + 130 - 120$
$= 220 - 120$
$= 100$
This is a 100-calorie gain.

47. $4 \cdot (15) + 3(-10) = 60 - 30 = 30$
There was \$30 in the account.

49. $\dfrac{5}{\cancel{8}} \cdot \overset{50}{\cancel{400}} = 250$

The value increased \$250.

51. $\dfrac{3}{\cancel{4}} \times \dfrac{\overset{3}{\cancel{12}}}{5} = \dfrac{9}{5}$

53. $\dfrac{5}{8} \div \dfrac{7}{16} = \dfrac{5}{\cancel{8}} \cdot \dfrac{\cancel{16}^{\,2}}{7} = \dfrac{10}{7}$

55. $0.32 \times 8 = 2.56$

57. $0.49 \div 7 = \dfrac{0.49}{7} = \dfrac{\cancel{49}^{\,7}}{\cancel{700}_{\,100}} = \dfrac{7}{100} = 0.07$

59. $\dfrac{0.64}{0.8} = \dfrac{\cancel{64}^{\,4}}{\cancel{80}_{\,5}} = \dfrac{4}{5} = 0.8$

61. $(+5) + 4(-2) = 5 - 8 = -3$

The valence of phosphate is –3.

63. $(+1) + (+5) + 3(-2) = 1 + 5 - 6 = 0$

The valence of sodium bromate is 0.

65. $2(+1) + 1(-2) = 2 - 2 = 0$

The valence of oxygen is 0.

67. Answer may vary.

Mastery Test 9.2

69. **a.** $(-10)^2 = (-10) \cdot (-10) = 100$

    b. $-10^2 = -(10 \cdot 10) = -100$

71. **a.** $42 \div 7 = 6$

    b. $\dfrac{-16}{4} = -4$

    c. $36 \div (-12) = -3$

    d. $\dfrac{-16}{-4} = 4$

Section 9.3 – The Rational Numbers

Problems

1. **a.** The additive inverse of $\dfrac{3}{4}$ is $-\dfrac{3}{4}$.

    b. The additive inverse of -5.1 is 5.1.

    c. The additive inverse of $-9\dfrac{1}{4}$ is $9\dfrac{1}{4}$.

    d. The additive inverse of 3.9 is -3.9.

2. **a.** $\left|-\dfrac{1}{8}\right| = \dfrac{1}{8}$ **b.** $|3.4| = 3.4$

    c. $\left|-3\dfrac{1}{4}\right| = 3\dfrac{1}{4}$ **d.** $|-8.2| = 8.2$

3. **a.** $-7.5 + 2.1 = -(7.5 - 2.1) = -5.4$

    b. $8.3 + (-9.7) = -(9.7 - 8.3) = -1.4$

    c. $-1.4 + (-6.1) = -(1.4 + 6.1) = -7.5$

4. **a.** $-\dfrac{5}{9} + \dfrac{7}{9} = +\left(\dfrac{7}{9} - \dfrac{5}{9}\right) = \dfrac{2}{9}$

    b. $\dfrac{3}{4} + \left(-\dfrac{5}{3}\right) = \dfrac{3 \cdot 3}{4 \cdot 3} + \left(-\dfrac{5 \cdot 4}{3 \cdot 4}\right)$

              $= \dfrac{9}{12} + \left(-\dfrac{20}{12}\right)$

              $= -\left(\dfrac{20}{12} - \dfrac{9}{12}\right) = -\dfrac{11}{12}$

5. **a.** $-3.8 - (-2.5) = -3.8 + 2.5$

                    $= -(3.8 - 2.5) = -1.3$

    b. $-4.7 - (-6.9) = -4.7 + 6.9$

                    $= +(6.9 - 4.7) = 2.2$

    c. $\dfrac{3}{8} - \left(-\dfrac{1}{8}\right) = \dfrac{3}{8} + \dfrac{1}{8} = \dfrac{4}{8} = \dfrac{1}{2}$

    d. $-\dfrac{7}{8} - \dfrac{5}{6} = -\left(\dfrac{7}{8} + \dfrac{5}{6}\right)$

                  $= -\left(\dfrac{7 \cdot 3}{8 \cdot 3} + \dfrac{5 \cdot 4}{6 \cdot 4}\right)$

                  $= -\left(\dfrac{21}{24} + \dfrac{20}{24}\right) = -\dfrac{41}{24}$

6. **a.** $-2.2 \cdot 3.2 = -7.04$

    b. $-1.3 \cdot (-4) = 5.2$

    c. $-\dfrac{3}{7} \cdot \left(-\dfrac{4}{5}\right) = \dfrac{12}{35}$

    d. $\dfrac{\cancel{6}^{\,2}}{7} \cdot \left(-\dfrac{2}{\cancel{3}}\right) = -\dfrac{4}{7}$

7. **a.** The reciprocal of $\dfrac{4}{5}$ is $\dfrac{5}{4}$.

    b. The reciprocal of $-\dfrac{7}{9}$ is $-\dfrac{9}{7}$.

8. a. $\dfrac{3}{5} \div \left(-\dfrac{4}{7}\right) = \dfrac{3}{5} \cdot \left(-\dfrac{7}{4}\right) = -\dfrac{21}{20}$

b. $-\dfrac{6}{7} \div \left(-\dfrac{3}{5}\right) = -\dfrac{\cancel{6}^{2}}{7} \cdot \left(-\dfrac{5}{\cancel{3}}\right) = \dfrac{10}{7}$

c. $-\dfrac{4}{5} \div \dfrac{8}{5} = -\dfrac{\cancel{4}}{\cancel{5}} \cdot \dfrac{\cancel{5}}{\cancel{8}_{2}} = -\dfrac{1}{2}$

9. a. $\dfrac{-3.6}{1.2} = \dfrac{-3.6 \cdot 10}{1.2 \cdot 10} = \dfrac{-36}{12} = -3$

b. $\dfrac{-3.1}{-12.4} = \dfrac{-3.1 \cdot 10}{-12.4 \cdot 10} = \dfrac{-31}{-124} = \dfrac{1}{4}$

c. $\dfrac{6.5}{-1.3} = \dfrac{6.5 \cdot 10}{-1.3 \cdot 10} = \dfrac{65}{-13} = -5$

10.

	a. $-\dfrac{2}{7}$	b. 3.4	c. 8	d. $\sqrt{81}$	e. $-2\dfrac{3}{4}$	f. $\sqrt{2}$	g. $0.\overline{36}$	h. 0
N			✓	✓				
W			✓	✓				✓
I			✓	✓				✓
Rat.	✓	✓	✓	✓	✓		✓	✓
Irr.						✓		
R	✓	✓	✓	✓	✓	✓	✓	✓

Exercises 9.3

1. The additive inverse of $\dfrac{7}{3}$ is $-\dfrac{7}{3}$.

3. The additive inverse of -6.4 is 6.4.

5. The additive inverse of $3\dfrac{1}{7}$ is $-3\dfrac{1}{7}$.

7. $\left|-\dfrac{4}{5}\right| = \dfrac{4}{5}$

9. $|-3.4| = 3.4$

11. $\left|1\dfrac{1}{2}\right| = 1\dfrac{1}{2}$

13. $-7.8 + 3.1 = -(7.8 - 3.1) = -4.7$

15. $3.2 + (-8.6) = 3.2 - 8.6$
$= -(8.6 - 3.2) = -5.4$

17. $-3.4 + (-5.2) = -3.4 - 5.2$
$= -(3.4 + 5.2) = -8.6$

19. $-\dfrac{2}{7} + \dfrac{5}{7} = +\left(\dfrac{5}{7} - \dfrac{2}{7}\right) = \dfrac{3}{7}$

21. $-\dfrac{3}{4} + \dfrac{1}{4} = -\left(\dfrac{3}{4} - \dfrac{1}{4}\right) = -\dfrac{2}{4} = -\dfrac{1}{2}$

23. $\dfrac{3}{4} + \left(-\dfrac{5}{6}\right) = \dfrac{3}{4} - \dfrac{5}{6}$
$= \dfrac{9}{12} - \dfrac{10}{12} = -\left(\dfrac{10}{12} - \dfrac{9}{12}\right) = -\dfrac{1}{12}$

25. $-\dfrac{1}{6} + \dfrac{3}{4} = -\dfrac{2}{12} + \dfrac{9}{12} = +\left(\dfrac{9}{12} - \dfrac{2}{12}\right) = \dfrac{7}{12}$

27. $-\dfrac{1}{3} + \left(-\dfrac{2}{7}\right) = -\dfrac{1}{3} - \dfrac{2}{7}$
$= -\dfrac{7}{21} - \dfrac{6}{21}$
$= -\left(\dfrac{7}{21} + \dfrac{6}{21}\right) = -\dfrac{13}{21}$

29. $-\dfrac{5}{6} + \left(-\dfrac{8}{9}\right) = -\dfrac{5}{6} - \dfrac{8}{9}$
$= -\dfrac{30}{36} - \dfrac{32}{36}$
$= -\left(\dfrac{30}{36} + \dfrac{32}{36}\right) = -\dfrac{62}{36} = -\dfrac{31}{18}$

31. $-3.8 - (-1.2) = -3.8 + 1.2$
$= -(3.8 - 1.2) = -2.6$

33. $-3.5 - (-8.7) = -3.5 + 8.7$
$= +(8.7 - 3.5) = 5.2$

35. $4.5 - 8.2 = -(8.2 - 4.5) = -3.7$

37. $\dfrac{3}{7} - \left(-\dfrac{1}{7}\right) = \dfrac{3}{7} + \dfrac{1}{7} = \dfrac{4}{7}$

39. $-\dfrac{5}{4} - \dfrac{7}{6} = -\dfrac{15}{12} - \dfrac{14}{12} = -\left(\dfrac{15}{12} + \dfrac{14}{12}\right) = -\dfrac{29}{12}$

41. $-2.2 \cdot 3.3 = -7.26$

43. $-1.3 \cdot (-2.2) = 2.86$

45. $\dfrac{5}{6}\cdot\left(-\dfrac{5}{7}\right)=-\dfrac{25}{42}$

47. $-\dfrac{\cancel{3}}{\cancel{8}}\cdot\left(-\dfrac{\cancel{6}}{\cancel{12}_4}\right)=\dfrac{1}{4}$

49. $-\dfrac{^3\cancel{6}}{\cancel{7}}\cdot\left(\dfrac{\cancel{35}^5}{\cancel{8}_4}\right)=-\dfrac{15}{4}$

51. $\dfrac{3}{5}\div\left(-\dfrac{4}{7}\right)=\dfrac{3}{5}\cdot\left(-\dfrac{7}{4}\right)=-\dfrac{21}{20}$

53. $-\dfrac{2}{3}\div\left(-\dfrac{7}{6}\right)=-\dfrac{2}{\cancel{3}}\cdot\left(-\dfrac{\cancel{6}^2}{7}\right)=\dfrac{4}{7}$

55. $-\dfrac{5}{8}\div\dfrac{7}{8}=-\dfrac{5}{\cancel{8}}\cdot\dfrac{\cancel{8}}{7}=-\dfrac{5}{7}$

57. $\dfrac{-3.1}{6.2}=\dfrac{-31}{62}=-\dfrac{1}{2}=-0.5$

59. $\dfrac{-1.6}{-9.6}=\dfrac{-16}{-96}=\dfrac{1}{6}$

	61.	**63.**	**65.**	**67.**
	$\sqrt{16}$	0	3	$0.\overline{68}$
N	✓		✓	
W	✓	✓	✓	
I	✓	✓	✓	
Rat.	✓	✓	✓	✓
Irr.				
R	✓	✓	✓	✓

69. $7\cdot9-5=63-5=58$

71. $6\div3-(3-5)=6\div3-(-2)=2+2=4$

73. Slightly wicked $=(0.54)\cdot(-2.5)=-1.35$

75. Extemely disgusting $=(1.45)\cdot(-2.1)$
$=-3.045$

77. Very good $=(1.75)\cdot(3.1)=5.425$

79. Answers may vary.

Mastery Test 9.3

81. The additive inverse of $\dfrac{13}{10}$ is $-\dfrac{13}{10}$.

83. The additive inverse of $-1\dfrac{1}{5}$ is $1\dfrac{1}{5}$.

85. $\left|-\dfrac{6}{17}\right|=\dfrac{6}{17}$

87. $\left|-7\dfrac{9}{11}\right|=7\dfrac{9}{11}$

89. $-4.6+4.3=-(4.6-4.3)=-0.3$

91. $-1.1+(-2.4)=-1.1-2.4$
$=-(1.1+2.4)=-3.5$

93. $\dfrac{1}{5}+\left(-\dfrac{1}{6}\right)=\dfrac{1}{5}-\dfrac{1}{6}=\dfrac{6}{30}-\dfrac{5}{30}=\dfrac{1}{30}$

95. $-7.8-1.7=-(7.8+1.7)=-9.5$

97. $-\dfrac{1}{8}-\dfrac{1}{6}=-\dfrac{3}{24}-\dfrac{4}{24}=-\left(\dfrac{3}{24}+\dfrac{4}{24}\right)=-\dfrac{7}{24}$

99. $-2.4\cdot(-2.6)=6.24$

101. $\dfrac{\cancel{8}}{_2\cancel{4}}\cdot\left(-\dfrac{\cancel{2}}{\cancel{8}}\right)=-\dfrac{1}{2}$

103. $\dfrac{3}{5}\div\dfrac{7}{15}=\dfrac{3}{\cancel{5}}\cdot\dfrac{\cancel{15}^3}{7}=\dfrac{9}{7}$

105. $-\dfrac{8}{7}\div\dfrac{4}{7}=-\dfrac{^2\cancel{8}}{\cancel{7}}\cdot\dfrac{\cancel{7}}{\cancel{4}}=-\dfrac{2}{1}=-2$

107. $\dfrac{-1.5}{-6}=\dfrac{-15}{-60}=\dfrac{1}{4}=0.25$

109.

	a. 3.2	**b.** $-\dfrac{8}{9}$	**c.** 9	**d.** $\sqrt{21}$
N			✓	
W			✓	
I			✓	
Rat.	✓	✓	✓	
Irr.				✓
R	✓	✓	✓	✓

Section 9.4 – Order of Operations

Problems

1. a. $-\dfrac{5}{7}\cdot 7 - 3 = -5 - 3 = -8$

b. $-20 + \dfrac{4}{9}\cdot 9 = -20 + 4 = -16$

2. a. $-64 \div \dfrac{8}{3} - (4+1) = -64 \div \dfrac{8}{3} - 5$
$$= -64 \cdot \dfrac{3}{8} - 5$$
$$= -24 - 5$$
$$= -29$$

b. $-27 \div 3^3 + 5 - 2 = -27 \div 27 + 5 - 2$
$$= -1 + 5 - 2$$
$$= 4 - 2$$
$$= 2$$

3. $-6 \div 3 \cdot 4 + 4(7-5) - 5 \cdot \dfrac{4}{5}$
$$= -6 \div 3 \cdot 4 + 4(2) - 5 \cdot \dfrac{4}{5}$$
$$= -2 \cdot 4 + 4(2) - 5 \cdot \dfrac{4}{5}$$
$$= -8 + 4(2) - 5 \cdot \dfrac{4}{5}$$
$$= -8 + 8 - 5 \cdot \dfrac{4}{5}$$
$$= -8 + 8 - 4$$
$$= 0 - 4$$
$$= -4$$

4. $-6^2 + \dfrac{2(6-2)}{2} + 15 \div 3$
$$= -6^2 + \dfrac{2(4)}{2} + 15 \div 3$$
$$= -36 + \dfrac{2(4)}{2} + 15 \div 3$$
$$= -36 + \dfrac{8}{2} + 15 \div 3$$
$$= -36 + 4 + 15 \div 3$$
$$= -36 + 4 + 5$$
$$= -32 + 5$$
$$= -27$$

Exercises 9.4

1. $\dfrac{-4}{5}\cdot 5 + 6 = -4 + 6 = 2$

3. $-7 + \dfrac{3}{2}\cdot 2 = -7 + 3 = -4$

5. $\dfrac{-7}{4}\cdot 8 - 3 = -14 - 3 = -17$

7. $20 - \dfrac{3}{5}\cdot 5 = 20 - 3 = 17$

9. $48 \div \dfrac{3}{4} - (3+2) = 48 \div \dfrac{3}{4} - 5$
$$= 48 \cdot \dfrac{4}{3} - 5 = 64 - 5 = 59$$

11. $3 \cdot 4 \div \frac{2}{3} + (6-2) = 3 \cdot 4 \div \frac{2}{3} + 4$

$$= 12 \div \frac{2}{3} + 4$$

$$= 12 \cdot \frac{3}{2} + 4$$

$$= 18 + 4$$

$$= 22$$

13. $-36 \div 3^2 + 4 - 1 = -36 \div 9 + 4 - 1$

$$= -4 + 4 - 1$$

$$= 0 - 1$$

$$= -1$$

15. $-8 \div 2^3 - 3 + 5 = -8 \div 8 - 3 + 5$

$$= -1 - 3 + 5$$

$$= -4 + 5$$

$$= 1$$

17. $-10 \div 5 \cdot 2 + 8 \cdot (6-4) - 3 \cdot 4$

$$= -10 \div 5 \cdot 2 + 8 \cdot (2) - 3 \cdot 4$$

$$= -2 \cdot 2 + 8 \cdot (2) - 3 \cdot 4$$

$$= -4 + 8 \cdot (2) - 3 \cdot 4$$

$$= -4 + 16 - 3 \cdot 4$$

$$= -4 + 16 - 12$$

$$= 12 - 12$$

$$= 0$$

19. $-4 \cdot 8 \div 2 - 3(4-1) + 9 \div 3$

$$= -4 \cdot 8 \div 2 - 3(3) + 9 \div 3$$

$$= -32 \div 2 - 3(3) + 9 \div 3$$

$$= -16 - 3(3) + 9 \div 3$$

$$= -16 - 9 + 9 \div 3$$

$$= -16 - 9 + 3$$

$$= -25 + 2$$

$$= -22$$

21. $-7^2 + \frac{3(8-4)}{4} + 10 \div 2 \cdot 3$

$$= -7^2 + \frac{3(4)}{4} + 10 \div 2 \cdot 3$$

$$= -49 + \frac{3(4)}{4} + 10 \div 2 \cdot 3$$

$$= -49 + \frac{12}{4} + 10 \div 2 \cdot 3$$

$$= -49 + 3 + 10 \div 2 \cdot 3$$

$$= -49 + 3 + 5 \cdot 3$$

$$= -49 + 3 + 15$$

$$= -46 + 15$$

$$= -31$$

23. $(-6)^2 \cdot 4 \div 4 - \frac{3(7-9)}{2} - 4 \cdot 3 \div 2^2$

$$= (-6)^2 \cdot 4 \div 4 - \frac{3(-2)}{2} - 4 \cdot 3 \div 2^2$$

$$= 36 \cdot 4 \div 4 - \frac{3(-2)}{2} - 4 \cdot 3 \div 4$$

$$= 144 \div 4 - \frac{3(-2)}{2} - 4 \cdot 3 \div 4$$

$$= 36 - \frac{3(-2)}{2} - 4 \cdot 3 \div 4$$

$$= 36 - \frac{(-6)}{2} - 4 \cdot 3 \div 4$$

$$= 36 + 3 - 4 \cdot 3 \div 4$$

$$= 36 + 3 - 12 \div 4$$

$$= 36 + 3 - 3$$

$$= 39 - 3$$

$$= 36$$

25. $\frac{R+M}{2} = \frac{92+82}{2} = \frac{174}{2} = 87$

The octane rating is 87.

27. a. $0.72(220 - A) = 0.72(220 - 20)$

$$= 0.72(200)$$

$$= 144$$

The minimum pulse rate for a 20-year old is 144.

b. $0.72(220 - 45) = 0.72(220 - 45)$

$$= 0.72(175)$$

$$= 126$$

The minimum pulse rate for a 45-year old is 126.

29. $5(7+2) = 5 \cdot 7 + 5 \cdot 2 = 35 + 10 = 45$;

$5(7+2) = 5(9) = 45$

31. $3(4+6) = 3 \cdot 4 + 3 \cdot 6 = 12 + 18 = 30$;

$3(4+6) = 3(10) = 30$

33. One possible answer: $3 + 2 \cdot 5 + 6$

35. One possible answer:
$1 + (2 - 3 - 4 - 5 + 6 + 7 + 8) \cdot 9$

37. No; answers may vary.

39. Answers may vary.

Mastery Test 9.4

41. $-64 \div 8 - (6 - 2) = -64 \div 8 - 4$
$$= -8 - 4$$
$$= -12$$

43. $\dfrac{-3}{4} \cdot 4 - 18 = -3 - 18 = -21$

45. $-12 \div 4 \cdot 2 + 2(5 - 3) - \dfrac{3}{4} \cdot 4$
$$= -12 \div 4 \cdot 2 + 2(2) - \dfrac{3}{4} \cdot 4$$
$$= -3 \cdot 2 + 2(2) - \dfrac{3}{4} \cdot 4$$
$$= -6 + 4 - 3$$
$$= -2 - 3$$
$$= -5$$

47. $[(205 - A) \cdot 7] \div 10 = [(205 - 35) \cdot 7] \div 10$
$$= [170 \cdot 7] \div 10$$
$$= 1190 \div 10 = 119$$

The ideal heart rate for a 35-year old is 119.

Review Exercises – Chapter 9

1. a. The additive inverse of 10 is –10.
 b. The additive inverse of 11 is –11.
 c. The additive inverse of 12 is –12.
 d. The additive inverse of 13 is –13.
 e. The additive inverse of 14 is –14.

2. a. The additive inverse of –9 is 9.
 b. The additive inverse of –8 is 8.
 c. The additive inverse of –7 is 7.
 d. The additive inverse of –6 is 6.
 e. The additive inverse of –5 is 5.

3. a. $|-7| = 7$
 b. $|8| = 8$
 c. $|-9| = 9$
 d. $|10| = 10$
 e. $|-11| = 11$

4. a. $-8 + (-5) = -8 - 5 = -(8 + 5) = -13$
 b. $-8 + (-4) = -8 - 4 = -(8 + 4) = -12$
 c. $-8 + (-3) = -8 - 3 = -(8 + 3) = -11$
 d. $-8 + (-2) = -8 - 2 = -(8 + 2) = -10$
 e. $-8 + (-1) = -8 - 1 = -(8 + 1) = -9$

5. a. $-12 + 5 = -(12 - 5) = -7$
 b. $-12 + 6 = -(12 - 6) = -6$
 c. $-12 + 7 = -(12 - 7) = -5$
 d. $-12 + 8 = -(12 - 8) = -4$
 e. $-12 + 9 = -(12 - 9) = -3$

6. a. $14 + (-4) = 14 - 4 = 10$
 b. $14 + (-5) = 14 - 5 = 9$
 c. $14 + (-6) = 14 - 6 = 8$
 d. $14 + (-7) = 14 - 7 = 7$
 e. $14 + (-8) = 14 - 8 = 6$

7. a. $-15 - 3 = -(15 + 3) = -18$
 b. $-15 - 4 = -(15 + 4) = -19$
 c. $-15 - 5 = -(15 + 5) = -20$
 d. $-15 - 6 = -(15 + 6) = -21$
 e. $-15 - 7 = -(15 + 7) = -22$

8. a. $-10 - (-2) = -10 + 2 = -(10 - 2) = -8$
 b. $-10 - (-3) = -10 + 3 = -(10 - 3) = -7$
 c. $-10 - (-1) = -10 + 1 = -(10 - 1) = -9$
 d. $-10 - (-4) = -10 + 4 = -(10 - 4) = -6$
 e. $-10 - (-5) = -10 + 5 = -(10 - 5) = -5$

9. a. $9 - 14 = -(14 - 9) = -5$
 b. $9 - 15 = -(15 - 9) = -6$
 c. $9 - 16 = -(16 - 9) = -7$
 d. $9 - 17 = -(17 - 9) = -8$
 e. $9 - 18 = -(18 - 9) = -9$

10. a. $-6 \cdot 4 = -24$ **b.** $-6 \cdot 5 = -30$
 c. $-6 \cdot 6 = -36$ **d.** $-6 \cdot 7 = -42$
 e. $-6 \cdot 8 = -48$

11. a. $-8 \cdot (-5) = 40$ **b.** $-8 \cdot (-6) = 48$
 c. $-8 \cdot (-7) = 56$ **d.** $-8 \cdot (-8) = 64$
 e. $-8 \cdot (-9) = 72$

12. a. $5 \cdot (-5) = -25$ **b.** $6 \cdot (-5) = -30$
   c. $7 \cdot (-5) = -35$ **d.** $8 \cdot (-5) = -40$
   e. $9 \cdot (-5) = -45$

13. a. $\dfrac{-58}{2} = -29$ **b.** $\dfrac{-48}{2} = -24$

   c. $\dfrac{-38}{2} = -19$ **d.** $\dfrac{-28}{2} = -14$

   e. $\dfrac{-18}{2} = -9$

14. a. $72 \div (-12) = -6$
   b. $72 \div (-18) = -4$
   c. $72 \div (-24) = -3$
   d. $72 \div (-36) = -2$
   e. $72 \div (-72) = -1$

15. a. $\dfrac{-15}{-5} = 3$ **b.** $\dfrac{-25}{-5} = 5$

   c. $\dfrac{-35}{-5} = 7$ **d.** $\dfrac{-45}{-5} = 9$

   e. $\dfrac{-55}{-5} = 11$

16. a. $(-4)^2 = 16$ **b.** $(-5)^2 = 25$
   c. $(-6)^2 = 36$ **d.** $(-7)^2 = 49$
   e. $(-8)^2 = 64$

17. a. $-4^2 = -16$ **b.** $-5^2 = -25$
   c. $-6^2 = -36$ **d.** $-7^2 = -49$
   e. $-8^2 = -64$

18. a. The additive inverse of $\dfrac{3}{11}$ is $-\dfrac{3}{11}$.

   b. The additive inverse of $\dfrac{4}{11}$ is $-\dfrac{4}{11}$.

   c. The additive inverse of $\dfrac{5}{11}$ is $-\dfrac{5}{11}$.

   d. The additive inverse of $\dfrac{6}{11}$ is $-\dfrac{6}{11}$.

   e. The additive inverse of $\dfrac{7}{11}$ is $-\dfrac{7}{11}$.

19. a. The additive inverse of -3.4 is 3.4.
   b. The additive inverse of -4.5 is 4.5.
   c. The additive inverse of -5.6 is 5.6.
   d. The additive inverse of -6.7 is 6.7.
   e. The additive inverse of -7.8 is 7.8.

20. a. The additive inverse of $-3\dfrac{1}{2}$ is $3\dfrac{1}{2}$.

   b. The additive inverse of $-4\dfrac{1}{2}$ is $4\dfrac{1}{2}$.

   c. The additive inverse of $-5\dfrac{1}{2}$ is $5\dfrac{1}{2}$.

   d. The additive inverse of $-6\dfrac{1}{2}$ is $6\dfrac{1}{2}$.

   e. The additive inverse of $-7\dfrac{1}{2}$ is $7\dfrac{1}{2}$.

21. a. $\left|-\dfrac{2}{11}\right| = \dfrac{2}{11}$ **b.** $\left|-\dfrac{3}{11}\right| = \dfrac{3}{11}$

   c. $\left|-\dfrac{4}{11}\right| = \dfrac{4}{11}$ **d.** $\left|\dfrac{5}{11}\right| = \dfrac{5}{11}$

   e. $\left|\dfrac{6}{11}\right| = \dfrac{6}{11}$

22. a. $\left|-3\dfrac{1}{4}\right| = 3\dfrac{1}{4}$ **b.** $\left|4\dfrac{1}{4}\right| = 4\dfrac{1}{4}$

   c. $\left|5\dfrac{1}{4}\right| = 5\dfrac{1}{4}$ **d.** $\left|-6\dfrac{1}{4}\right| = 6\dfrac{1}{4}$

   e. $\left|-7\dfrac{1}{4}\right| = 7\dfrac{1}{4}$

23. a. $|-5.1| = 5.1$ **b.** $|6.2| = 6.2$
   c. $|-7.3| = 7.3$ **d.** $|8.4| = 8.4$
   e. $|-9.5| = 9.5$

24. a. $-8.7 + 3.1 = -(8.7 - 3.1) = -5.6$
   b. $-8.7 + 3.2 = -(8.7 - 3.2) = -5.5$
   c. $-8.7 + 3.3 = -(8.7 - 3.3) = -5.4$
   d. $-8.7 + 3.4 = -(8.7 - 3.4) = -5.3$
   e. $-8.7 + 3.5 = -(8.7 - 3.5) = -5.2$

25. a. $6.2 + (-9.3) = -(9.3 - 6.2) = -3.1$
   b. $6.2 + (-9.4) = -(9.4 - 6.2) = -3.2$
   c. $6.2 + (-9.5) = -(9.5 - 6.2) = -3.3$
   d. $6.2 + (-9.6) = -(9.6 - 6.2) = -3.4$
   e. $6.2 + (-9.7) = -(9.7 - 6.2) = -3.5$

26. a. $-2.1 + (-3.2) = -2.1 - 3.2$
$= -(2.1 + 3.2) = -5.3$

   b. $-2.1 + (-3.3) = -2.1 - 3.3$
$= -(2.1 + 3.3) = -5.4$

   c. $-2.1 + (-3.4) = -2.1 - 3.4$
$= -(2.1 + 3.4) = -5.5$

   d. $-2.1 + (-3.5) = -2.1 - 3.5$
$= -(2.1 + 3.5) = -5.6$

   e. $-2.1 + (-3.6) = -2.1 - 3.6$
$= -(2.1 + 3.6) = -5.7$

27. a. $-\dfrac{3}{11} + \dfrac{5}{11} = \dfrac{2}{11}$

   b. $-\dfrac{3}{11} + \dfrac{6}{11} = \dfrac{3}{11}$

   c. $-\dfrac{3}{11} + \dfrac{7}{11} = \dfrac{4}{11}$

   d. $-\dfrac{3}{11} + \dfrac{8}{11} = \dfrac{5}{11}$

   e. $-\dfrac{3}{11} + \dfrac{9}{11} = \dfrac{6}{11}$

28. a. $\dfrac{1}{5} + \left(-\dfrac{2}{9}\right) = \dfrac{1}{5} - \dfrac{2}{9} = \dfrac{9}{45} - \dfrac{10}{45} = -\dfrac{1}{45}$

   b. $\dfrac{1}{5} + \left(-\dfrac{4}{9}\right) = \dfrac{1}{5} - \dfrac{4}{9} = \dfrac{9}{45} - \dfrac{20}{45} = -\dfrac{11}{45}$

   c. $\dfrac{1}{5} + \left(-\dfrac{5}{9}\right) = \dfrac{1}{5} - \dfrac{5}{9} = \dfrac{9}{45} - \dfrac{25}{45} = -\dfrac{16}{45}$

   d. $\dfrac{1}{5} + \left(-\dfrac{7}{9}\right) = \dfrac{1}{5} - \dfrac{7}{9} = \dfrac{9}{45} - \dfrac{35}{45} = -\dfrac{26}{45}$

   e. $\dfrac{1}{5} + \left(-\dfrac{8}{9}\right) = \dfrac{1}{5} - \dfrac{8}{9} = \dfrac{9}{45} - \dfrac{40}{45} = -\dfrac{31}{45}$

29. a. $-5.9 - (-3.1) = -5.9 + 3.1$
$= -(5.9 - 3.1) = -2.8$

   b. $-5.9 - (-3.2) = -5.9 + 3.2$
$= -(5.9 - 3.2) = -2.7$

   c. $-5.9 - (-3.3) = -5.9 + 3.3$
$= -(5.9 - 3.3) = -2.6$

   d. $-5.9 - (-3.4) = -5.9 + 3.4$
$= -(5.9 - 3.4) = -2.5$

   e. $-5.9 - (-3.5) = -5.9 + 3.5$
$= -(5.9 - 3.5) = -2.4$

30. a. $-3.2 - (-7.5) = -3.2 + 7.5$
$= +(7.5 - 3.2) = 4.3$

   b. $-3.2 - (-7.6) = -3.2 + 7.6$
$= +(7.6 - 3.2) = 4.4$

   c. $-3.2 - (-7.7) = -3.2 + 7.7$
$= +(7.7 - 3.2) = 4.5$

   d. $-3.2 - (-7.8) = -3.2 + 7.8$
$= +(7.8 - 3.2) = 4.6$

   e. $-3.2 - (-7.9) = -3.2 + 7.9$
$= +(7.9 - 3.2) = 4.7$

31. a. $\dfrac{2}{11} - \left(-\dfrac{3}{11}\right) = \dfrac{2}{11} + \dfrac{3}{11} = \dfrac{5}{11}$

   b. $\dfrac{2}{11} - \left(-\dfrac{4}{11}\right) = \dfrac{2}{11} + \dfrac{4}{11} = \dfrac{6}{11}$

   c. $\dfrac{2}{11} - \left(-\dfrac{5}{11}\right) = \dfrac{2}{11} + \dfrac{5}{11} = \dfrac{7}{11}$

   d. $\dfrac{2}{11} - \left(-\dfrac{6}{11}\right) = \dfrac{2}{11} + \dfrac{6}{11} = \dfrac{8}{11}$

   e. $\dfrac{2}{11} - \left(-\dfrac{7}{11}\right) = \dfrac{2}{11} + \dfrac{7}{11} = \dfrac{9}{11}$

32. a. $-\dfrac{5}{6} - \dfrac{4}{3} = -\dfrac{5}{6} - \dfrac{8}{6} = -\left(\dfrac{5}{6} + \dfrac{8}{6}\right) = -\dfrac{13}{6}$

   b. $-\dfrac{5}{6} - \dfrac{5}{3} = -\dfrac{5}{6} - \dfrac{10}{6}$
$= -\left(\dfrac{5}{6} + \dfrac{10}{6}\right) = -\dfrac{15}{6} = -\dfrac{5}{2}$

   c. $-\dfrac{5}{6} - \dfrac{7}{3} = -\dfrac{5}{6} - \dfrac{14}{6} = -\left(\dfrac{5}{6} + \dfrac{14}{6}\right) = -\dfrac{19}{6}$

   d. $-\dfrac{5}{6} - \dfrac{8}{3} = -\dfrac{5}{6} - \dfrac{16}{6}$
$= -\left(\dfrac{5}{6} + \dfrac{16}{6}\right) = -\dfrac{21}{6} = -\dfrac{7}{2}$

   e. $-\dfrac{5}{6} - \dfrac{2}{3} = -\dfrac{5}{6} - \dfrac{4}{6}$
$= -\left(\dfrac{5}{6} + \dfrac{4}{6}\right) = -\dfrac{9}{6} = -\dfrac{3}{2}$

33. a. $-3.1 \cdot 4.2 = -13.02$

   b. $-3.1 \cdot 4.3 = -13.33$

   c. $-3.1 \cdot 4.4 = -13.64$

   d. $-3.1 \cdot 4.5 = -13.95$

   e. $-3.1 \cdot 4.6 = -14.26$

34. a. $-3.1 \cdot (-2.1) = 6.51$
    b. $-3.1 \cdot (-2.2) = 6.82$
    c. $-3.1 \cdot (-2.3) = 7.13$
    d. $-3.1 \cdot (-2.4) = 7.44$
    e. $-3.1 \cdot (-2.5) = 7.75$

35. a. $-\dfrac{2}{3} \cdot \left(-\dfrac{2}{3}\right) = \dfrac{4}{9}$
    b. $-\dfrac{2}{3} \cdot \left(-\dfrac{4}{3}\right) = \dfrac{8}{9}$
    c. $-\dfrac{2}{3} \cdot \left(-\dfrac{5}{3}\right) = \dfrac{10}{9}$
    d. $-\dfrac{2}{3} \cdot \left(-\dfrac{7}{3}\right) = \dfrac{14}{9}$
    e. $-\dfrac{2}{3} \cdot \left(-\dfrac{8}{3}\right) = \dfrac{16}{9}$

36. a. $\dfrac{5}{\cancel{2}} \cdot \left(-\dfrac{\cancel{2}}{7}\right) = -\dfrac{5}{7}$
    b. $\dfrac{5}{2} \cdot \left(-\dfrac{3}{7}\right) = -\dfrac{15}{14}$
    c. $\dfrac{5}{\cancel{2}} \cdot \left(-\dfrac{\cancel{4}^{\,2}}{7}\right) = -\dfrac{10}{7}$
    d. $\dfrac{5}{\cancel{2}} \cdot \left(-\dfrac{\cancel{6}^{\,3}}{7}\right) = -\dfrac{15}{7}$
    e. $\dfrac{5}{\cancel{2}} \cdot \left(-\dfrac{\cancel{8}^{\,4}}{7}\right) = -\dfrac{20}{7}$

37. a. The reciprocal of $-\dfrac{2}{11}$ is $-\dfrac{11}{2}$.
    b. The reciprocal of $-\dfrac{3}{11}$ is $-\dfrac{11}{3}$.
    c. The reciprocal of $-\dfrac{4}{11}$ is $-\dfrac{11}{4}$.
    d. The reciprocal of $-\dfrac{5}{11}$ is $-\dfrac{11}{5}$.
    e. The reciprocal of $-\dfrac{6}{11}$ is $-\dfrac{11}{6}$.

38. a. $\dfrac{1}{5} \div \left(-\dfrac{1}{7}\right) = \dfrac{1}{5} \cdot \left(-\dfrac{7}{1}\right) = -\dfrac{7}{5}$
    b. $\dfrac{1}{5} \div \left(-\dfrac{2}{7}\right) = \dfrac{1}{5} \cdot \left(-\dfrac{7}{2}\right) = -\dfrac{7}{10}$

c. $\dfrac{1}{5} \div \left(-\dfrac{3}{7}\right) = \dfrac{1}{5} \cdot \left(-\dfrac{7}{3}\right) = -\dfrac{7}{15}$
d. $\dfrac{1}{5} \div \left(-\dfrac{4}{7}\right) = \dfrac{1}{5} \cdot \left(-\dfrac{7}{4}\right) = -\dfrac{7}{20}$
e. $\dfrac{1}{5} \div \left(-\dfrac{5}{7}\right) = \dfrac{1}{5} \cdot \left(-\dfrac{7}{5}\right) = -\dfrac{7}{25}$

39. a. $-\dfrac{5}{2} \div \left(-\dfrac{1}{4}\right) = -\dfrac{5}{\cancel{2}} \cdot \left(-\dfrac{\cancel{4}^{\,2}}{1}\right) = 10$
    b. $-\dfrac{5}{2} \div \left(-\dfrac{1}{6}\right) = -\dfrac{5}{\cancel{2}} \cdot \left(-\dfrac{\cancel{6}^{\,3}}{1}\right) = 15$
    c. $-\dfrac{5}{2} \div \left(-\dfrac{1}{8}\right) = -\dfrac{5}{\cancel{2}} \cdot \left(-\dfrac{\cancel{8}^{\,4}}{1}\right) = 20$
    d. $-\dfrac{5}{2} \div \left(-\dfrac{1}{10}\right) = -\dfrac{5}{\cancel{2}} \cdot \left(-\dfrac{\cancel{10}^{\,5}}{1}\right) = 25$
    e. $-\dfrac{5}{2} \div \left(-\dfrac{1}{12}\right) = -\dfrac{5}{\cancel{2}} \cdot \left(-\dfrac{\cancel{12}^{\,6}}{1}\right) = 30$

40. a. $-\dfrac{2}{11} \div \dfrac{3}{11} = -\dfrac{2}{\cancel{11}} \cdot \dfrac{\cancel{11}}{3} = -\dfrac{2}{3}$
    b. $-\dfrac{2}{11} \div \dfrac{4}{11} = -\dfrac{\cancel{2}}{\cancel{11}} \cdot \dfrac{\cancel{11}}{\cancel{4}_{2}} = -\dfrac{1}{2}$
    c. $-\dfrac{2}{11} \div \dfrac{5}{11} = -\dfrac{2}{\cancel{11}} \cdot \dfrac{\cancel{11}}{5} = -\dfrac{2}{5}$
    d. $-\dfrac{2}{11} \div \dfrac{6}{11} = -\dfrac{\cancel{2}}{\cancel{11}} \cdot \dfrac{\cancel{11}}{\cancel{6}_{3}} = -\dfrac{1}{3}$
    e. $-\dfrac{2}{11} \div \dfrac{7}{11} = -\dfrac{2}{\cancel{11}} \cdot \dfrac{\cancel{11}}{7} = -\dfrac{2}{7}$

41. a. $\dfrac{-2.2}{1.1} = \dfrac{-22}{11} = -2$
    b. $\dfrac{-3.3}{1.1} = \dfrac{-33}{11} = -3$
    c. $\dfrac{-4.4}{1.1} = \dfrac{-44}{11} = -4$
    d. $\dfrac{-5.5}{1.1} = \dfrac{-55}{11} = -5$
    e. $\dfrac{-6.6}{1.1} = \dfrac{-66}{11} = -6$

42. a. $\dfrac{-1.1}{-2.2} = \dfrac{-11}{-22} = \dfrac{1}{2}$

b. $\dfrac{-1.1}{-3.3} = \dfrac{-11}{-33} = \dfrac{1}{3}$

c. $\dfrac{-1.1}{-4.4} = \dfrac{-11}{-44} = \dfrac{1}{4}$

d. $\dfrac{-1.1}{-5.5} = \dfrac{-11}{-55} = \dfrac{1}{5}$

e. $\dfrac{-1.1}{-6.6} = \dfrac{-11}{-66} = \dfrac{1}{6}$

43. a. $\dfrac{2.2}{-1.1} = \dfrac{22}{-11} = -2$

b. $\dfrac{3.3}{-1.1} = \dfrac{33}{-11} = -3$

c. $\dfrac{4.4}{-1.1} = \dfrac{44}{-11} = -4$

d. $\dfrac{5.5}{-1.1} = \dfrac{55}{-11} = -5$

e. $\dfrac{6.6}{-1.1} = \dfrac{66}{-11} = -6$

44.

	a. 3.7	**b.** $\sqrt{121}$	**c.** $0.\overline{56}$	**d.** $-\dfrac{8}{9}$	**e.** $\sqrt{21}$
N		✓			
W		✓			
I		✓			
Rat.	✓	✓	✓	✓	
Irr.					✓
R	✓	✓	✓	✓	✓

45. a. $-\dfrac{7}{2} \cdot 2 - 5 = -7 - 5 = -(7+5) = -12$

b. $-\dfrac{7}{3} \cdot 3 - 6 = -7 - 6 = -(7+6) = -13$

c. $-\dfrac{7}{4} \cdot 4 - 7 = -7 - 7 = -(7+7) = -14$

d. $-\dfrac{7}{5} \cdot 5 - 8 = -7 - 8 = -(7+8) = -15$

e. $-\dfrac{7}{6} \cdot 6 - 9 = -7 - 9 = -(7+9) = -16$

46. a. $-30 \div \dfrac{3}{4} - (3+4) = -30 \div \dfrac{3}{4} - 7$

$= -30 \cdot \dfrac{4}{3} - 7$

$= -40 - 7 = -47$

b. $-30 \div \dfrac{3}{5} - (3+4) = -30 \div \dfrac{3}{5} - 7$

$= -30 \cdot \dfrac{5}{3} - 7$

$= -50 - 7 = -57$

c. $-30 \div \dfrac{3}{7} - (3+4) = -30 \div \dfrac{3}{7} - 7$

$= -30 \cdot \dfrac{7}{3} - 7$

$= -70 - 7 = -77$

d. $-30 \div \dfrac{3}{8} - (3+4) = -30 \div \dfrac{3}{8} - 7$

$= -30 \cdot \dfrac{8}{3} - 7$

$= -80 - 7 = -87$

e. $-30 \div \dfrac{3}{9} - (3+4) = -30 \div \dfrac{3}{9} - 7$

$= -30 \cdot \dfrac{9}{3} - 7$

$= -90 - 7 = -97$

47. a. $-64 \div 4 \cdot 2 + 3(5-3) - 4 \cdot \dfrac{3}{4}$

$= -64 \div 4 \cdot 2 + 3(2) - 4 \cdot \dfrac{3}{4}$

$= -16 \cdot 2 + 3(2) - 4 \cdot \dfrac{3}{4}$

$= -32 + 3(2) - 4 \cdot \dfrac{3}{4}$

$= -32 + 6 - 4 \cdot \dfrac{3}{4}$

$= -32 + 6 - 3$

$= -26 - 3 = -29$

b. $-64 \div 8 \cdot 2 + 3(5-3) - 8 \cdot \dfrac{3}{8}$

$= -64 \div 8 \cdot 2 + 3(2) - 8 \cdot \dfrac{3}{8}$

$= -8 \cdot 2 + 3(2) - 8 \cdot \dfrac{3}{8}$

$= -16 + 3(2) - 8 \cdot \dfrac{3}{8}$

$= -16 + 6 - 8 \cdot \dfrac{3}{8}$

$= -16 + 6 - 3$

$= -10 - 3 = -13$

c. $-64 \div 16 \cdot 2 + 3(5-3) - 16 \cdot \frac{3}{16}$

$= -64 \div 16 \cdot 2 + 3(2) - 16 \cdot \frac{3}{16}$

$= -4 \cdot 2 + 3(2) - 16 \cdot \frac{3}{16}$

$= -8 + 3(2) - 16 \cdot \frac{3}{16}$

$= -8 + 6 - 16 \cdot \frac{3}{16}$

$= -8 + 6 - 3$

$= -2 - 3 = -5$

d. $-64 \div 32 \cdot 2 + 3(5-3) - 32 \cdot \frac{3}{32}$

$= -64 \div 32 \cdot 2 + 3(2) - 32 \cdot \frac{3}{32}$

$= -2 \cdot 2 + 3(2) - 32 \cdot \frac{3}{32}$

$= -4 + 3(2) - 32 \cdot \frac{3}{32}$

$= -4 + 6 - 32 \cdot \frac{3}{32}$

$= -4 + 6 - 3$

$= 2 - 3 = -1$

e. $-64 \div 64 \cdot 2 + 3(5-3) - 64 \cdot \frac{3}{64}$

$= -64 \div 64 \cdot 2 + 3(2) - 64 \cdot \frac{3}{64}$

$= -1 \cdot 2 + 3(2) - 64 \cdot \frac{3}{64}$

$= -2 + 3(2) - 64 \cdot \frac{3}{64}$

$= -2 + 6 - 64 \cdot \frac{3}{64}$

$= -2 + 6 - 3$

$= 4 - 3 = 1$

48. a. $-6^2 + \frac{3(4-8)}{2} + 10 \div 5$

$= -6^2 + \frac{3(-4)}{2} + 10 \div 5$

$= -36 + \frac{3(-4)}{2} + 10 \div 5$

$= -36 + \frac{(-12)}{2} + 10 \div 5$

$= -36 - 6 + 10 \div 5$

$= -36 - 6 + 2$

$= -42 + 2 = -40$

b. $-7^2 + \frac{3(4-8)}{2} + 20 \div 5$

$= -7^2 + \frac{3(-4)}{2} + 20 \div 5$

$= -49 + \frac{3(-4)}{2} + 20 \div 5$

$= -49 + \frac{(-12)}{2} + 20 \div 5$

$= -49 - 6 + 20 \div 5$

$= -49 - 6 + 4$

$= -55 + 4 = -51$

c. $-7^2 + \frac{3(4-8)}{2} + 30 \div 5$

$= -7^2 + \frac{3(-4)}{2} + 30 \div 5$

$= -49 + \frac{3(-4)}{2} + 30 \div 5$

$= -49 + \frac{(-12)}{2} + 30 \div 5$

$= -49 - 6 + 30 \div 5$

$= -49 - 6 + 6$

$= -55 + 6 = -49$

d. $-8^2 + \frac{3(4-8)}{2} + 40 \div 5$

$= -8^2 + \frac{3(-4)}{2} + 40 \div 5$

$= -64 + \frac{3(-4)}{2} + 40 \div 5$

$= -64 + \frac{(-12)}{2} + 40 \div 5$

$= -64 - 6 + 40 \div 5$

$= -64 - 6 + 8$

$= -70 + 8 = -62$

e. $-9^2 + \frac{3(4-8)}{2} + 50 \div 5$

$= -9^2 + \frac{3(-4)}{2} + 50 \div 5$

$= -81 + \frac{3(-4)}{2} + 50 \div 5$

$= -81 + \frac{(-12)}{2} + 50 \div 5$

$= -81 - 6 + 50 \div 5$

$= -81 - 6 + 10$

$= -87 + 10 = -77$

Cumulative Review Chapters 1–9

1. $9 \div 3 \cdot 3 + 5 - 4 = 3 \cdot 3 + 5 - 4$
$$= 9 + 5 - 4$$
$$= 14 - 4$$
$$= 10$$

2. $44\underline{9}.851 \rightarrow 450$

3. $90 \div 0.13 = \dfrac{90}{0.13} = \dfrac{9000}{13}$

$$= 13\overline{)9000.000} \approx 692.31$$

$$\begin{array}{r}
692.307 \\
\hline
78 \\
\hline
120 \\
117 \\
\hline
30 \\
26 \\
\hline
4\,0 \\
3\,9 \\
\hline
100 \\
91 \\
\hline
9
\end{array}$$

4. $7.2 = 0.9y$

$$\dfrac{7.2}{0.9} = \dfrac{\cancel{0.9}\,y}{\cancel{0.9}}$$

$$\dfrac{72}{9} = y$$

$$8 = y$$

5. $\quad 6 = \dfrac{z}{6.6}$

$$6.6 \times 6 = z$$

$$39.6 = z$$

6. $\dfrac{c}{4} = \dfrac{4}{64}$

$$64c = 4 \cdot 4$$

$$c = \dfrac{16}{64} = \dfrac{1}{4}$$

7. $23\% = 0.23$

8. Let P = the percentage.
$$0.80 \times 40 = P$$
$$32 = P$$
Thus, 80% of 40 is 32.

9. Let R = the percent.
$$R \times 4 = 2$$
$$R = \dfrac{2}{4} = \dfrac{1}{2} = 0.5 = 50\%$$
Thus, 50% of 4 is 2.

10. Let B = the number.
$$30 = 0.40 \times B$$
$$\dfrac{30}{0.40} = B \quad \text{so} \quad B = \dfrac{300}{4} = 75$$
Thus, 30 is 40% of 75.

11. $I = P \times R \times T = 900 \times 0.045 \times 4 = \$162$

12. Agriculture

13.

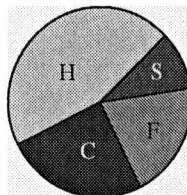

14. 4% of the families in Chicago have incomes between $50,000 and $79,999.

15. $77° - 75° = 2°$. The average temperature is about 2° higher in June than in May.

16. $30 + 17 = 47\%$ of these hours is in art and English combined.

17. The mode is 6, since it occurs most often.

18. The sum of the 11 numbers is 187. Thus the mean is $\dfrac{187}{11} = 17$.

19. Arrange the numbers in order:
2, 2, 2, 2, 3, 4, 10, 14, 19, 23, 29.
Since there are an odd number of values, the median is the middle number. Thus the median is 4.

20. 9 yd $= 9 \cdot (3 \text{ ft}) = 27 (12 \text{ in.}) = 324$ in.

21. 16 in. $= 16 \text{ in.} \cdot \dfrac{1 \text{ ft}}{12 \text{ in.}} = \dfrac{16}{12} \text{ ft} = 1\dfrac{1}{3}$ ft

22. $17 \text{ ft} = 17 \cancel{\text{ft}} \cdot \dfrac{1 \text{ yd}}{3 \cancel{\text{ft}}} = \dfrac{17}{3} \text{ yd} = 5\dfrac{2}{3} \text{ yd}$

23. $6 \text{ ft} = 6 \cdot (12 \text{ in.}) = 72 \text{ in.}$

24. $170 \text{ m} = 1700 \text{ dm}$

25. $P = 2 \cdot (4.4) + 2 \cdot (2.3)$
$= 8.8 + 4.6$
$= 13.4 \text{ in.}$

26. $d = 2 \cdot r = 2 \cdot 8 \text{ ft} = 16 \text{ ft};$
$C = \pi \cdot d = 3.14 \cdot (16 \text{ ft}) = 50.24 \text{ ft}$

27. $A = \dfrac{1}{2} \cdot b \cdot h$
$= \dfrac{1}{\cancel{2}} \cdot (\overset{8}{\cancel{16}} \text{ cm}) \cdot (20 \text{ cm}) = 160 \text{ cm}^2$

28. $A = \pi r^2$
$= 3.14 \cdot (6 \text{ cm})^2$
$= 3.14 \cdot (36) \text{ cm}^2 = 113.04 \text{ cm}^2$

29. $A = \dfrac{1}{2} \cdot h \cdot (a + b)$
$= \dfrac{1}{2} \cdot (9 \text{ in.}) \cdot (24 \text{ in.} + 34 \text{ in.})$
$= \dfrac{1}{\cancel{2}} \cdot (9 \text{ in.}) \cdot (\overset{29}{\cancel{58}} \text{ in.}) = 261 \text{ in.}^2$

30. $V = L \cdot W \cdot H$
$= (4 \text{ cm}) \cdot (5 \text{ cm}) \cdot (10 \text{ cm}) = 200 \text{ cm}^3$

31. $V = \dfrac{4}{3} \cdot \pi r^3$
$= \dfrac{4}{3} \cdot (3.14) \cdot (9 \text{ in.})^3$
$= \dfrac{4}{\cancel{3}} \cdot (3.14) \cdot \overset{243}{\cancel{729}} \text{ in.}^3 = 3052.08 \text{ in.}^3$

32. Obtuse

33. $m\angle A + m\angle B + m\angle C = 180°$
$30° + 25° + m\angle C = 180°$
$55° + m\angle C = 180°$
$m\angle C = 180° - 55° = 125°$

34. We need a number whose square is 4900. The number is 70. Thus $\sqrt{4900} = 70$.

35. $a^2 + b^2 = c^2$
$8^2 + 6^2 = c^2$
$64 + 36 = c^2$
$100 = c^2$
$c = \sqrt{100} = 10$
The hypotenuse is 10.

36. The additive inverse of $-4\dfrac{3}{4}$ is $4\dfrac{3}{4}$.

37. $\left| -9.8 \right| = 9.8$

38. $-7.2 + (-2.6) = -7.2 - 2.6$
$= -(7.2 + 2.6) = -9.8$

39. $\dfrac{1}{5} + \left(-\dfrac{1}{7} \right) = \dfrac{1}{5} - \dfrac{1}{7} = \dfrac{7}{35} - \dfrac{5}{35} = \dfrac{2}{35}$

40. $-4.2 - (-8.5) = -4.2 + 8.5$
$= +(8.5 - 4.2) = 4.3$

41. $-\dfrac{1}{5} - \dfrac{1}{6} = -\left(\dfrac{1}{5} + \dfrac{1}{6} \right) = -\left(\dfrac{6}{30} + \dfrac{5}{30} \right) = -\dfrac{11}{30}$

42. $-\dfrac{\cancel{2}}{\cancel{8}} \cdot \dfrac{\cancel{8}}{\cancel{9}_3} = -\dfrac{1}{3}$

43. $-\dfrac{9}{8} \div \dfrac{5}{8} = -\dfrac{9}{\cancel{8}} \cdot \dfrac{\cancel{8}}{5} = -\dfrac{9}{5}$

44. $(-9)^2 = (-9) \cdot (-9) = 81$

Chapter 10

Introduction to Algebra

Section 10.1 – Introduction to Algebra

Problems

1. The terms are $8z^2$, $-5z$, and 2.

2. a. $3(a+5b) = 3 \cdot a + 3 \cdot 5b = 3a + 15b$
   b. $6(4a-5b) = 6 \cdot 4a - 6 \cdot 5b = 24a - 30b$
   c. $-2(a-3b+c)$
   $= -2 \cdot 2 - (-2) \cdot 3b + (-2) \cdot c$
   $= -4 + 6b - 2c$

3. a. $5a - 15b = 5 \cdot a - 5 \cdot 3b = 5(a-3b)$
   b. $ab + ac - ad = a \cdot b + a \cdot c - a \cdot d$
   $= a(b+c-d)$
   c. $6a - 12b + 18c = 6 \cdot a - 6 \cdot 2b + 6 \cdot 3c$
   $= 6(a - 2b + 3c)$

4. a. $6a - 2a = (6-2)a = 8a$
   b. $5a + 7b - a + 2b = 5a - a + 7b + 2b$
   $= (5-1)a + (7+2)b$
   $= 4a + 9b$
   c. $0.3a + 0.21b - 0.8a + 0.32b$
   $= 0.3a - 0.8a + 0.21b + 0.32b$
   $= (0.3 - 0.8)a + (0.21 + 0.32)b$
   $= -0.5a + 0.53b$
   d. $\dfrac{1}{5}a + \dfrac{4}{7}b + \dfrac{2}{5}a - \dfrac{1}{7}b$
   $= \dfrac{1}{5}a + \dfrac{2}{5}a + \dfrac{4}{7} - \dfrac{1}{7}b$
   $= \left(\dfrac{1}{5} + \dfrac{2}{5}\right)a + \left(\dfrac{4}{7} - \dfrac{1}{7}\right)b$
   $= \dfrac{3}{5}a + \dfrac{3}{7}b$

5. a. $(4a+5b) - (a+3b) = 4a + 5b - a - 3b$
   $= 3a + 2b$
   b. $(6a+5b) - (8a-2b) = 6a + 5b - 8a + 2b$
   $= -2a + 7b$

6. a. $-(-2x+5y) - 6x = 2x - 5y - 6x$
   $= -4x - 5y$
   b. $-(-2x^2 - 3y) + 3x = 2x^2 + 3y + 3x$

7. a. $a + b = 4 + (-9) = 4 - 9 = -5$
   b. $5a + b = 5(3) + 5 = 20$
   c. $-(a+b) = -(6+3) = -9$

Exercises 10.1

1. $2x$; -3

3. $-3x^2$; $0.5x$; -6

5. $\dfrac{1}{5}x$; $-\dfrac{3}{4}y$; $\dfrac{1}{8}z$

7. $3(2x+y) = 3 \cdot 2x + 3 \cdot y = 6x + 3y$

9. $3(a-b) = 3 \cdot a - 3 \cdot b = 3a - 3b$

11. $-5(2x-y) = -5 \cdot 2x - (-5) \cdot y = -10x + 5y$

13. $8(3x^2 + 2) = 8 \cdot 3x^2 + 8 \cdot 2 = 24x^2 + 16$

15. $-6(2x^2 - 3) = -6 \cdot 2x^2 - (-6) \cdot 3$
   $= -12x^2 + 18$

17. $3(2x^2 + 3x + 5) = 3 \cdot 2x^2 + 3 \cdot 3x + 3 \cdot 5$
   $= 6x^2 + 9x + 15$

19. $-5(3x^2 - 2x - 3)$
   $= -5 \cdot 3x^2 - (-5) \cdot 2x - (-5) \cdot 3$
   $= -15x^2 + 10x + 15$

21. $0.5(x+y-2) = 0.5 \cdot x + 0.5 \cdot y - 0.5 \cdot 2$
$\qquad = 0.5x + 0.5y - 1$

23. $\dfrac{6}{5}(a-b+5) = \dfrac{6}{5} \cdot a + \dfrac{6}{5} \cdot b + \dfrac{6}{5} \cdot 5$
$\qquad\qquad = \dfrac{6}{5}a + \dfrac{6}{5}b + 6$

25. $-2(x-y+4) = -2 \cdot x - (-2) \cdot y + (-2) \cdot 4$
$\qquad\qquad = -2x + 2y - 8$

27. $-0.3(x+y-6)$
$\quad = -0.3 \cdot x + (-0.3) \cdot y - (-0.3) \cdot 6$
$\quad = -0.3x - 0.3y + 1.8$

29. $-\dfrac{5}{2}(a-2b+c-1)$
$\quad = -\dfrac{5}{2} \cdot a - \left(-\dfrac{5}{2}\right) \cdot 2b + \left(-\dfrac{5}{2}\right) \cdot c - \left(-\dfrac{5}{2}\right) \cdot 1$
$\quad = -\dfrac{5}{2}a + 5b - \dfrac{5}{2}c + \dfrac{5}{2}$

31. $3x + 15 = 3 \cdot x + 3 \cdot 5 = 3(x+5)$

33. $9y - 18 = 9 \cdot y - 9 \cdot 2 = 9(y-2)$

35. $-5y + 20 = -5y - (-20)$
$\qquad\qquad = -5 \cdot y - (-5) \cdot 4$
$\qquad\qquad = -5(y-4)$

37. $-3x - 27 = -3x + (-27)$
$\qquad\qquad = -3 \cdot x + (-3) \cdot 9$
$\qquad\qquad = -3(x+9)$

39. $bx - by + bz = b \cdot x - b \cdot y + b \cdot z$
$\qquad\qquad\quad = b(x-y+z)$

41. $8a + 2a = (8+2)a = 10a$

43. $4x + 2x = (4+2)x = 6x$

45. $8a^2 + 2a^2 = (8+2)a^2 = 10a^2$

47. $13x - 2x = (13-2)x = 11x$

49. $17y^2 - 12y^2 = (17-12)y^2 = 5y^2$

51. $7x + 3y - 2x = 7x - 2x + 3y$
$\qquad\qquad\quad = (7-2)x + 3y$
$\qquad\qquad\quad = 5x + 3y$

53. $13x + 5 - 2y - 3x - 9 - y$
$\quad = 13x - 3x - 2y - y + 5 - 9$
$\quad = (13-3)x + (-2-1)y + (5-9)$
$\quad = 10x + (-3)y + (-4)$
$\quad = 10x - 3y - 4$

55. $3.9a + 4.5b - 3 - 1.5a - 7.5b - 4$
$\quad = 3.9a - 1.5a + 4.5b - 7.5b - 3 - 4$
$\quad = (3.9-1.5)a + (4.5-7.5)b - 7$
$\quad = 2.4a + (-3)b - 7$
$\quad = 2.4a - 3b - 7$

57. $\dfrac{1}{7}a - \dfrac{1}{5}b + \dfrac{3}{7}a - \dfrac{3}{5}b$
$\quad = \dfrac{1}{7}a + \dfrac{3}{7}a - \dfrac{1}{5}b - \dfrac{3}{5}b$
$\quad = \left(\dfrac{1}{7} + \dfrac{3}{7}\right)a + \left(-\dfrac{1}{5} - \dfrac{3}{5}\right)b$
$\quad = \dfrac{4}{7}a + \left(-\dfrac{4}{5}\right)b$
$\quad = \dfrac{4}{7}a - \dfrac{4}{5}b$

59. $-\dfrac{1}{8}x - \dfrac{3}{11}b + \dfrac{3}{8}x - \dfrac{2}{11}b$
$\quad = -\dfrac{1}{8}x + \dfrac{3}{8}x - \dfrac{3}{11}b - \dfrac{2}{11}b$
$\quad = \left(-\dfrac{1}{8} + \dfrac{3}{8}\right)x + \left(-\dfrac{3}{11} - \dfrac{2}{11}\right)b$
$\quad = \dfrac{2}{8}x + \left(-\dfrac{5}{11}\right)b$
$\quad = \dfrac{1}{4}x - \dfrac{5}{11}b$

61. $-(-a+5b) - (7a+8b) = a - 5b - 7a - 8b$
$\qquad\qquad\qquad\qquad\quad = -6a - 13b$

63. $-(-2a+b) - (4b+a) = 2a - b - 4b - a$
$\qquad\qquad\qquad\qquad\quad = a - 5b$

65. $-(-b^2 - 2a) + (3a - 4b^2)$
$\quad = b^2 + 2a + 3a - 4b^2$
$\quad = -3b^2 + 5a \ \text{ or } \ 5a - 3b^2$

67. $-(-a^2 - 3b) + (2a^2 - 5b)$
$= a^2 + 3b + 2a^2 - 5b$
$= 3a^2 - 2b$

69. $-(-2b^2 - 3a) - (6b^2 - 5a)$
$= 2b^2 + 3a - 6b^2 + 5a$
$= -4b^2 + 8a$

71. $m + n = 7 + (-9) = 7 - 9 = -2$

73. $-(u + v) = -(5 + 3) = -8$

75. $3x^2 y = 3(3)^3(5) = 3 \cdot 9 \cdot 5 = 27 \cdot 5 = 135$

77. $\dfrac{x}{3} + x = \dfrac{6}{3} + 6 = 2 + 6 = 8$

79. $LWH = (2)(3.2)(5) = 32$

81. $3^2 \cdot 5^2 \cdot 7^0 = (3 \cdot 3) \cdot (5 \cdot 5) \cdot 1 = 9 \cdot 25 = 225$

83. $10^3 \cdot 3^0 \cdot 5^2 \cdot 8^1 = (10 \cdot 10 \cdot 10) \cdot 1 \cdot (5 \cdot 5) \cdot 8$
$= 1000 \cdot 25 \cdot 8$
$= 200,000$

85. Answers may vary.

87. Answers may vary.

Mastery Test 10.1

89. a. $(4x + 3y) - (x + 2y) = 4x + 3y - x - 2y$
$= 3x + y$

 b. $(5x + 3y) - (7x - 2y) = 5x + 3y - 7x + 2y$
$= -2x + 5y$

91. a. $7a - 3a = (7 - 3)a = 4a$

 b. $5a + 3b - a + 3b = 5a - a + 3b + 3b$
$= 4a + 6b$

 c. $\dfrac{2}{9}a + \dfrac{5}{7}b + \dfrac{3}{9}a - \dfrac{4}{7}b$
$= \left(\dfrac{2}{9} + \dfrac{3}{9}\right)a + \left(\dfrac{5}{7} - \dfrac{4}{7}\right)b$
$= \dfrac{5}{9}a + \dfrac{1}{7}b$

93. a. $3(a + 2b) = 3 \cdot a + 3 \cdot 2b = 3a + 6b$

 b. $5(2a - 3b) = 5 \cdot 2a - 5 \cdot 3b = 10a - 15b$

 c. $-4(a - 3b + c)$
$= -4 \cdot a - (-4) \cdot 3b + (-4) \cdot c$
$= -4a + 12b - 4c$

Section 10.2 – The Algebra of Exponents

Problems

1. a. $5^{-2} = \dfrac{1}{5^2} = \dfrac{1}{5 \cdot 5} = \dfrac{1}{25}$

 b. $3^{-3} = \dfrac{1}{3^3} = \dfrac{1}{3 \cdot 3 \cdot 3} = \dfrac{1}{27}$

2. a. $\dfrac{1}{7^4} = 7^{-4}$ **b.** $\dfrac{1}{6^5} = 6^{-5}$

3. a. $2^5 \cdot 2^{-3} = 2^{5+(-3)} = 2^2 = 4$
 b. $x^5 \cdot x^3 = x^{5+3} = x^8$
 c. $3^4 \cdot 3^{-7} = 3^{4+(-7)} = 3^{-3} = \dfrac{1}{3^3} = \dfrac{1}{27}$
 d. $x^3 \cdot x^{-4} = x^{3+(-4)} = x^{-1} = \dfrac{1}{x}$
 e. $y^3 \cdot y^{-3} = y^{3+(-3)} = y^0 = 1$

4. a. $\dfrac{7^2}{7^{-3}} = 7^{2-(-3)} = 7^{2+3} = 7^5 = 16,807$

 b. $\dfrac{y}{y^6} = y^{1-6} = y^{-5} = \dfrac{1}{y^5}$

 c. $\dfrac{x^{-3}}{x^{-3}} = x^{-3-(-3)} = x^{-3+3} = x^0 = 1$

 d. $\dfrac{z^{-4}}{z^{-5}} = z^{-4-(-5)} = z^{-4+5} = z^1 = z$

5. a. $(5^3)^2 = 5^{3 \cdot 2} = 5^6 = 15,625$

 b. $(x^{-3})^4 = x^{-3 \cdot 4} = x^{-12} = \dfrac{1}{x^{12}}$

 c. $(y^3)^{-6} = y^{3(-6)} = y^{-18} = \dfrac{1}{y^{18}}$

 d. $(z^{-3})^{-5} = z^{-3(-5)} = z^{15}$

6. a. $(x^2 y^{-2})^3 = (x^2)^3 (y^{-2})^3 = x^6 y^{-6} = \dfrac{x^6}{y^6}$

  b. $(x^{-3} y^2)^3 = (x^{-3})^3 (y^2)^3 = x^{-9} y^6 = \dfrac{y^6}{x^9}$

  c. $(x^3 y^2)^{-2} = (x^3)^{-2} (y^2)^{-2}$
  $\phantom{(x^3 y^2)^{-2}} = x^{-6} y^{-4} = \dfrac{1}{x^6 y^4}$

7. $A = P(1+r)^n$
  $ = 500(1+0.10)^2$
  $ = 500(1.10)^2 = 500(1.21) = \$605$

Exercises 10.2

1. a. $4^{-2} = \dfrac{1}{4^2} = \dfrac{1}{4 \cdot 4} = \dfrac{1}{16}$

  b. $x^{-2} = \dfrac{1}{x^2} = \dfrac{1}{x \cdot x} = \dfrac{1}{x^2}$

3. a. $5^{-3} = \dfrac{1}{5^3} = \dfrac{1}{5 \cdot 5 \cdot 5} = \dfrac{1}{125}$

  b. $y^{-3} = \dfrac{1}{y^3}$

5. a. $3^{-4} = \dfrac{1}{3^4} = \dfrac{1}{3 \cdot 3 \cdot 3 \cdot 3} = \dfrac{1}{81}$

  b. $z^{-4} = \dfrac{1}{z^4}$

7. a. $\dfrac{1}{2^3} = 2^{-3}$ **b.** $\dfrac{1}{x^3} = x^{-3}$

9. a. $\dfrac{1}{4^5} = 4^{-5}$ **b.** $\dfrac{1}{y^5} = y^{-5}$

11. a. $\dfrac{1}{3^5} = 3^{-5}$ **b.** $\dfrac{1}{z^5} = z^{-5}$

13. a. $3^2 \cdot 3^3 = 3^{2+3} = 3^5 = 243$
  b. $x^5 \cdot x^8 = x^{5+8} = x^{13}$

15. a. $2^{-5} \cdot 2^7 = 2^{-5+7} = 2^2 = 4$
  b. $y^{-3} \cdot y^8 = y^{-3+8} = y^5$

17. a. $4^{-6} \cdot 4^4 = 4^{-6+4} = 4^{-2} = \dfrac{1}{4^2} = \dfrac{1}{16}$

  b. $x^{-7} \cdot x^3 = x^{-7+3} = x^{-4} = \dfrac{1}{x^4}$

19. a. $6^{-1} \cdot 6^{-2} = 6^{-1+(-2)} = 6^{-3} = \dfrac{1}{6^3} = \dfrac{1}{216}$

  b. $y^{-3} \cdot y^{-4} = y^{-3+(-4)} = y^{-7} = \dfrac{1}{y^7}$

21. a. $2^{-4} \cdot 2^{-2} = 2^{-4+(-2)} = 2^{-6} = \dfrac{1}{2^6} = \dfrac{1}{64}$

  b. $x^{-3} \cdot x^{-7} = x^{-3+(-7)} = x^{-10} = \dfrac{1}{x^{10}}$

23. $x^6 \cdot x^{-4} = x^{6+(-4)} = x^2$

25. $y^{-3} \cdot y^5 = y^{-3+5} = y^2$

27. $a^3 \cdot a^{-8} = a^{3+(-8)} = a^{-5} = \dfrac{1}{a^5}$

29. $x^{-5} \cdot x^3 = x^{-5+3} = x^{-2} = \dfrac{1}{x^2}$

31. $x \cdot x^{-3} = x^{1+(-3)} = x^{-2} = \dfrac{1}{x^2}$

33. $a^{-2} \cdot a^{-3} = a^{-2+(-3)} = a^{-5} = \dfrac{1}{a^5}$

35. $b^{-3} \cdot b^3 = b^{-3+3} = b^0 = 1$

37. $\dfrac{3^4}{3^{-1}} = 3^{4-(-1)} = 3^5 = 243$

39. $\dfrac{4^{-1}}{4^2} = 4^{-1-2} = 4^{-3} = \dfrac{1}{4^3} = \dfrac{1}{64}$

41. $\dfrac{y}{y^3} = y^{1-3} = y^{-2} = \dfrac{1}{y^2}$

43. $\dfrac{x}{x^{-2}} = x^{1-(-2)} = x^3$

45. $\dfrac{x^{-3}}{x^{-1}} = x^{-3-(-1)} = x^{-3+1} = x^{-2} = \dfrac{1}{x^2}$

47. $\dfrac{x^{-3}}{x^4} = x^{-3-4} = x^{-7} = \dfrac{1}{x^7}$

49. $\dfrac{x^{-2}}{x^{-5}} = x^{-2-(-5)} = x^{-2+5} = x^3$

51. $(3^2)^2 = 3^{2\cdot2} = 3^4 = 81$

53. $(3^{-1})^2 = 3^{-1\cdot2} = 3^{-2} = \dfrac{1}{3^2} = \dfrac{1}{9}$

55. $(2^{-2})^{-3} = 2^{-2(-3)} = 2^6 = 64$

57. $(3^2)^{-1} = 3^{2(-1)} = 3^{-2} = \dfrac{1}{3^2} = \dfrac{1}{9}$

59. $(x^3)^{-3} = x^{3(-3)} = x^{-9} = \dfrac{1}{x^9}$

61. $(y^{-3})^2 = y^{-3\cdot2} = y^{-6} = \dfrac{1}{y^6}$

63. $(a^{-2})^{-3} = a^{-2(-3)} = a^6$

65. $(x^3 y^{-2})^3 = (x^3)^3 (y^{-2})^3 = x^9 y^{-6} = \dfrac{x^9}{y^6}$

67. $(x^{-2} y^3)^2 = (x^{-2})^2 (y^3)^2 = x^{-4} y^6 = \dfrac{y^6}{x^4}$

69. $(x^3 y^2)^{-3} = (x^3)^{-3} (y^2)^{-3} = x^{-9} y^{-6} = \dfrac{1}{x^9 y^6}$

71. $(x^{-6} y^{-3})^2 = (x^{-6})^2 (y^{-3})^2 = x^{-12} y^{-6} = \dfrac{1}{x^{12} y^6}$

73. $(x^{-4} y^{-4})^{-3} = (x^{-4})^{-3} (y^{-4})^{-3} = x^{12} y^{12}$

75. $A = P(1+r)^n$
$\quad = 1000(1 + 0.08)^3$
$\quad = 1000(1.08)^3$
$\quad = 1000(1.259712)$
$\quad = \$1259.71$

77. $A = P(1+r)^n$
$\quad = 1000(1 + 0.10)^4$
$\quad = 1000(1.10)^4$
$\quad = 1000(1.4641)$
$\quad = \$1464.10$

79. $(1.31 \times 8.7) \times (10^5 \times 10^4)$
$\quad = 11.397 \times 10^9$
$\quad = 11.397 \times 1,000,000,000$
$\quad = 11,397,000,000$ km

81. $7.31 \times 10^1 = 7.31 \times 10 = 73.1$

83. $7.31 \times 10^4 = 7.31 \times (10 \cdot 10 \cdot 10 \cdot 10)$
$\quad = 7.31 \times 10,000$
$\quad = 73,100$

85. a. $(11,111)^2 = 123,454,321$

    b. $(111,111)^2 = 12,345,654,321$

87. a. $1 + 3 + 5 + 7 + 9 = 5^2$

    b. $1 + 3 + 5 + 7 + 9 + 11 = 6^2$

89. Answers may vary.

91. Answers may vary.

Mastery Test 10.2

93. a. $(2^2)^3 = 2^{2\cdot3} = 2^6 = 64$

    b. $(x^{-3})^2 = x^{-3\cdot2} = x^{-6} = \dfrac{1}{x^6}$

    c. $(a^3)^{-5} = y^{3(-5)} = y^{-15} = \dfrac{1}{y^{15}}$

    d. $(b^{-2})^{-5} = b^{-2(-5)} = b^{10}$

95. a. $(a^3 b^{-2})^4 = (a^3)^4 (b^{-2})^4 = a^{12} b^{-8} = \dfrac{a^{12}}{b^8}$

    b. $(a^{-4} b^4)^3 = (a^{-4})^3 (b^4)^3 = a^{-12} b^{12} = \dfrac{b^{12}}{a^{12}}$

    c. $(a^{-3} b^4)^{-2} = (a^{-3})^{-2} (b^4)^{-2} = a^6 b^{-8} = \dfrac{a^6}{b^8}$

97. a. $\dfrac{1}{6^3} = 6^{-3}$

    b. $\dfrac{1}{c^4} = c^{-4}$

Section 10.3 – Scientific Notation

Problems

1. $239{,}000 = 2.39 \times 10^5;$
$0.123456 = 1.23456 \times 10^{-1}$

2. $4.08 \times 10^5 = 408{,}000;$
$3.125 \times 10^{-4} = 0.0003125$

3. a. $(6 \times 10^4) \times (2.2 \times 10^3)$
$= (6 \times 2.2) \times (10^4 \times 10^3)$
$= (13.2) \times 10^{4+3}$
$= (1.32 \times 10) \times 10^7$
$= 1.32 \times 10^8$

    b. $(4.1 \times 10^2) \times (3 \times 10^{-5})$
$= (4.1 \times 3) \times (10^2 \times 10^{-5})$
$= (12.3) \times 10^{2+(-5)}$
$= (1.23 \times 10) \times 10^{-3}$
$= 1.23 \times 10^{-2}$

4. $(2.52 \times 10^{-2}) \div (4.2 \times 10^{-3})$
$= \dfrac{2.52}{4.2} \times \dfrac{10^{-2}}{10^{-3}}$
$= 0.6 \times 10^{-2-(-3)}$
$= (6 \times 10^{-1}) \times 10^1$
$= 6 \times 10^0$
$= 6$

5. $\dfrac{2.8 \times 10^8}{1.4 \times 10^5} = \dfrac{2.8}{1.4} \times \dfrac{10^8}{10^5}$
$= 2 \times 10^{8-5}$
$= 2 \times 10^3 \text{ hours}$

Exercises 10.3

1. $68{,}000{,}000 = 6.8 \times 10^7$

3. $293{,}000{,}000 = 2.93 \times 10^8$

5. $1{,}900{,}000{,}000 = 1.9 \times 10^9$

7. $0.00024 = 2.4 \times 10^{-4}$

9. $0.000000002 = 2 \times 10^{-9}$

11. $2.35 \times 10^2 = 235$

13. $8 \times 10^6 = 8{,}000{,}000$

15. $6.8 \times 10^9 = 6{,}800{,}000{,}000$

17. $2.3 \times 10^{-1} = 0.23$

19. $2.5 \times 10^{-4} = 0.00025$

21. $(3 \times 10^4) \times (5 \times 10^5) = (3 \times 5) \times (10^4 \times 10^5)$
$= 15 \times 10^{4+5}$
$= (1.5 \times 10) \times 10^9$
$= 1.5 \times 10^{10}$

23. $(6 \times 10^{-3}) \times (5.1 \times 10^6)$
$= (6 \times 5.1) \times (10^{-3} \times 10^6)$
$= 30.6 \times 10^3$
$= (3.06 \times 10) \times 10^3$
$= 3.06 \times 10^4$

25. $(4 \times 10^{-2}) \times (3.1 \times 10^{-3})$
$= (4 \times 3.1) \times (10^{-2} \times 10^{-3})$
$= 12.4 \times 10^{-5}$
$= (1.24 \times 10) \times 10^{-5}$
$= 1.24 \times 10^{-4}$

27. $\dfrac{4.2 \times 10^5}{2.1 \times 10^2} = \dfrac{4.2}{2.1} \times \dfrac{10^5}{10^2} = 2 \times 10^{5-2} = 2 \times 10^3$

29. $\dfrac{2.2 \times 10^4}{8.8 \times 10^6} = \dfrac{2.2}{8.8} \times \dfrac{10^4}{10^6}$
$= 0.25 \times 10^{4-6}$
$= (2.5 \times 10^{-1}) \times 10^{-2}$
$= 2.5 \times 10^{-3}$

31. a. $(1.6 \times 10^2) \times (2.9 \times 10^8)$
$= (1.6 \times 2.9) \times (10^2 \times 10^8)$
$= 4.64 \times 10^{2+8}$
$= 4.64 \times 10^{10}$ lb
  b. $4.64 \times 10^{10} = 46,400,000,000$ lb

33. $\dfrac{(2.5 \times 10^8) \times (2 \times 10^3)}{(2.9 \times 10^8) \times (3.6 \times 10^2)}$

$= \dfrac{(2.5 \times 2) \times (10^8 \times 10^3)}{(2.9 \times 3.6) \times (10^8 \times 10^2)}$

$= \dfrac{5 \times 10^{11}}{10.44 \times 10^{10}}$

$= \dfrac{5}{10.44} \times 10^{11-10}$

$= 0.4789272 \times 10$

$= 4.79$ lb/day

35. $\dfrac{4.7 \times 10^9}{235} = \dfrac{4.7}{235} \times 10^9$
$= 0.02 \times 10^9$
$= 2 \times 10^7$ kilocalories

37. $x + \dfrac{3}{5} = \dfrac{1}{2}$
$x = \dfrac{1}{2} - \dfrac{3}{5} = \dfrac{5}{10} - \dfrac{6}{10} = -\dfrac{1}{10}$

39. $x + 0.7 = 0.2$
$x = 0.2 - 0.7 = -0.5$

41. $-0.3x = 0.9$
$x = \dfrac{0.9}{-0.3} = \dfrac{9}{-3} = -3$

43. $299,792,458 = 2.99792458 \times 10^8$ m/sec

45. $(2.06 \times 10^5) \times (1.5 \times 10^8)$
$= (2.06 \times 1.5) \times (10^5 \times 10^8)$
$= 3.09 \times 10^{5+8}$
$= 3.09 \times 10^{13}$ kilometers

47. $\dfrac{3.09 \times 10^{13}}{9.46 \times 10^{12}} = \dfrac{3.09}{9.46} \times \dfrac{10^{13}}{10^{12}}$
$= 0.3266385 \times 10^{13-12}$
$= 0.3266385 \times 10$
$= 3.266385$
$= 3.27$

49. Answers may vary.

51. -9.97×10^{-6}

53. $1.23 \quad -7$

Mastery Test 10.3

55. $\dfrac{2.48 \times 10^{-2}}{6.2 \times 10^{-4}} = \dfrac{2.48}{6.2} \times \dfrac{10^{-2}}{10^{-4}}$
$= 0.4 \times 10^{-2-(-4)}$
$= 0.4 \times 10^2$
$= 40$

57. $250,000 = 2.5 \times 10^5$ years

59. $4.5 \times 10^9 = 4,500,000,000$ years

Section 10.4 – Solving Linear Equations

Problems

1. a.
$$x - 6 = -8$$
$$x - 6 + 6 = -8 + 6$$
$$x = -2$$
Check: $x - 6 \overset{?}{=} -8$
$$-2 - 6 = -8$$
$$-8 = -8 \text{ True}$$
Thus the solution is $x = -2$.

b.
$$y + \frac{1}{5} = \frac{1}{2}$$
$$y + \frac{1}{5} - \frac{1}{5} = \frac{1}{2} - \frac{1}{5}$$
$$y = \frac{5}{10} - \frac{2}{10} = \frac{3}{10}$$
Check: $y + \frac{1}{5} \overset{?}{=} \frac{1}{2}$
$$\frac{3}{10} + \frac{1}{5} = \frac{1}{2}$$
$$\frac{1}{2} = \frac{1}{2} \text{ True}$$
Thus the solution is $y = \frac{3}{10}$.

2. a.
$$-x = 8$$
$$-1 \cdot (-1x) = -1 \cdot 8$$
$$x = -8$$
Check: $-x \overset{?}{=} 8$
$$-(-8) = 8$$
$$8 = 8 \text{ True}$$
Thus the solution is $x = -8$.

b.
$$\frac{-y}{4} = 6$$
$$4 \cdot \frac{-y}{4} = 6 \cdot 4$$
$$-y = 24$$
$$-1 \cdot (-y) = -1 \cdot 24$$
$$y = -24$$
Check: $\frac{-y}{4} \overset{?}{=} 6$
$$\frac{-(-24)}{4} = 6$$
$$6 = 6 \text{ True}$$
Thus the solution is $y = -24$.

c.
$$-1.2z = 4.8$$
$$\frac{-1.2z}{-1.2} = \frac{4.8}{-1.2}$$
$$z = -4$$
Check: $-1.2z \overset{?}{=} 4.8$
$$-1.2(-4) = 4.8$$
$$4.8 = 4.8 \text{ True}$$
Thus the solution is $z = -4$.

3.
$$-2x + 6 = 9$$
$$-2x + 6 - 6 = 9 - 3$$
$$-2x = 3$$
$$\frac{-2x}{-2} = \frac{3}{-2}$$
$$x = -\frac{3}{2}$$
Check: $-2x + 6 \overset{?}{=} 9$
$$-2\left(-\frac{3}{2}\right) + 6 = 9$$
$$3 + 6 = 9$$
$$9 = 9 \text{ True}$$
Thus the solution is $x = -\frac{3}{2}$.

4.
$$3x - 5 = 7$$
$$3x - 5 + 5 = 7 + 5$$
$$3x = 12$$
$$\frac{3x}{3} = \frac{12}{3}$$
$$x = 4$$
Check: $3x - 5 \overset{?}{=} 7$
$$3(4) - 5 = 7$$
$$7 = 7 \text{ True}$$
Thus the solution is $x = 4$.

5.
$$3x + 2.4 = x + 8.6$$
$$3x + 2.4 - 2.4 = x + 8.6 - 2.4$$
$$3x = x + 6.2$$
$$3x - x = x - x + 6.2$$
$$2x = 6.2$$
$$\frac{2x}{2} = \frac{6.2}{2}$$
$$x = 3.1$$

Check: $3x + 2.4 \overset{?}{=} x + 8.6$

$3(3.1) + 2.4 = 3.1 + 8.6$

$11.7 = 11.7$ True

Thus the solution is $x = 3.1$.

6. $5(x + 2) = 2x + 19$

$5x + 10 = 2x + 19$

$5x + 10 - 10 = 2x + 19 - 10$

$5x = 2x + 9$

$5x - 2x = 2x - 2x + 9$

$3x = 9$

$\dfrac{3x}{3} = \dfrac{9}{3}$

$x = 3$

Check: $5(x + 2) \overset{?}{=} 2x + 19$

$5(3 + 2) = 2(3) + 19$

$25 = 25$ True

Thus the solution is $x = 3$.

7. $10 = 2(x + 3)$

$10 = 2x + 6$

$10 - 6 = 2x + 6 - 6$

$4 = 2x$

$\dfrac{4}{2} = \dfrac{2x}{2}$

$2 = x$

Check: $10 \overset{?}{=} 2(x + 3)$

$10 = 2(2 + 3)$

$10 = 10$ True

Thus the solution is $x = 2$.

8. a. $\dfrac{3}{4} x = -2$

$\dfrac{4}{3} \cdot \dfrac{3}{4} x = \dfrac{4}{3} \cdot (-2)$

$x = -\dfrac{8}{3}$

Check: $\dfrac{3}{4} x \overset{?}{=} -2$

$\dfrac{3}{4}\left(-\dfrac{8}{3}\right) = -2$

$-2 = -2$ True

Thus the solution is $x = -\dfrac{8}{3}$.

b. $\dfrac{1}{3} x - \dfrac{1}{4} + \dfrac{3}{4} x = \dfrac{1}{3} + 2x$

$12 \cdot \left(\dfrac{1}{3} x - \dfrac{1}{4} + \dfrac{3}{4} x\right) = 12 \cdot \left(\dfrac{1}{3} + 2x\right)$

$4x - 3 + 9x = 4 + 24x$

$13x - 3 = 4 + 24x$

$13x - 3 - 4 = 4 - 4 + 24x$

$13x - 7 = 24x$

$13x - 13x - 7 = 24x - 13x$

$-7 = 11x$

$\dfrac{-7}{11} = \dfrac{11x}{11}$

$x = -\dfrac{7}{11}$

The check is left for you. Thus the solution is $x = -\dfrac{7}{11}$.

Exercises 10.4

1. $x - 2 = -4$

$x - 2 + 2 = -4 + 2$

$x = -2$

3. $-\dfrac{3}{4} = y - \dfrac{1}{4}$

$-\dfrac{3}{4} + \dfrac{1}{4} = y - \dfrac{1}{4} + \dfrac{1}{4}$

$-\dfrac{2}{4} = y$

$y = -\dfrac{1}{2}$

5. $x - 1.9 = -8.9$

$x - 1.9 + 1.9 = -8.9 + 1.9$

$x = -7$

7. $\dfrac{1}{2} = y + \dfrac{1}{5}$

$\dfrac{1}{2} - \dfrac{1}{5} = y + \dfrac{1}{5} - \dfrac{1}{5}$

$\dfrac{5}{10} - \dfrac{2}{10} = y$

$y = \dfrac{3}{10}$

9.
$$x + 3.8 = 9.9$$
$$x + 3.8 - 3.8 = 9.9 - 3.8$$
$$x = 6.1$$

11.
$$-x = 8$$
$$-1 \cdot (-x) = -1 \cdot (8)$$
$$x = -8$$

13.
$$-y = -\frac{1}{5}$$
$$-1 \cdot (-y) = -1 \cdot \left(-\frac{1}{5}\right)$$
$$y = \frac{1}{5}$$

15.
$$-x = 2.3$$
$$-1 \cdot (-x) = -1 \cdot (2.3)$$
$$x = -2.3$$

17.
$$\frac{-x}{2} = 8$$
$$2 \cdot \frac{-x}{2} = 2 \cdot 8$$
$$-x = 16$$
$$-1 \cdot (-x) = -1 \cdot 16$$
$$x = -16$$

19.
$$3.1 = \frac{-y}{4}$$
$$4 \cdot (3.1) = 4 \cdot \frac{-y}{4}$$
$$12.4 = -y$$
$$-1 \cdot (12.4) = -1 \cdot (-y)$$
$$-12.4 = y$$

21.
$$-2.1z = 4.2$$
$$\frac{-2.1z}{-2.1} = \frac{4.2}{-2.1}$$
$$z = -2$$

23.
$$-1.2 = 2.4x$$
$$\frac{-1.2}{2.4} = \frac{2.4x}{2.4}$$
$$-\frac{12}{24} = x$$
$$x = -\frac{1}{2}$$

25.
$$3.6y = 4.8$$
$$y = \frac{4.8}{3.6} = \frac{48}{36} = \frac{4}{3}$$
$$y = \frac{4}{3} \text{ or } 1\frac{1}{3}$$

27.
$$2x + 7.1 = 9.3$$
$$2x + 7.1 - 7.1 = 9.3 - 7.1$$
$$2x = 2.2$$
$$\frac{2x}{2} = \frac{2.2}{2}$$
$$x = 1.1$$

29.
$$6.5 = 3y + 3.2$$
$$6.5 - 3.2 = 3y + 3.2 - 3.2$$
$$3.3 = 3y$$
$$\frac{3.3}{3} = \frac{3y}{3}$$
$$y = 1.1$$

31.
$$\frac{3}{5}x - \frac{6}{5} = \frac{4}{5}$$
$$\frac{3}{5}x - \frac{6}{5} + \frac{6}{5} = \frac{4}{5} + \frac{6}{5}$$
$$\frac{3}{5}x = 2$$
$$\frac{5}{3} \cdot \frac{3}{5}x = \frac{5}{3} \cdot 2$$
$$x = \frac{10}{3}$$

33.
$$\frac{1}{3} = \frac{5}{2}x - \frac{8}{3}$$
$$\frac{8}{3} + \frac{1}{3} = \frac{5}{2}x - \frac{8}{3} + \frac{8}{3}$$
$$3 = \frac{5}{2}x$$
$$\frac{2}{5} \cdot 3 = \frac{2}{5} \cdot \frac{5}{2}x$$
$$x = \frac{6}{5}$$

35.
$$3.5x + 5 = 1.5x + 7$$
$$3.5x + 5 - 5 = 1.5x + 7 - 5$$
$$3.5x = 1.5x + 2$$
$$3.5x - 1.5x = 1.5x - 1.5x + 2$$
$$2x = 2$$
$$x = 1$$

37.
$$21 - 1.5y = 6.5y + 5$$
$$21 - 5 - 1.5y = 6.5y + 5 - 5$$
$$16 - 1.5y = 6.5y$$
$$16 - 1.5y + 1.5y = 6.5y + 1.5y$$
$$16 = 8y$$
$$2 = y$$

39.
$$3(y + 2) = -24$$
$$3y + 6 = -24$$
$$3y + 6 - 6 = -24 - 6$$
$$3y = -30$$
$$y = -10$$

41.
$$6(x - 1) = x + 6$$
$$6x - 6 = x + 6$$
$$6x - 6 + 6 = x + 6 + 6$$
$$6x = x + 12$$
$$6x - x = x - x + 12$$
$$5x = 12$$
$$x = \frac{12}{5}$$

43.
$$y + 6 = -2(y + 2)$$
$$y + 6 = -2y - 4$$
$$y + 6 - 6 = -2y - 4 - 6$$
$$y = -2y - 10$$
$$y + 2y = -2y + 2y - 10$$
$$3y = -10$$
$$y = -\frac{10}{3}$$

45.
$$1 - y = 5(y - 1)$$
$$1 - y = 5y - 5$$
$$1 + 5 - y = 5y - 5 + 5$$
$$6 - y = 5y$$
$$6 - y + y = 5y + y$$
$$6 = 6y$$
$$1 = y$$

47.
$$3x + 1 = 2(x + 1)$$
$$3x + 1 = 2x + 2$$
$$3x + 1 - 1 = 2x + 2 - 1$$
$$3x = 2x + 1$$
$$3x - 2x = 2x - 2x + 1$$
$$x = 1$$

49.
$$6x + 2 = -6x + 2$$
$$6x + 2 - 2 = -6x + 2 - 2$$
$$6x = -6x$$
$$6x + 6x = -6x + 6x$$
$$12x = 0$$
$$x = 0$$

51.
$$\frac{3}{5}x = -4$$
$$\frac{5}{3} \cdot \frac{3}{5}x = \frac{5}{3} \cdot (-4)$$
$$x = -\frac{20}{3}$$

53.
$$-\frac{3}{8}y = 2$$
$$-\frac{8}{3} \cdot \left(-\frac{3}{8}\right)y = -\frac{8}{3} \cdot 2$$
$$y = -\frac{16}{3}$$

55.
$$-\frac{2}{5}x = -3$$
$$-\frac{5}{2} \cdot \left(-\frac{2}{5}\right)x = -\frac{5}{2} \cdot (-3)$$
$$x = \frac{15}{2}$$

57.
$$\frac{5}{6}y - \frac{1}{4} = \frac{1}{2} - \frac{2}{3}y$$
$$12 \cdot \left(\frac{5}{6}y - \frac{1}{4}\right) = 12 \cdot \left(\frac{1}{2} - \frac{2}{3}y\right)$$
$$10y - 3 = 6 - 8y$$
$$10y - 3 + 3 = 6 + 3 - 8y$$
$$10y = 9 - 8y$$
$$10y + 8y = 9 - 8y + 8y$$
$$18y = 9$$
$$y = \frac{1}{2}$$

59.
$$\frac{3}{2}y - \frac{1}{3} = \frac{5}{4}y + \frac{1}{8}$$
$$24 \cdot \left(\frac{3}{2}y - \frac{1}{3}\right) = 24 \cdot \left(\frac{5}{4}y + \frac{1}{8}\right)$$
$$36y - 8 = 30y + 3$$
$$36y - 8 + 8 = 30y + 3 + 8$$
$$36y = 30y + 11$$
$$36y - 30y = 30y - 30y + 11$$
$$6y = 11$$
$$y = \frac{11}{6}$$

61. $\dfrac{2}{3} \cdot 1000 = \dfrac{2}{3} \cdot \dfrac{1000}{1} = \dfrac{2000}{3}$

63. $67.50 \div 0.25 = \dfrac{67.50}{0.25} = \dfrac{6750}{25} = 270$

65.
$$S + C = T$$
$$S + 66 = 176$$
$$S + 66 - 66 = 176 - 66$$
$$S = 110$$
The person's salary was $110.

67.
$$D - W = B$$
$$308 - W = 186$$
$$308 - 308 - W = 186 - 308$$
$$-W = -122$$
$$-1 \cdot (-W) = -1 \cdot (-122)$$
$$W = 122$$
The person withdrew $122.

69.
$$W = 5H - 190$$
$$160 = 5H - 190$$
$$160 + 190 = 5H - 190 + 190$$
$$350 = 5H$$
$$\dfrac{350}{5} = \dfrac{5H}{5}$$
$$70 = H$$
His height should be 70 inches.

71. Answers may vary.

73.
$$x - 3 = -4$$
$$x - 3 + 3 = -4 + 3$$
$$x = -1$$

75.
$$-y = 7$$
$$-1 \cdot (-y) = -1 \cdot 7$$
$$y = -7$$

77.
$$-3.3z = 9.9$$
$$\dfrac{-3.3z}{-3.3} = \dfrac{9.9}{-3.3}$$
$$z = \dfrac{99}{-33} = -3$$

79.
$$3x - 6 = 9$$
$$3x - 6 + 6 = 9 + 6$$
$$3x = 15$$
$$\dfrac{3x}{3} = \dfrac{15}{3}$$
$$x = 5$$

81.
$$3(x + 1) = 2x + 7$$
$$3x + 3 = 2x + 7$$
$$3x + 3 - 3 = 2x + 7 - 3$$
$$3x = 2x + 4$$
$$3x - 2x = 2x - 2x + 4$$
$$x = 4$$

83.
$$\dfrac{2}{3}x = -6$$
$$\dfrac{3}{2} \cdot \dfrac{2}{3}x = \dfrac{3}{2} \cdot (-6)$$
$$x = -9$$

Section 10.5 – Applications: Word Problems

Problems

1. Let w = his weight on the moon.
His wt. on earth = 6 × his wt. on moon
$$162 = 6w$$
$$\dfrac{162}{6} = \dfrac{6w}{6}$$
$$27 = w$$
$$w = 27$$
Verify: $162 = 6 \cdot 27$ is correct.
He weighs 27 pounds on the moon.

2. Let w = weight at the beginning.
His weight decreased 357 lb to 130 lb
$$w - 357 = 130$$
$$w - 357 + 357 = 130 + 357$$
$$w = 487$$
Verify: $487 - 357 = 130$ is correct.
He weighed 487 pounds at the beginning of his diet.

3. Let a = amount of water in the bucket.

$\frac{8}{9}$ of water amt. is the amt. of oxygen

$$\frac{8}{9}a = 368$$
$$\frac{9}{8} \cdot \frac{8}{9}a = \frac{9}{8} \cdot 368$$
$$a = \frac{3312}{8}$$
$$a = 414$$

Verify: $\frac{8}{\cancel{9}} \cdot \overset{46}{\cancel{414}} = 368$ ✓

There are 414 g of water in the bucket.

4.
$$C = 15 + 0.20m$$
$$65 = 15 + 0.20m$$
$$65 - 15 = 15 - 15 + 0.20m$$
$$50 = 0.20m$$
$$\frac{50}{0.20} = m$$
$$\frac{500}{2} = m$$
$$m = 250$$

Verify: $15 + 0.20(250) = 15 + 50 = 65$ ✓

The person traveled 250 miles.

5. Let R = the rate.
$$I = PRT$$
$$600 = 10,000 \cdot R \cdot \frac{1}{2}$$
$$600 = 5000R$$
$$\frac{6\cancel{00}}{50\cancel{00}} = R \text{ so } R = \frac{6}{50} = 0.12 = 12\%$$

Verify: $10,000(0.12)\left(\frac{1}{2}\right) = 600$ ✓

The certificate paid 12% simple interest.

6. Let n = the number.
$$6 + 3n = 6n$$
$$6 + 3n - 3n = 6n - 3n$$
$$6 = 3n$$
$$\frac{6}{3} = n$$
$$n = 2$$

Verify: $6 + 3(2) = 6(2)$
$6 + 6 = 12$ is correct.

The number is 2.

7. Let h = no. of calories in the hamburger and f = no. of calories in the french fries. Now $2h = f + 140$ so $f = 2h - 140$.
$$h + f = 820$$
$$h + (2h - 140) = 820$$
$$3h - 140 = 820$$
$$3h = 960$$
$$h = 320$$
and $f = 2 \cdot 320 - 140 = 640 - 140 = 500$.

Verify: $320 + 500 = 820$ ✓

There are 320 calories in the hamburger and 500 calories in the french fries.

Exercises 10.5

1. Let I = amount of interest earned.
Balance = $300 plus interest
$$318 = 300 + I$$
$$318 - 300 = 300 - 300 + I$$
$$18 = I$$

Verify: $318 = 300 + 18$ is correct.

She earned $18 in interest during the year.

3. Let v = number of votes the loser had.
The winner had 632 more than the loser
$$10,839 = v + 632$$
$$10,839 - 632 = v + 632 - 632$$
$$10,207 = v$$

Verify: $10,839 = 10,207 + 632$ is correct.

The loser had 10,207 votes.

5. Let v = number of votes the winner got.
Loser had 849 votes less than the winner
$$9347 = v - 849$$
$$9347 + 849 = v - 849 + 849$$
$$10,196 = v$$

Verify: $9347 = 10,196 - 849$ is correct.

The winner got 10,196 votes.

7. Let s = selling price.
Profit = selling price minus cost
$$11 = s - 57$$
$$11 + 57 = s - 57 + 57$$
$$68 = s$$

Verify: $11 = 68 - 57$ is correct.

The selling price is $68.

9. Let p = number of pints pumped at rest.
   Exercise number is four times normal rate
   $$40 = 4p$$
   $$\frac{40}{4} = p$$
   $$p = 10$$
   Verify: $40 = 4 \cdot 10$ is correct.
   The heart pumps 10 pints of blood per minute when at rest.

11. Let b = number of breaths when at rest.
   Exercise number is five times normal rate
   $$60 = 5b$$
   $$\frac{60}{5} = b$$
   $$b = 12$$
   Verify: $60 = 5 \cdot 12$ is correct.
   A person should take 12 breaths per minute when at rest.

13. Let c = capacity of the tank.
   Amt. in tank is $\frac{2}{3}$ of its capacity.
   $$18 = \frac{2}{3}c$$
   $$\frac{3}{2} \cdot 18 = \frac{3}{2} \cdot \frac{2}{3}c$$
   $$c = 27$$
   Verify: $18 = \frac{2}{\cancel{3}} \cdot \overset{9}{\cancel{27}} = 18$ ✓

   The tank's capacity is 27 gallons.

15. Let h = height of St. Peter's dome
   Village dome is $\frac{1}{16}$ of St. Peter's dome
   $$43.75 = \frac{1}{16}h$$
   $$16 \cdot (43.75) = 16 \cdot \frac{1}{16}h$$
   $$700 = h$$
   Verify: $43.75 = \frac{1}{16} \cdot 700 = \frac{700}{16} = 43.75$ ✓
   The dome of St. Peter's Basilica is 700 feet high.

17. $P/E = \dfrac{M}{E}$
   $$12 = \frac{M}{3}$$
   $$3 \cdot 12 = M$$
   $$M = 36$$
   The market value should be \$36.

19. Let d = no. of dialects he can translate.
   and l = no. of languages.
   Now $l = d + 42$;
   $$d + l = 186$$
   $$d + (d + 42) = 186$$
   $$2d + 42 = 186$$
   $$2d = 144$$
   $$d = 72; \quad l = 72 + 42 = 114$$
   Verify: $72 + 114 = 186$ ✓
   He can translate 72 dialects.

21. Let R = the rate.
   $$I = PRT$$
   $$80 = 1000 \cdot R \cdot 1$$
   $$80 = 1000R$$
   $$\frac{8\cancel{0}}{100\cancel{0}} = R \text{ so } R = \frac{8}{100} = 8\%$$
   Verify: $1000(0.08)(1) = 80$ ✓
   The certificate paid 8% simple interest.

23. Let R = the rate.
   $$I = PRT$$
   $$240 = 3000 \cdot R \cdot 1$$
   $$240 = 3000R$$
   $$\frac{24\cancel{0}}{300\cancel{0}} = R \text{ so } R = \frac{24}{300} = \frac{2}{25} = 0.08\%$$
   Verify: $3000(0.08)(1) = 240$ ✓
   The bonds paid 8% simple interest.

25. Let R = the rate.
   $$I = PRT$$
   $$588 = 1400 \cdot R \cdot 2$$
   $$588 = 2800R$$
   $$\frac{588}{2800} = R \text{ so } R = \frac{588}{2800} = \frac{21}{100} = 21\%$$
   Verify: $1400(0.21)(2) = 588$ ✓
   The loan company is charging 21% simple interest.

27. Let n = the number.
$$2n + 11 = 19$$
$$2n = 8$$
$$n = 4$$
The number is 4.

29. Let n = the number.
$$7n + 6 = 69$$
$$7n = 63$$
$$n = 9$$
The number is 9.

31. $\quad 2(1) + 2(Cr) + 7(-2) = 0$
$$2 + 2(Cr) - 14 = 0$$
$$-12 + 2(Cr) = 0$$
$$2(Cr) = 12$$
$$Cr = +6$$

33. $\quad 1(P) + 4(-2) = 0$
$$P - 8 = 0$$
$$P = +8$$

35. Answers may vary.

37. Answers may vary.

Mastery Test 10.5

39. Let n = the number.
$$2n + 12 = 5n$$
$$12 = 3n$$
$$4 = n$$
The number is 4.

41. Let m = the miles traveled.
The cost is $30 plus $0.25 per mi traveled
$$110 = 30 + 0.25m$$
$$80 = 0.25m$$
$$\frac{80}{0.25} = m$$
$$\frac{8000}{25} = m$$
$$320 = m$$
Verify:
$$110 = 30 + 0.25(320) = 30 + 80 = 110 \checkmark$$
There were 320 miles traveled.

43. Let w = its weight on Earth.
Its wt. on earth = 6 × his wt. on moon
$$w = 6 \cdot 30$$
$$w = 180$$
It weight would be 180 pounds on Earth.

Review Exercises – Chapter 10

1. a. $6x^3; \; -3x^2; \; 7x; \; -7$

   b. $5x^3; \; -4x^2; \; 6x; \; -6$

   c. $4x^3; \; -5x^2; \; 5x; \; -5$

   d. $3x^3; \; -6x^2; \; 4x; \; -4$

   e. $2x^3; \; -7x^2; \; 3x; \; -3$

2. a. $2(x + 2y) = 2 \cdot x + 2 \cdot 2y = 2x + 4y$

   b. $3(x + 2y) = 3 \cdot x + 3 \cdot 2y = 3x + 6y$

   c. $4(x + 2y) = 4 \cdot x + 4 \cdot 2y = 4x + 8y$

   d. $5(x + 2y) = 5 \cdot x + 5 \cdot 2y = 5x + 10y$

   e. $6(x + 2y) = 6 \cdot x + 6 \cdot 2y = 6x + 12y$

3. a. $-4(x - 2y + z)$
$$= -4 \cdot x - (-4) \cdot 2y + (-4) \cdot z$$
$$= -4x + 8y - 4z$$

   b. $-5(x - 2y + z)$
$$= -5 \cdot x - (-5) \cdot 2y + (-5) \cdot z$$
$$= -5x + 10y - 5z$$

   c. $-6(x - 2y + z)$
$$= -6 \cdot x - (-6) \cdot 2y + (-6) \cdot z$$
$$= -6x + 12y - 6z$$

   d. $-7(x - 2y + z)$
$$= -7 \cdot x - (-7) \cdot 2y + (-7) \cdot z$$
$$= -7x + 14y - 7z$$

   e. $-8(x - 2y + z)$
$$= -8 \cdot x - (-8) \cdot 2y + (-8) \cdot z$$
$$= -8x + 16y - 8z$$

4. a. $3x - 12y = 3 \cdot x - 3 \cdot 4y = 3(x - 4y)$

   b. $3x - 15y = 3 \cdot x - 3 \cdot 5y = 3(x - 5y)$

   c. $3x - 18y = 3 \cdot x - 3 \cdot 6y = 3(x - 6y)$

   d. $3x - 21y = 3 \cdot x - 3 \cdot 7y = 3(x - 7y)$

   e. $3x - 24y = 3 \cdot x - 3 \cdot 8y = 3(x - 8y)$

5. a. $40x - 50y - 60z$
$= 10 \cdot 4x - 10 \cdot 5y - 10 \cdot 6z$
$= 10(4x - 5y - 6z)$

b. $30x - 40y - 50z$
$= 10 \cdot 3x - 10 \cdot 4y - 10 \cdot 5z$
$= 10(3x - 4y - 5z)$

c. $20x - 30y - 40z$
$= 10 \cdot 2x - 10 \cdot 3y - 10 \cdot 4z$
$= 10(2x - 3y - 4z)$

d. $10x - 20y - 30z$
$= 10 \cdot x - 10 \cdot 2y - 10 \cdot 3z$
$= 10(x - 2y - 3z)$

e. $10x - 10y - 20z$
$= 10 \cdot x - 10 \cdot y - 10 \cdot 2z$
$= 10(x - y - 2z)$

6. a. $3x + 5y - x + 4y = 3x - x + 5y + 4y$
$\qquad = 2x + 9y$

b. $4x + 5y - x + 3y = 4x - x + 5y + 3y$
$\qquad = 3x + 8y$

c. $5x + 5y - x + 2y = 5x - x + 5y + 2y$
$\qquad = 4x + 7y$

d. $6x + 5y - 2x + 4y = 6x - 2x + 5y + 4y$
$\qquad = 4x + 9y$

e. $7x + 5y - 3x + 4y = 7x - 3x + 5y + 4y$
$\qquad = 4x + 9y$

7. a. $0.3x + 0.31y - 0.5x + 0.23y$
$= 0.3x - 0.5x + 0.31y + 0.23y$
$= -0.2x + 0.54y$

b. $0.3x + 0.32y - 0.6x + 0.23y$
$= 0.3x - 0.6x + 0.32y + 0.23y$
$= -0.3x + 0.55y$

c. $0.3x + 0.33y - 0.7x + 0.23y$
$= 0.3x - 0.7x + 0.33y + 0.23y$
$= -0.4x + 0.56y$

d. $0.3x + 0.34y - 0.8x + 0.23y$
$= 0.3x - 0.8x + 0.34y + 0.23y$
$= -0.5x + 0.57y$

e. $0.3x + 0.35y - 0.9x + 0.23y$
$= 0.3x - 0.9x + 0.35y + 0.23y$
$= -0.6x + 0.58y$

8. a. $\dfrac{1}{11}x + \dfrac{9}{13}y + \dfrac{3}{11}x - \dfrac{2}{13}y$

$= \dfrac{1}{11}x + \dfrac{3}{11}x + \dfrac{9}{13}y - \dfrac{2}{13}y$

$= \dfrac{4}{11}x + \dfrac{7}{13}y$

b. $\dfrac{2}{11}x + \dfrac{8}{13}y + \dfrac{3}{11}x - \dfrac{2}{13}y$

$= \dfrac{2}{11}x + \dfrac{3}{11}x + \dfrac{8}{13}y - \dfrac{2}{13}y$

$= \dfrac{5}{11}x + \dfrac{6}{13}y$

c. $\dfrac{3}{11}x + \dfrac{7}{13}y + \dfrac{3}{11}x - \dfrac{2}{13}y$

$= \dfrac{3}{11}x + \dfrac{3}{11}x + \dfrac{7}{13}y - \dfrac{2}{13}y$

$= \dfrac{6}{11}x + \dfrac{5}{13}y$

d. $\dfrac{4}{11}x + \dfrac{6}{13}y + \dfrac{3}{11}x - \dfrac{2}{13}y$

$= \dfrac{4}{11}x + \dfrac{3}{11}x + \dfrac{6}{13}y - \dfrac{2}{13}y$

$= \dfrac{7}{11}x + \dfrac{4}{13}y$

e. $\dfrac{5}{11}x + \dfrac{5}{13}y + \dfrac{3}{11}x - \dfrac{2}{13}y$

$= \dfrac{5}{11}x + \dfrac{3}{11}x + \dfrac{5}{13}y - \dfrac{2}{13}y$

$= \dfrac{8}{11}x + \dfrac{3}{13}y$

9. a. $(4a + 5b) - (a + 2b) = 4a + 5b - a - 2b$
$\qquad = 3a + 3b$

b. $(5a + 5b) - (a + 2b) = 5a + 5b - a - 2b$
$\qquad = 4a + 3b$

c. $(6a + 5b) - (a - 2b) = 6a + 5b - a + 2b$
$\qquad = 5a + 7b$

d. $(7a + 5b) - (-a + 2b) = 7a + 5b + a - 2b$
$\qquad = 8a + 3b$

e. $(8a + 5b) - (a - 3b) = 8a + 5b - a + 3b$
$\qquad = 7a + 8b$

10. a. $2^{-2} = \dfrac{1}{2^2} = \dfrac{1}{4}$ **b.** $2^{-3} = \dfrac{1}{2^3} = \dfrac{1}{8}$

c. $2^{-4} = \dfrac{1}{2^4} = \dfrac{1}{16}$ **d.** $2^{-5} = \dfrac{1}{2^5} = \dfrac{1}{32}$

e. $2^{-6} = \dfrac{1}{2^6} = \dfrac{1}{64}$

11. a. $\dfrac{1}{2^3} = 2^{-3}$ **b.** $\dfrac{1}{2^4} = 2^{-4}$

c. $\dfrac{1}{2^5} = 2^{-5}$ **d.** $\dfrac{1}{2^6} = 2^{-6}$

e. $\dfrac{1}{2^7} = 2^{-7}$

12. a. $x^3 \cdot x^4 = x^{3+4} = x^7$

b. $x^3 \cdot x^5 = x^{3+5} = x^8$

c. $x^3 \cdot x^6 = x^{3+6} = x^9$

d. $x^3 \cdot x^7 = x^{3+7} = x^{10}$

e. $x^3 \cdot x^8 = x^{3+8} = x^{11}$

13. a. $2^3 \cdot 2^{-5} = 2^{3+(-5)} = 2^{-2} = \dfrac{1}{2^2} = \dfrac{1}{4}$

b. $2^3 \cdot 2^{-6} = 2^{3+(-6)} = 2^{-3} = \dfrac{1}{2^3} = \dfrac{1}{8}$

c. $2^3 \cdot 2^{-7} = 2^{3+(-7)} = 2^{-4} = \dfrac{1}{2^4} = \dfrac{1}{16}$

d. $2^3 \cdot 2^{-8} = 2^{3+(-8)} = 2^{-5} = \dfrac{1}{2^5} = \dfrac{1}{32}$

e. $2^3 \cdot 2^{-9} = 2^{3+(-9)} = 2^{-6} = \dfrac{1}{2^6} = \dfrac{1}{64}$

14. a. $y^3 \cdot y^{-5} = y^{3+(-5)} = y^{-2} = \dfrac{1}{y^2}$

b. $y^3 \cdot y^{-6} = y^{3+(-6)} = y^{-3} = \dfrac{1}{y^3}$

c. $y^3 \cdot y^{-7} = y^{3+(-7)} = y^{-4} = \dfrac{1}{y^4}$

d. $y^3 \cdot y^{-8} = y^{3+(-8)} = y^{-5} = \dfrac{1}{y^5}$

e. $y^3 \cdot y^{-9} = y^{3+(-9)} = y^{-6} = \dfrac{1}{y^6}$

15. a. $\dfrac{2^4}{2^{-2}} = 2^{4-(-2)} = 2^6 = 64$

b. $\dfrac{2^5}{2^{-2}} = 2^{5-(-2)} = 2^7 = 128$

c. $\dfrac{2^6}{2^{-2}} = 2^{6-(-2)} = 2^8 = 256$

d. $\dfrac{2^7}{2^{-2}} = 2^{7-(-2)} = 2^9 = 512$

e. $\dfrac{2^8}{2^{-2}} = 2^{8-(-2)} = 2^{10} = 1024$

16. a. $\dfrac{x^2}{x^5} = x^{2-5} = x^{-3} = \dfrac{1}{x^3}$

b. $\dfrac{x^2}{x^6} = x^{2-6} = x^{-4} = \dfrac{1}{x^4}$

c. $\dfrac{x^2}{x^7} = x^{2-7} = x^{-5} = \dfrac{1}{x^5}$

d. $\dfrac{x^2}{x^8} = x^{2-8} = x^{-6} = \dfrac{1}{x^6}$

e. $\dfrac{x^2}{x^9} = x^{2-9} = x^{-7} = \dfrac{1}{x^7}$

17. a. $\dfrac{y^{-2}}{y^{-4}} = y^{-2-(-4)} = y^{-2+4} = y^2$

b. $\dfrac{y^{-2}}{y^{-5}} = y^{-2-(-5)} = y^{-2+5} = y^3$

c. $\dfrac{y^{-2}}{y^{-6}} = y^{-2-(-6)} = y^{-2+6} = y^4$

d. $\dfrac{y^{-2}}{y^{-7}} = y^{-2-(-7)} = y^{-2+7} = y^5$

e. $\dfrac{y^{-2}}{y^{-8}} = y^{-2-(-8)} = y^{-2+8} = y^6$

18. a. $(2^2)^1 = 2^{2(1)} = 2^2 = 4$

b. $(2^2)^2 = 2^{2(2)} = 2^4 = 16$

c. $(2^2)^3 = 2^{2(3)} = 2^6 = 64$

d. $(2^2)^4 = 2^{2(4)} = 2^8 = 256$

e. $(2^2)^5 = 2^{2(5)} = 2^{10} = 1024$

19. a. $(y^4)^{-2} = y^{4(-2)} = y^{-8} = \dfrac{1}{y^8}$

b. $(y^4)^{-3} = y^{4(-3)} = y^{-12} = \dfrac{1}{y^{12}}$

c. $(y^4)^{-4} = y^{4(-4)} = y^{-16} = \dfrac{1}{y^{16}}$

d. $(y^4)^{-5} = y^{4(-5)} = y^{-20} = \dfrac{1}{y^{20}}$

e. $(y^4)^{-6} = y^{4(-6)} = y^{-24} = \dfrac{1}{y^{24}}$

20. a. $(z^{-3})^{-2} = z^{-3(-2)} = z^6$

    b. $(z^{-3})^{-3} = z^{-3(-3)} = z^9$

    c. $(z^{-3})^{-4} = z^{-3(-4)} = z^{12}$

    d. $(z^{-3})^{-5} = z^{-3(-5)} = z^{15}$

    e. $(z^{-3})^{-6} = z^{-3(-6)} = z^{18}$

21. a. $(x^3 y^{-2})^2 = (x^3)^2 (y^{-2})^2 = x^6 y^{-4} = \dfrac{x^6}{y^4}$

    b. $(x^3 y^{-2})^3 = (x^3)^3 (y^{-2})^3 = x^9 y^{-6} = \dfrac{x^9}{y^6}$

    c. $(x^3 y^{-2})^4 = (x^3)^4 (y^{-2})^4 = x^{12} y^{-8} = \dfrac{x^{12}}{y^8}$

    d. $(x^3 y^{-2})^5 = (x^3)^5 (y^{-2})^5 = x^{15} y^{-10} = \dfrac{x^{15}}{y^{10}}$

    e. $(x^3 y^{-2})^6 = (x^3)^6 (y^{-2})^6 = x^{18} y^{-12} = \dfrac{x^{18}}{y^{12}}$

22. a. $(x^{-2} y^3)^{-2} = (x^{-2})^{-2} (y^3)^{-2} = x^4 y^{-6} = \dfrac{x^4}{y^6}$

    b. $(x^{-2} y^3)^{-3} = (x^{-2})^{-3} (y^3)^{-3} = x^6 y^{-9} = \dfrac{x^6}{y^9}$

    c. $(x^{-2} y^3)^{-4} = (x^{-2})^{-4} (y^3)^{-4} = x^8 y^{-12} = \dfrac{x^8}{y^{12}}$

    d. $(x^{-2} y^3)^{-5} = (x^{-2})^{-5} (y^3)^{-5} = x^{10} y^{-15} = \dfrac{x^{10}}{y^{15}}$

    e. $(x^{-2} y^3)^{-6} = (x^{-2})^{-6} (y^3)^{-6} = x^{12} y^{-18} = \dfrac{x^{12}}{y^{18}}$

23. a. $44,000,000 = 4.4 \times 10^7$

    b. $450,000,000 = 4.5 \times 10^8$

    c. $4,600,000 = 4.6 \times 10^6$

    d. $47,000 = 4.7 \times 10^4$

    e. $48,000,000 = 4.8 \times 10^7$

24. a. $0.0014 = 1.4 \times 10^{-3}$

    b. $0.00015 = 1.5 \times 10^{-4}$

    c. $0.000016 = 1.6 \times 10^{-5}$

    d. $0.0000017 = 1.7 \times 10^{-6}$

    e. $0.00000018 = 1.8 \times 10^{-7}$

25. a. $7.83 \times 10^3 = 7830$

    b. $6.83 \times 10^4 = 68,300$

    c. $5.83 \times 10^5 = 583,000$

    d. $4.83 \times 10^6 = 4,830,000$

    e. $3.83 \times 10^7 = 38,300,000$

26. a. $8.4 \times 10^{-2} = 0.084$

    b. $7.4 \times 10^{-3} = 0.0074$

    c. $6.4 \times 10^{-4} = 0.00064$

    d. $5.4 \times 10^{-5} = 0.000054$

    e. $4.4 \times 10^{-6} = 0.0000044$

27. a. $(2 \times 10^2) \times (1.1 \times 10^3)$
$= (2 \times 1.1) \times (10^2 \times 10^3)$
$= 2.2 \times 10^5$

    b. $(3 \times 10^2) \times (3.1 \times 10^4)$
$= (3 \times 3.1) \times (10^2 \times 10^4)$
$= 9.3 \times 10^6$

    c. $(4 \times 10^2) \times (3.1 \times 10^5)$
$= (4 \times 3.1) \times (10^2 \times 10^5)$
$= 12.4 \times 10^7$
$= (1.24 \times 10) \times 10^7$
$= 1.24 \times 10^8$

    d. $(5 \times 10^2) \times (3.1 \times 10^6)$
$= (5 \times 3.1) \times (10^2 \times 10^6)$
$= 15.5 \times 10^8$
$= (1.55 \times 10) \times 10^8$
$= 1.55 \times 10^9$

    e. $(6 \times 10^2) \times (3.1 \times 10^6)$
$= (6 \times 3.1) \times (10^2 \times 10^6)$
$= 18.6 \times 10^8$
$= (1.86 \times 10) \times 10^8$
$= 1.86 \times 10^9$

28. a. $(1.1 \times 10^3) \times (2 \times 10^{-5})$
$= (1.1 \times 2) \times (10^3 \times 10^{-5})$
$= 2.2 \times 10^{-2}$

    b. $(1.1 \times 10^4) \times (3 \times 10^{-5})$
$= (1.1 \times 3) \times (10^4 \times 10^{-5})$
$= 3.3 \times 10^{-1}$

c. $(1.1 \times 10^3) \times (4 \times 10^{-6})$

$= (1.1 \times 4) \times (10^3 \times 10^{-6})$

$= 4.4 \times 10^{-3}$

d. $(1.1 \times 10^3) \times (5 \times 10^{-7})$

$= (1.1 \times 5) \times (10^3 \times 10^{-7})$

$= 5.5 \times 10^{-4}$

e. $(1.1 \times 10^4) \times (6 \times 10^{-8})$

$= (1.1 \times 6) \times (10^4 \times 10^{-8})$

$= 6.6 \times 10^{-4}$

29. a. $(1.15 \times 10^{-3}) \div (2.3 \times 10^{-4})$

$= \dfrac{1.15 \times 10^{-3}}{2.3 \times 10^{-4}}$

$= \dfrac{1.15}{2.3} \times \dfrac{10^{-3}}{10^{-4}}$

$= 0.5 \times 10^{-3-(-4)}$

$= (5 \times 10^{-1}) \times 10^1$

$= 5 \times 10^0$

$= 5 \times 1 = 5$

b. $(1.38 \times 10^{-2}) \div (2.3 \times 10^{-4})$

$= \dfrac{1.38 \times 10^{-2}}{2.3 \times 10^{-4}}$

$= \dfrac{1.15}{2.3} \times \dfrac{10^{-2}}{10^{-4}}$

$= 0.6 \times 10^{-2-(-4)}$

$= (6 \times 10^{-1}) \times 10^2$

$= 6 \times 10^1$

c. $(1.61 \times 10^{-3}) \div (2.3 \times 10^{-5})$

$= \dfrac{1.61 \times 10^{-3}}{2.3 \times 10^{-5}}$

$= \dfrac{1.61}{2.3} \times \dfrac{10^{-3}}{10^{-5}}$

$= 0.7 \times 10^{-3-(-5)}$

$= (7 \times 10^{-1}) \times 10^2$

$= 7 \times 10^1$

d. $(1.84 \times 10^{-4}) \div (2.3 \times 10^{-4})$

$= \dfrac{1.84 \times 10^{-4}}{2.3 \times 10^{-4}}$

$= \dfrac{1.84}{2.3} \times \dfrac{10^{-4}}{10^{-4}}$

$= 0.8 \times 1$

$= 8 \times 10^{-1}$

e. $(2.07 \times 10^{-3}) \div (2.3 \times 10^{-4})$

$= \dfrac{2.07 \times 10^{-3}}{2.3 \times 10^{-4}}$

$= \dfrac{2.07}{2.3} \times \dfrac{10^{-3}}{10^{-4}}$

$= 0.9 \times 10^{-3-(-4)}$

$= (9 \times 10^{-1}) \times 10^1$

$= 9 \times 10^0$

$= 9 \times 1 = 9$

30. a. $\quad y + \dfrac{1}{3} = \dfrac{1}{2}$

$6 \cdot \left(y + \dfrac{1}{3} \right) = 6 \cdot \dfrac{1}{2}$

$6y + 2 = 3$

$6y = 1$

$y = \dfrac{1}{6}$

b. $\quad y + \dfrac{1}{4} = \dfrac{1}{2}$

$4 \cdot \left(y + \dfrac{1}{4} \right) = 4 \cdot \dfrac{1}{2}$

$4y + 1 = 2$

$4y = 1$

$y = \dfrac{1}{4}$

c. $\quad y + \dfrac{1}{5} = \dfrac{1}{2}$

$10 \cdot \left(y + \dfrac{1}{5} \right) = 10 \cdot \dfrac{1}{2}$

$10y + 2 = 5$

$10y = 3$

$y = \dfrac{3}{10}$

d. $\quad y + \dfrac{1}{6} = \dfrac{1}{2}$

$6 \cdot \left(y + \dfrac{1}{6} \right) = 6 \cdot \dfrac{1}{2}$

$6y + 1 = 3$

$6y = 2$

$y = \dfrac{2}{6} = \dfrac{1}{3}$

e. $14 \cdot \left(y + \dfrac{1}{7} \right) = 14 \cdot \dfrac{1}{2}$

$14y + 2 = 7$

$14y = 5$

$y = \dfrac{5}{14}$

31. a. $\dfrac{-y}{3} = 2$ **b.** $\dfrac{-y}{3} = 3$

$3 \cdot \dfrac{-y}{3} = 3 \cdot 2$ $3 \cdot \dfrac{-y}{3} = 3 \cdot 3$

$-y = 6$ $-y = 9$

$y = -6$ $y = -9$

c. $\dfrac{-y}{3} = 4$ **d.** $\dfrac{-y}{3} = 5$

$3 \cdot \dfrac{-y}{3} = 3 \cdot 4$ $3 \cdot \dfrac{-y}{3} = 3 \cdot 5$

$-y = 12$ $-y = 15$

$y = -12$ $y = -15$

e. $\dfrac{-y}{3} = 6$

$3 \cdot \dfrac{-y}{3} = 3 \cdot 6$

$-y = 18$

$y = -18$

32. a. $-1.2z = 2.4$ **b.** $-1.2z = 3.6$

$\dfrac{-1.2z}{-1.2} = \dfrac{2.4}{-1.2}$ $\dfrac{-1.2z}{-1.2} = \dfrac{3.6}{-1.2}$

$z = \dfrac{24}{-12}$ $z = \dfrac{36}{-12}$

$z = -2$ $z = -3$

c. $-1.2z = 4.8$ **d.** $-1.2z = 7.2$

$\dfrac{-1.2z}{-1.2} = \dfrac{4.8}{-1.2}$ $\dfrac{-1.2z}{-1.2} = \dfrac{7.2}{-1.2}$

$z = \dfrac{48}{-12}$ $z = \dfrac{72}{-12}$

$z = -4$ $z = -6$

e. $-1.2z = 8.4$

$\dfrac{-1.2z}{-1.2} = \dfrac{8.4}{-1.2}$

$z = \dfrac{84}{-12}$

$z = -7$

33. a. $-3x + 5 = 10$ **b.** $-3x + 5 = 13$

$-3x = 5$ $-3x = 8$

$x = -\dfrac{5}{3}$ $x = -\dfrac{8}{3}$

c. $-3x + 5 = 15$ **d.** $-3x + 5 = 18$

$-3x = 10$ $-3x = 13$

$x = -\dfrac{10}{3}$ $x = -\dfrac{13}{3}$

e. $-3x + 5 = 21$

$-3x = 16$

$x = -\dfrac{16}{3}$

34. a. $2x - 6 = 10$ **b.** $2x - 6 = 12$

$2x = 16$ $2x = 18$

$x = 8$ $x = 9$

c. $2x - 6 = 14$ **d.** $2x - 6 = 16$

$2x = 20$ $2x = 22$

$x = 10$ $x = 11$

e. $2x - 6 = 18$

$2x = 24$

$x = 12$

35. a. $4x + 5.4 = 2x + 9.6$

$4x = 2x + 4.2$

$2x = 4.2$

$x = 2.1$

b. $5x + 5.4 = 3x + 9.6$

$5x = 3x + 4.2$

$2x = 4.2$

$x = 2.1$

c. $6x + 5.4 = 4x + 9.6$

$6x = 4x + 4.2$

$2x = 4.2$

$x = 2.1$

d. $7x + 5.4 = 5x + 9.6$

$7x = 5x + 4.2$

$2x = 4.2$

$x = 2.1$

e. $8x + 5.4 = 6x + 9.6$

$8x = 6x + 4.2$

$2x = 4.2$

$x = 2.1$

36. a. $36 = 3(x + 1)$ **b.** $27 = 3(x + 1)$

$36 = 3x + 3$ $27 = 3x + 3$

$33 = 3x$ $24 = 3x$

$11 = x$ $8 = x$

c. $24 = 3(x + 1)$ **d.** $21 = 3(x + 1)$

$24 = 3x + 3$ $21 = 3x + 3$

$21 = 3x$ $18 = 3x$

$7 = x$ $6 = x$

e. $18 = 3(x + 1)$

$18 = 3x + 3$

$15 = 3x$

$5 = x$

37. a. $\dfrac{2}{7}x = -3$

$\dfrac{7}{2} \cdot \dfrac{2}{7}x = \dfrac{7}{2} \cdot (-3)$

$x = -\dfrac{21}{2}$

b. $\dfrac{2}{7}x = -5$

$\dfrac{7}{2} \cdot \dfrac{2}{7}x = \dfrac{7}{2} \cdot (-5)$

$x = -\dfrac{35}{2}$

c. $\dfrac{2}{7}x = -7$

$\dfrac{7}{2} \cdot \dfrac{2}{7}x = \dfrac{7}{2} \cdot (-7)$

$x = -\dfrac{49}{2}$

d. $\dfrac{2}{7}x = -9$

$\dfrac{7}{2} \cdot \dfrac{2}{7}x = \dfrac{7}{2} \cdot (-9)$

$x = -\dfrac{63}{2}$

e. $\dfrac{2}{7}x = -11$

$\dfrac{7}{2} \cdot \dfrac{2}{7}x = \dfrac{7}{2} \cdot (-11)$

$x = -\dfrac{77}{2}$

c. $\dfrac{1}{4}x - \dfrac{3}{8} + \dfrac{7}{2}x = \dfrac{1}{4} + 4x$

$8 \cdot \left(\dfrac{1}{4}x - \dfrac{3}{8} + \dfrac{7}{2}x \right) = 8 \cdot \left(\dfrac{1}{4} + 4x \right)$

$2x - 3 + 28x = 2 + 32x$

$30x - 3 = 2 + 32x$

$30x - 5 = 32x$

$-5 = 2x$

$x = -\dfrac{5}{2}$

d. $\dfrac{1}{4}x - \dfrac{3}{8} + \dfrac{9}{2}x = \dfrac{1}{4} + 5x$

$8 \cdot \left(\dfrac{1}{4}x - \dfrac{3}{8} + \dfrac{9}{2}x \right) = 8 \cdot \left(\dfrac{1}{4} + 5x \right)$

$2x - 3 + 36x = 2 + 40x$

$38x - 3 = 2 + 40x$

$38x - 5 = 40x$

$-5 = 2x$

$x = -\dfrac{5}{2}$

e. $\dfrac{1}{4}x - \dfrac{3}{8} + \dfrac{11}{2}x = \dfrac{1}{4} + 6x$

$8 \cdot \left(\dfrac{1}{4}x - \dfrac{3}{8} + \dfrac{11}{2}x \right) = 8 \cdot \left(\dfrac{1}{4} + 6x \right)$

$2x - 3 + 44x = 2 + 48x$

$46x - 3 = 2 + 48x$

$46x - 5 = 48x$

$-5 = 2x$

$x = -\dfrac{5}{2}$

38. a. $\dfrac{1}{4}x - \dfrac{3}{8} + \dfrac{3}{2}x = \dfrac{1}{4} + 2x$

$8 \cdot \left(\dfrac{1}{4}x - \dfrac{3}{8} + \dfrac{3}{2}x \right) = 8 \cdot \left(\dfrac{1}{4} + 2x \right)$

$2x - 3 + 12x = 2 + 16x$

$14x - 3 = 2 + 16x$

$14x - 5 = 16x$

$-5 = 2x$

$x = -\dfrac{5}{2}$

b. $\dfrac{1}{4}x - \dfrac{3}{8} + \dfrac{5}{2}x = \dfrac{1}{4} + 3x$

$8 \cdot \left(\dfrac{1}{4}x - \dfrac{3}{8} + \dfrac{5}{2}x \right) = 8 \cdot \left(\dfrac{1}{4} + 3x \right)$

$2x - 3 + 20x = 2 + 24x$

$22x - 3 = 2 + 24x$

$22x - 5 = 24x$

$-5 = 2x$

$x = -\dfrac{5}{2}$

39. a. $C = 30 + 0.20m$

$70.40 = 30 + 0.20m$

$40.40 = 0.20m$

$\dfrac{40.40}{0.20} = m$

$m = 202$ miles

b. $C = 30 + 0.20m$

$70.60 = 30 + 0.20m$

$40.60 = 0.20m$

$\dfrac{40.60}{0.20} = m$

$m = 203$ miles

c. $C = 30 + 0.20m$

$70.80 = 30 + 0.20m$

$40.80 = 0.20m$

$\dfrac{40.80}{0.20} = m$

$m = 204$ miles

d. $C = 30 + 0.20m$
$71.20 = 30 + 0.20m$
$41.20 = 0.20m$
$\dfrac{41.20}{0.20} = m$
$m = 206$ miles

e. $C = 30 + 0.20m$
$71.40 = 30 + 0.20m$
$41.40 = 0.20m$
$\dfrac{41.40}{0.20} = m$
$m = 207$ miles

40. a. $I = PRT$
$30 = 1000 \cdot R \cdot \dfrac{1}{2}$
$30 = 500R$
$\dfrac{3\cancel{0}}{50\cancel{0}} = R$ so $R = \dfrac{3}{50} = 0.06 = 6\%$

b. $I = PRT$
$40 = 1000 \cdot R \cdot \dfrac{1}{2}$
$40 = 500R$
$\dfrac{4\cancel{0}}{50\cancel{0}} = R$ so $R = \dfrac{4}{50} = 0.08 = 8\%$

c. $I = PRT$
$50 = 1000 \cdot R \cdot \dfrac{1}{2}$
$50 = 500R$
$\dfrac{5\cancel{0}}{50\cancel{0}} = R$ so $R = \dfrac{5}{50} = 0.10 = 10\%$

d. $I = PRT$
$60 = 1000 \cdot R \cdot \dfrac{1}{2}$
$60 = 500R$
$\dfrac{6\cancel{0}}{50\cancel{0}} = R$ so $R = \dfrac{6}{50} = 0.12 = 12\%$

e. $I = PRT$
$70 = 1000 \cdot R \cdot \dfrac{1}{2}$
$70 = 500R$
$\dfrac{7\cancel{0}}{50\cancel{0}} = R$ so $R = \dfrac{7}{50} = 0.14 = 14\%$

Cumulative Review Chapters 1–10

1. $49 \div 7 \cdot 7 + 8 - 5 = 7 \cdot 7 + 8 - 5$
$= 49 + 8 - 5$
$= 57 - 5$
$= 52$

2. $80 \div 0.13 = \dfrac{80}{0.13} = \dfrac{8000}{13}$

$$= 13\overline{)8000.000} \approx 615.38$$

$$
\begin{array}{r}
615.384 \\
13\overline{)8000.000} \\
\underline{78} \\
20 \\
\underline{13} \\
70 \\
\underline{65} \\
5\,0 \\
3\,9 \\
\overline{1\,10} \\
1\,04 \\
\overline{60} \\
\underline{52} \\
8
\end{array}
$$

3. $23\% = 0.23$

4.   Let P = the percentage.
$0.90 \times 40 = P$
$36 = P$
Thus, 90% of 40 is 36.

5.   Let R = the percent.
$R \times 8 = 4$
$R = \dfrac{4}{8} = 0.5 = 50\%$
Thus, 50% of 8 is 4.

6.   Let B = the number.
$6 = 0.30 \times B$
$\dfrac{6}{0.30} = B$
$\dfrac{60}{3} = B$
$20 = B$
Thus, 6 is 30% of 20.

7. $I = PRT = 300 \times 0.075 \times 5 = \$112.50$

8. 1% of the families in Portland have incomes $120,000 and over.

9. The percent of these hours that is in art and math together is 25% + 15% = 40%.

10. The mode is 5, since it is the value that occurs most often.

11. Mean $= \dfrac{2+16+1+28+1+30+1+25}{8}$

$= \dfrac{104}{8}$

$= 13$

12. Arrange the numbers in order:
5, 6, 7, 9, 10, 14, 14, 14, 29.
Since there are an odd number of values, the median is the middle number. Thus, the median is 10.

13. 4 yd $= 4 \cdot (3 \text{ ft}) = 12 \cdot (12 \text{ in.}) = 144 \text{ in.}$

14. 17 in. $= 17 \text{ in.} \cdot \dfrac{1 \text{ ft}}{12 \text{ in.}} = \dfrac{17}{12} \text{ ft} = 1\dfrac{5}{12} \text{ ft}$

15. 17 ft $= 17 \text{ ft} \cdot \dfrac{1 \text{ yd}}{3 \text{ ft}} = \dfrac{17}{3} \text{ yd} = 5\dfrac{2}{3} \text{ yd}$

16. 7 ft $= 7 \cdot (12 \text{ in.}) = 84 \text{ in.}$

17. 240 m = 2400 dm

18. $P = 2 \cdot (4.8 \text{ cm}) + 2 \cdot (2.8 \text{ cm})$
$= 9.6 \text{ cm} + 5.6 \text{ cm}$
$= 15.2 \text{ cm}$

19. $d = 2 \cdot r = 2 \cdot 9 \text{ in.} = 18 \text{ in.};$
$C = \pi \cdot d = 3.14 \cdot (18 \text{ in.}) = 56.52 \text{ in.}$

20. $A = \dfrac{1}{2} \cdot b \cdot h$

$= \dfrac{1}{2} \cdot (\overset{5}{\cancel{10}} \text{ cm}) \cdot (16 \text{ cm})$

$= 80 \text{ cm}^2$

21. $A = \pi r^2$
$= 3.14 \cdot (3 \text{ cm})^2$
$= 3.14 \cdot (9) \text{ cm}^2$
$= 28.26 \text{ cm}^2$

22. $V = LWH$
$= (5 \text{ cm}) \cdot (7 \text{ cm}) \cdot (3 \text{ cm}) = 105 \text{ cm}^3$

23. $V = \dfrac{4}{3} \pi r^3$

$= \dfrac{4}{3} \cdot (3.14) \cdot (12 \text{ in.})^3$

$= \dfrac{4}{\cancel{3}} \cdot (3.14) \cdot \overset{576}{\cancel{1728}} \text{ in.}^3$

$= 7234.56 \text{ in.}^3$

24. $\angle \alpha$ is a Right angle.

25. $m\angle A + m\angle B + m\angle C = 180°$
$35° + 25° + m\angle C = 180°$
$60° + m\angle C = 180°$
$m\angle C = 120°$

26. We need a number that, when squared, we get 8100. The number is 90. Thus
$\sqrt{8100} = 90$.

27. $a^2 + b^2 = c^2$
$9^2 + 12^2 = c^2$
$81 + 144 = c^2$
$225 = c^2$
$c = \sqrt{225} = 15$
The hypotenuse is 15.

28. The additive inverse of $-3\dfrac{4}{5}$ is $3\dfrac{4}{5}$.

29. $\left|-3.5\right| = 3.5$

30. $-4.7 + (-6.8) = -4.7 - 6.8$
$= -(4.7 + 6.8) = -11.5$

31. $\dfrac{2}{5} + \left(-\dfrac{1}{8}\right) = \dfrac{2}{5} - \dfrac{1}{8} = \dfrac{16}{40} - \dfrac{5}{40} = \dfrac{11}{40}$

32. $-3.8 - (-4.1) = -3.8 + 4.1$
$$= +(4.1 - 3.8) = 0.3$$

33. $-\dfrac{2}{5} - \dfrac{2}{7} = -\left(\dfrac{2}{5} + \dfrac{2}{7}\right) = -\left(\dfrac{14}{35} + \dfrac{10}{35}\right) = -\dfrac{24}{35}$

34. $-\dfrac{\cancel{7}}{8} \cdot \dfrac{9}{\cancel{7}} = -\dfrac{9}{8}$

35. $-\dfrac{10}{7} \div \dfrac{3}{7} = -\dfrac{10}{\cancel{7}} \cdot \dfrac{\cancel{7}}{3} = -\dfrac{10}{3}$

36. $(-7)^2 = (-7)(-7) = 49$

37. $5(5x - 8y) = 5 \cdot 4x - 5 \cdot 8y = 20x - 40y$

38. $60x - 30y - 40z$
$$= 10 \cdot 6x - 10 \cdot 3y - 10 \cdot 4z$$
$$= 10(6x - 3y - 4z)$$

39. $(2x + y) - (7x - 2y) = 2x + y - 7x + 2y$
$$= 2x - 7x + y + 2y$$
$$= -5x + 3y$$

40. $9^{-2} = \dfrac{1}{9^2} = \dfrac{1}{81}$

41. $x^{-2} \cdot x^{-5} = x^{-2+(-5)} = x^{-7} = \dfrac{1}{x^7}$

42. $(3x^3 y^{-2})^2 = 3^2 (x^3)^2 (y^{-2})^2$
$$= 9x^6 y^{-4}$$
$$= \dfrac{9x^6}{y^4}$$

43. $0.000000024 = 2.4 \times 10^{-8}$

44. $2.82 \times 10^7 = 28,200,000$

45. $(2.695 \times 10^{-3}) \div (7.7 \times 10^4)$
$$= \dfrac{2.695 \times 10^{-3}}{7.7 \times 10^4}$$
$$= \dfrac{2.695}{7.7} \times \dfrac{10^{-3}}{10^4}$$
$$= 0.35 \times 10^{-3-4}$$
$$= (3.5 \times 10^{-1}) \times 10^{-7}$$
$$= 3.5 \times 10^{-8}$$

46. $\dfrac{5}{7}x = -35$
$$\dfrac{7}{5} \cdot \dfrac{5}{7}x = \dfrac{7}{5} \cdot (-35)$$
$$x = -49$$

47. $9x - 10 = 8$
$$9x = 18$$
$$x = 2$$

48. $2x - 4 = x + 3$
$$2x = x + 7$$
$$x = 7$$

49. $16 = 4(x - 9)$
$$16 = 4x - 36$$
$$52 = 4x$$
$$13 = x$$